Dynamical Systems

An International Symposium

Volume 1

SOLOMON LEFSCHETZ
1884–1972

Dynamical Systems

An International Symposium
Volume 1

Edited by

Lamberto Cesari

Department of Mathematics
University of Michigan
Ann Arbor, Michigan

Jack K. Hale

Joseph P. LaSalle

Lefschetz Center for Dynamical Systems
Division of Applied Mathematics
Brown University
Providence, Rhode Island

ACADEMIC PRESS   New York   San Francisco   London   1976

A Subsidiary of Harcourt Brace Jovanovich, Publishers

ACADEMIC PRESS, INC.
111 Fifth Avenue, New York, New York 10003

United Kingdom Edition published by
ACADEMIC PRESS, INC. (LONDON) LTD.
24/28 Oval Road, London NW1

Library of Congress Cataloging in Publication Data
Main entry under title:

Dynamical systems.

    Includes bibliographies and index.
    1.   Differential equations—Congresses.   2.   Topolog-
ical dynamics—Congresses.   3.   Differentiable dynami-
cal systems—Congresses.   I.   Cesari, Lamberto.
II.   Hale, Jack K.   III.   LaSalle, Joseph P.
QA371.D9   515   75-13095
ISBN 0−12−164901−6 (v. 1)

It was Solomon Lefschetz who made the subject of differential equations both respectable and lively in this country, and who through his projects at Princeton and RIAS and his association with the Center for Dynamical Systems at Brown made it possible with his boundless enthusiasm, inspiration, and guidance for many young people to establish deep roots in the subject. We will remember always his love of life, humor, self-discipline, courage, wisdom, vigor, incessant curiosity, towering intellect, and creative genius.

J. P. LaSalle, IEEE Memorial, 1973

Contents

Chapter 1   QUALITATIVE THEORY

CHARLES CONLEY

MICHAEL SHUB

Chapter 2   GENERAL THEORY

LAMBERTO CESARI

JEAN MAWHIN

PAUL H. RABINOWITZ

WILLIAM T. REID

WENDELL H. FLEMING AND C. P. TSAI

Chapter 7   CONTROL THEORY

List of Contributors

Numbers in parentheses indicate the pages on which the authors' contributions begin.

THOMAS S. ANGELL (*311*), Department of Mathematics, University of Delaware, Newark, Delaware

H. T. BANKS (*287*), Lefschetz Center for Dynamical Systems, Division of Applied Mathematics, Brown University, Providence, Rhode Island

JOHN A. BURNS* (*287*), Lefschetz Center for Dynamical Systems, Division of Applied Mathematics, Brown University, Providence, Rhode Island

LAMBERTO CESARI (*29, 251*), Department of Mathematics, University of Michigan, Ann Arbor, Michigan

NATHANIEL CHAFEE (*263*), School of Mathematics, Georgia Institute of Technology, Atlanta, Georgia

WILLIAM C. CHEWNING† (*303*), Department of Mathematics and Computer Science, University of South Carolina, Columbia, South Carolina

A. K. CHOUDHURY (*317*), Department of Electrical Engineering, Howard University, Washington, D.C.

E. N. CHUKWU (*307*), Department of Mathematics, The Cleveland State University, Cleveland, Ohio

CHARLES CONLEY (*1*), Department of Mathematics, University of Wisconsin, Madison, Wisconsin

ROBERTO CONTI (*283*), Istituto Matematico U. Dini, Università di Firenze, Florence, Italy

MICHAEL G. CRANDALL (*131*), Department of Mathematics, University of California, Los Angeles, California, and Mathematics Research Center, University of Wisconsin, Madison, Wisconsin

J. P. FINK (*267*), Department of Mathematics, University of Pittsburgh, Pittsburgh, Pennsylvania

WENDELL H. FLEMING (*103*), Lefschetz Center for Dynamical Systems, Division of Applied Mathematics, Brown University, Providence, Rhode Island

MARVIN I. FREEDMAN (*325*), Department of Mathematics, Boston University, Boston, Massachusetts

---

* Present address: Department of Mathematics, Virginia Polytechnic Institute and State University, Blacksburg, Virginia.
† Deceased.

O. HÁJEK (*307*), Department of Mathematics, Case Western Reserve University, Cleveland, Ohio

JACK K. HALE (*179*), Lefschetz Center for Dynamical Systems, Division of Applied Mathematics, Brown University, Providence, Rhode Island

WILLIAM S. HALL (*267*), Department of Mathematics, University of Pittsburgh, Pittsburgh, Pennsylvania

A. R. HAUSRATH (*267*), Department of Mathematics, University of Pittsburgh, Pittsburgh, Pennsylvania

MARC Q. JACOBS (*297*), Department of Mathematics, University of Missouri, Columbia, Missouri

JAMES L. KAPLAN (*325*), Department of Mathematics, Boston University, Boston, Massachusetts

KLAUS KIRCHGÄSSNER (*115*), Mathematisches Institut A der Universität Stuttgart, Stuttgart, Germany

WERNER KRABS (*291*), Fachbereich Mathematik der Technischen Hochschule, Darmstadt, West Germany

C. E. LANGENHOP (*297*), Department of Mathematics, Southern Illinois University, Carbondale, Illinois

JOSEPH P. LASALLE (*211*), Lefschetz Center for Dynamical Systems, Division of Applied Mathematics, Brown University, Providence, Rhode Island

JEAN MAWHIN (*51*), Université de Louvain, Institut Mathématique, Louvain-la-Neuve, Belgium

RICHARD K. MILLER (*223*), Mathematics Department, Iowa State University, Ames, Iowa

WALDYR M. OLIVA (*195*), Instituto de Matemática e Estatistica, Universidade de São Paulo, Sao Paulo, Brazil

PAUL H. RABINOWITZ (*83*), Department of Mathematics, University of Wisconsin, Madison, Wisconsin

WILLIAM T. REID (*97*), Department of Mathematics, The University of Oklahoma, Norman, Oklahoma

RUSSELL D. RUPP (*331*), Department of Mathematics, State University of New York, Albany, New York

JÜRGEN SCHEURLE (*115*), Mathematisches Institut A der Universität Stuttgart, Stuttgart, Germany

THOMAS I. SEIDMAN (*273*), Department of Mathematics, University of Maryland, Baltimore County, Baltimore, Maryland

GEORGE R. SELL (*223*), School of Mathematics, University of Minnesota, Minneapolis, Minnesota

MICHAEL SHUB (*13*), Department of Mathematics, Queens College, Flushing, New York

J. T. STUART (*277*), Mathematics Department, Imperial College, London, England

LUC TARTAR* (*167*), University of Paris, Paris, France

C. P. TSAI (*103*), Lefschetz Center for Dynamical Systems, Division of Applied Mathematics, Brown University, Providence, Rhode Island

---

* Present address: Université Paris-Sud Mathematiques, Orsay, Cedex, France.

Preface

The International Symposium on Dynamical Systems, of which this volume is the proceedings, was held at Brown University, August 12–16, 1974 and was the formal occasion for dedicating the Lefschetz Center for Dynamical Systems to the memory of Solomon Lefschetz. The central theme of the symposium was the manner in which the theory of dynamical systems continues to permeate current research in ordinary and functional differential equations, and how this approach and the techniques of ordinary differential equations have begun to influence in a significant way research on certain types of partial differential equations and evolutionary equations in general. This volume provides an exposition of recent advances, present status, and prospects for future research and applications.

The editors and the Lefschetz Center for Dynamical Systems wish to thank the Air Force Office of Scientific Research, the Army Research Office (Durham), the National Science Foundation, the Office of Naval Research, and Brown University for the generous support that made this symposium possible. The editors were responsible for the program, and we wish here to express on behalf of all the participants our appreciation to H. Thomas Banks, Ettore F. Infante, and Constantine Dafermos for their planning and organization of the meeting.

SOLOMON LEFSCHETZ: A Memorial Address

Solomon Lefschetz died not quite two years ago. This September will be the 90th anniversay of his birth. All of us here know of Solomon Lefschetz and many of us were fortunate to have known him well.

He was born in Moscow on September 3, 1884 but returned shortly thereafter with his mother to Paris. There is not much known of his early life in Paris, and I never heard him speak about his youth except to say he did not learn Russian until he was in his teens and how glad he was to get his engineering degree so that he could forget all the uninteresting things he had been forced to learn. He received his engineering degree in 1905 from the École Centrale in Paris. Shortly after receiving his degree he left for the United States, worked for a while at the Baldwin Locomotive works in Philadelphia as an engineering apprentice, and in 1907 accepted a position as a power engineer with Westinghouse in Pittsburgh. In 1910 he lost both of his arms below his elbows in a high-voltage accident. While recovering from this accident he decided that he could no longer be an engineer. He himself at a later date said simply: "For six years I was an engineer. I soon realized my true path was not engineering but mathematics." He then pointed out that two of his professors at the École Centrale were Emile Picard and Paul Appel. "Each had written a three-volume treatise: *Analysis* (Picard) and *Analytical Mechanics* (Appel). I plunged into these and gave myself a self-taught graduate course. What with a strong French training in the equivalent of an undergraduate course, I was all set."

And so he was. A year later he received his Ph.D. in mathematics from nearby Clark University (Worcester, Massachusetts). He received an assistantship (soon changed to an instructorship) at the University of Nebraska. He was there for two years. In 1912 he become an American citizen. In 1913 he married Miss Alice Berg Hayes, a fellow student at Clark. About Mrs. Lefschetz and Clark, Lefschetz writes: "there was fortunately a first-rate librarian, Dr. L. N. Wilson, and a well-kept mathematical library. Just two of us enjoyed it—my fellow graduate student and future wife, and myself. I took advantage of the library to learn about a number of highly interesting new fields, notably about the superb Italian school of algebraic geometry."

Mrs. Lefschetz was a perfect wife and companion for Lefschetz. She, too, had strong convictions and an independent mind. She remained much in the background. It was most delightful to spend hours in conversation with her and Lefschetz. I once took my young son with me on a visit with the Lefschetz's at 11 Lake Lane in Princeton. I had warned him to be on his best behavior and to remember that the Lefschetz's were old and not used to children. We were there for almost four hours, and my son was as quiet as I have ever seen him. As soon as we left and were in the car by ourselves, he said to me indignantly, "They are not old!" During the time that Lefschetz was traveling by plane, he and Mrs. Lefschetz never flew together, even though they had no dependents. I once asked Lefschetz about this. He said, "She stays home to clean the house. She's afraid someone will break in while we are gone and find the place dirty."

Lefschetz left Nebraska after two years and was an instructor at the University of Kansas from 1913 to 1916, an associate professor from 1916 to 1919, and a professor from 1923 to 1925. Lefschetz in describing his research at Kansas on algebraic curves on hyperelliptic surfaces writes: "This [research] launched me into Poincaré-type topology, the 1919 Bordin Prize of the Paris Academy, and in 1924 Princeton! . . . The immèdiate effect of the prize was the Kansas promotion (January 1920) to associate professor plus a schedule reduction." Until that time he had been teaching eighteen hours or more a week. He later speaks wistfully of "the mathematical calm of Nebraska–Kansas which I had so enjoyed without realizing it."

Well, he did go to Princeton in 1924 and, as we know, eventually became one of the most original and influential and well-recognized mathematicians of his generation.

During World War II and at a time when he was 60 years old and had already made superb contributions to two fields of mathematics, he turned his attention to the subject of differential equations. His interest in and his promotion of differential equations lasted throughout the remaining 28 years of his life. I visited Lefschetz a few months before he died. He was first interested in everything that was going on at Brown, both mathematically and otherwise. Next, he wanted me to take him out to dinner.

He writes: "Then came World War II and I turned my attention to differential equations. With Office of Naval Research backing (1946–1955), I conducted a seminar on the subject from which there emanated a number of really capable fellows, also a book: *Differential Equations: Geometric Theory* (1957)." The acme of understatement.

The year that Lefschetz' interest in differential equations began was the year of the death of George David Birkhoff (1884–1944). Lefschetz and Birkhoff were contemporaries (they were the same age) but were less than friendly. It was only much later (around 1965) that Lefschetz finally recognized that, after Poincaré and Liapunov, Birkhoff was one of the cofounders of the geometric theory of differential equations. This was after the work of Moser, and after I had established a relationship between Birkhoff's limit sets and Liapunov stability theory. William Hodges points out, in relationship to this, an interesting aspect of Lefschetz' character. "He [Lefschetz] asserted (with much truth) 'he made up his mind in a flash and found his reasons later.' Naturally, he made mistakes this way, but once he was really convinced that he was wrong he could be extremely generous."

But to return to 1944, the kindest word for the state of differential equations within mathematics in the United States is, perhaps, dormant. At the 1965 International Symposium on Differential Equations and Dynamical Systems in Puerto Rico, of which this symposium is the successor, Lefschetz in his talk (after claiming that as an "amateur" he may have "a broader view of the field of differential equations than its professionals!") said:

"The study of geometric differential equations as a chapter of differential equations was really founded by Henri Poincaré in his classical "Memoirs" on curves defined by a differential equation (1881). The cofounders include Liapunov, who in his great "Memoire" created a true theory of dynamic stability, and G. D. Birkhoff, who is responsible for so many new concepts in dynamics.

"Curiously, until recently the immense contributions of the founders were practically ignored by the mathematical public in general and especially by applied mathematicians [he is speaking of the United States]. . . . The reason is not far to seek . . . almost universal linearization prevailed." The major interest among electrical engineers in the United States during the period before World War II was in the development of long distance telephony and radio communication. The problems were linearized, no differential equations were required, and the mathematics exploited was the theory of functions of a complex variable. During this time one might add that differential equations was one of the poorest taught and most boring of our undergraduate courses in mathematics. Some students who studied topology not only did not know what a differential equation was, they did not even know the meaning of a derivative.

Well, I do not need to relate to this audience what happened to the field of differential equations in this country following Lefschetz. I will, however, allow myself one remark. It is interesting (and quite useless) to speculate about what might have happened had Lefschetz remained an industrial engineer. I feel sure of only one thing. Short of an amendment to the Constitution he would not have become President of the United States. But having left engineering and turned to mathematics, he made in turn a profound contribution to engineering. He was responsible for introducing us (and the whole Western world) to two of the mainstreams of modern control theory: Liapunov's theory of stability and the mathematical theory of optimal control. Of the latter, he was, through his students and "a number of really capable fellows" who came through his Princeton project, instrumental in its founding.

After his retirement from Princeton in 1953, he had also to retire as director of his differential equations project (a university rule) and during the next five years his project was gradually phased out. The project which ran from 1946 through 1959 operated initially on an annual grant of $25,000 from ONR, later reduced to $20,000 a year and shared for a few years with the Air Research and Development Command.

After attempts by Lefschetz to found a research institute at other universities failed, he received *carte blanche* from the Martin Company in November 1957 to form a mathematics center at their Research Institute for Advanced Study, Baltimore, and was charged with making "this Center an outstanding example of its kind in the world." Lefschetz was then almost 70 years old. Within three years Lefschetz had done this, and his group had achieved an international reputation. Lefschetz writes of this as follows: "Very shortly we became known. A considerable number of good differential equationists visited us, and some few were invited for a year or so. After some six years it was necessary to transfer elsewhere. This operation . . . resulted in our becoming part of the Division of Applied Mathematics at Brown University as the "Center for Dynamical Systems" with LaSalle as director and myself as (once weekly) visiting professor. At Brown our general relationship was excellent." As with much of Lefschetz' mathematics, one would have to dig deeply to comprehend fully what lies behind what he has written. Suffice it to say that we are still grateful to the Martin Company (now Martin–Marrietta) for the opportunity RIAS offered us and for the generous support while we were there from AFOSR, ARO,

NASA, and ONR. Because of an administrative ruling (not a part of Congress' basic law in establishing NSF) we were not eligible at RIAS for NSF support. For this and other reasons, the support of basic research outside universities is difficult and unstable in this country, and we had a strong desire to train young people. Lefschetz was instrumental in our selecting Brown as a place to establish what is now his Center.

We are lineal descendants of Lefschetz' Princeton differential equations project, and we hope that this Center will be a lasting memorial to him and that it will remind everyone who works and visits here, or knows of us, that we reside eternally in his debt.

In final tribute, I quote from my memorial to Solomon Lefschetz written for the *IEEE Transactions on Automatic Control*:

> It was Solomon Lefschetz who made the subject of differential equations both respectable and lively [in this country] and who through his projects at Princeton and RIAS and his association with the Center for Dynamical Systems at Brown made it possible with his boundless enthusiasm, inspiration, and guidance for many young people to establish deep roots in the subject. We will remember always his love of life, humor, self-discipline, courage, wisdom, vigor, incessant curiosity, towering intellect, and creative genius.

At Princeton the students in their songs about faculty had the following verse:

> Here's to Lefschetz (Solomon L.)
> Who's as argumentative as hell,
> When he's at last beneath the sod
> Then he'll start to heckle God.

I feel sure that Lefschetz and God are getting along very well.

Joseph P. LaSalle, August 1974

Contents of Volume 2

Chapter 1: QUALITATIVE THEORY

Some Aspects of the Qualitative Theory of Differential Equations

CHARLES CONLEY
Department of Mathematics
University of Wisconsin, Madison, Wisconsin

In these remarks some problems of a qualitative nature are discussed. A salient feature of the problems is that they do not require any difficult computations; a definition of *qualitative* will not be given. All the examples here concern the existence of orbits that are situated in some special way; in a general sense they are boundary-value problems.

In Fig. 1 a "tube" with square cross section is pictured. There is supposed to be a differential equation whose solutions behave in the following way: on the top, bottom, and left end they cross into the tube; on the sides and the right end they cross out of the tube. In particular, the orbit through any boundary point leaves the tube in one or the other (or both) time directions—in the terms of T. Wazewski each boundary point is a point of strict exit or entrance. Consequently, the mapping that assigns to each point of the tube its first point of exit is continuous, as is the analogous mapping for the backward flow; the mapping from entering to exiting points is a homeomorphism wherever it is defined.

In the present case it is assumed that there is no invariant set in the tube—thus every point is mapped both to the exit set and the entrance set under the above mappings.

An orbit segment will be said to cross the tube if it runs from the left end to the right end without leaving the tube. The first question concerns the existence of these. The answer is that the set of such orbits meets the left end in a set that connects the top and bottom of the tube and likewise meets the right end in a set that connects the sides of the tube.

The main part of the argument is this: consider any arc in the left end that connects the sides; the ends go out (immediately) from different sides. Since the arc goes out through an arc in the exit set, some points of it must leave through the right end.

1

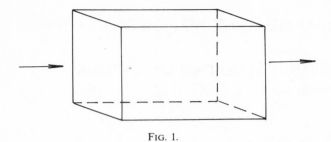

FIG. 1.

Here is an "application": consider the equations

$$\dot{x} = x + f(x, y, t), \qquad \dot{y} = -y + g(x, y, t), \qquad \dot{t} = 1,$$

with the hypothesis that $(f^2 + g^2)/(x^2 + y^2) \to 0$ as $x^2 + y^2 \to +\infty$ uniformly in $t \in (-\infty, \infty)$. The question concerns the existence of bounded solutions.

The hypothesis implies there is an infinitely long tube of the form $\{(x, y, t) \,|\, |x|\,|y| \le C\}$, where C is a large enough constant. An argument similar to that above implies that the set of orbits that stay in the right half of this tube (from $t = 0$) meets the $t = 0$ section in a set connecting the top and bottom. Likewise, this section contains a set connecting the sides such that the left half orbit through points in this set stays in the left half tube. Since these two connecting sets must intersect, bounded orbits exist.

A sequence of tubes will be said to be correctly connected if the right end of the kth overlaps the left end of the $(k + 1)$st as shown in Fig. 2. If a finite sequence of tubes is correctly connected there are orbit segments that run from the left end of the first to the right end of the last passing through the tubes in sequence and contained in their union. The proof is based on the observation that any arc in the left end of the kth tube that connects the sides contains a subarc that is carried to the right end so that its image connects the sides of the $(k + 1)$st tube.

Given a bi-infinite sequence of correctly connected tubes such that the time to cross any one is more than (say) 1, it follows that there are orbits (at least one) that are contained in the union of the tubes and pass through them in sequence. If the sequence of tubes is periodic, then there must be a periodic orbit in the union of the tubes. The argument in this case needs an additional ingredient.

These remarks apply to compact hyperbolic invariant sets of smooth

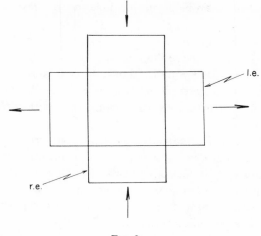

FIG. 2.

flows. The application requires the following definition: a finite sequence of orbit segments is called an ε, t-chain from x to y if (1) the segments are each of time length at least t, (2) x is within $\varepsilon/2$ of the initial point of the first arc, (3) the terminal point of the kth arc is within ε of the initial point of the $(k+1)$st arc, and (4) the terminal point of the last arc is within $\varepsilon/2$ of y. The distances are chosen so that ε, t-chains from x to y and y to z can be put together to make one from x to z.

If H is a compact hyperbolic set, then for sufficiently small ε there corresponds to any ε, t-chain (in H) from x to y a sequence of tubes, one for each segment in the chain, which are correctly connected and which approximately contain their corresponding orbit segments. Furthermore, the tubes have roughly the same cross-sectional diameter and this can be chosen to go to zero with ε independently of the ε, t-chain.

This consequence is fairly evident from the definition of hyperbolic: namely, that the tangent flow restricted to the tangent bundle over H is the sum of three invariant subbundles, the first spanned by contracting directions (vertical in Fig. 1), the second by the flow direction (end to end), and the last by expanding directions (side to side).

Suppose now that for some $x \in H$ there is an ε, t-chain from x back to x all of whose segments lie in H. Then (if ε is small enough) there is a periodic orbit passing near x (it may not be in H though). If an ε-pseudo-orbit means a bi-infinite ε, t-chain, then: uniformly close to an ε pseudo-orbit in H there lies a real orbit. For flows a more careful statement concerning time parametrization could be given; for diffeomorphisms the problem does not arise.

Actually, in this setting, bi-infinite sequences of tubes determine unique orbits (compare to the classical theorem for the first set of equations when the partials of f and g are small); thus one can guess at the structural stability statement for hyperbolic invariant sets. The sequence of tubes is "stable" under C^1 perturbations and defines the orbit correspondence at least.

Here is a relevant problem: suppose the nonwandering set (of a smooth flow on a compact manifold) is a hyperbolic invariant set. Does it follow that it is the closure of the periodic orbits? For "proof": first, every point x of the nonwandering set is chain-recurrent, meaning that for all ε, $t > 0$, there is an ε, t-chain from x back to itself (only one segment is required). As above, there is a periodic orbit near x. This "proof" is wrong because the segments in the chain may not lie in the nonwandering set.

Avoiding this problem, define the chain-recurrent set of a flow to be the set of chain-recurrent points—it is a closed invariant set containing the nonwandering set. In contrast to the case of the nonwandering set, however, if one restricts the flow to its chain-recurrent set R, then the chain-recurrent set of this new flow is R itself. This means that for $x \in R$, the chains from x to x can be chosen to lie in R. Thus, if the chain-recurrent set is hyperbolic, the periodic orbits are dense.

An axiom A flow is one such that the nonwandering set is hyperbolic and also the closure of the periodic points. A reasonable conjecture is as follows: A flow satisfies Axiom A plus the strong transversality condition if and only if the chain-recurrent set is hyperbolic.

One other remark about H: if $x \in H$ is chain-recurrent but not periodic, then Smale's shift automorphism is embedded near x; one can use the tube construction together with two distinct periodic orbits near x to construct a mapping with all the essential features of Smale's horseshoe mapping.

A problem in which something close to these tubes plays a role is the following: Does the equation below admit a nontrivial standing wave solution?

$$p_t = p_{xx} + s(x)p(1 - p).$$

This is really an ordinary differential equations problem; standing wave means $p_t = 0$. Here x runs from $-\infty$ to $+\infty$. Given that $s(x)$ approaches a negative limit at $-\infty$ and a positive limit at $+\infty$, one proves that there is a standing wave that goes to 0 at $-\infty$ and to 1 at $+\infty$, and otherwise lies between 0 and 1.

Writing the equation for standing waves as a system, one has

$$\dot{p} = q, \qquad \dot{q} = -s(x)p(1-p).$$

Choose x_1 such that $s(x)$ is close to $\delta > 0$ for $x \geq x_1$. Now consider the set ("tube") $\{p, q, x \,|\, x \geq x_1 ; 0 \leq p \leq 1 ; q \geq 0\}$ (cf. Fig. 3). Since $\delta(x) > 0$ for $x \geq x_1$, one finds that the face $\{p = 0 ; q > 0\}$ consists of strict entrance points, while the faces $\{0 < p < 1 ; q = 0\}$ and $\{p = 1 ; q > 0\}$ consist of strict exit points. With the same type of argument as applied before one sees that any arc from the horizontal exit face to the vertical one must contain a point that stays in the tube as x increases. A little analysis shows that orbits through such points satisfy the right-hand boundary condition (one uses the fact that $s(x)$ is not integrable near $+\infty$). Finally, an estimate on the behavior of orbits with an initially large q value shows that the set of orbits satisfying the right-hand boundary condition meets the left end of the tube in a set that connects the entering face to the point $p = 1$, $q = 0$ (cf. Fig. 3).

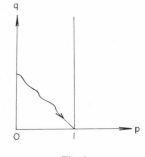

Fig. 3.

Reasoning in a similar way with x_0 such that $s(x)$ is close to $-\Delta < 0$ for $x \leq x_0$, one obtains an analogous picture (Fig. 4) of the set of points with $x = x_0$ that are on orbits satisfying the left-hand boundary condition.

To span the gap from x_0 to x_1 one uses the slab $\{0 \leq p \leq 1, x_0 \leq x \leq x_1\}$, which also has the essential property that exit and entrance points are strict. One finds the arc of Fig. 4 is carried so that it must intersect that of Fig. 3. In this example not every boundary point of the tubes used is an entrance or exit point, but those that are are strict, which is the essential thing.

The aim now is to lead up to another kind of "boundary-value" problem; however, the route will be somewhat roundabout in order that some vaguely relevant facts can be brought into brief focus.

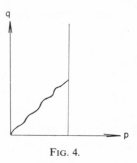

FIG. 4.

An attractor means here a set that is the ω-limit set of some compact neighborhood of itself; this is more restrictive than saying it contains the limit set of each point in some neighborhood of itself.

Define a "block" to be a compact set each one of whose boundary points is a point of strict exit or strict entrance. The definition of an attractor is equivalent to saying it is the maximal (inclusion) invariant set in some block all of whose boundary points are entering points. For example, the empty set and the whole space (if compact) and any closed–open subset of the space are (trivial) attractors—they serve as their own blocks.

A filtration (of a flow on a compact space) means a finite increasing sequence of attractors the last of which is the whole space. Thinking of the use of filtration in topology, perhaps one should say instead that it is an increasing sequence of blocks for these attractors, but the attractors themselves are more relevant here.

Each attractor has a dual repeller (i.e., attractor for the reverse flow). Namely, it is the maximal invariant set in the complement of any block for the attractor.

The Morse sets of a filtration $\mathscr{F} = \{\phi = A_0, A_1, \ldots, A_n = X\}$ are the sets $M_k = A_k \cap A_{k-1}^*$ $(k = 1, \ldots, n)$, where A_{k-1}^* is the repeller dual to A_{k-1}.

Here is the theorem: Let $\mathscr{M}(\mathscr{F})$ be the union of the Morse sets of $\mathscr{F}$. Then the chain-recurrent set of a flow on X is the intersection over all filtrations $\mathscr{F}$ of the corresponding sets $\mathscr{M}(\mathscr{F})$. In particular, every point of X is chain-recurrent if and only if the flow has no nontrivial attractors. This implies the existence of a "Liapounoff function" that is constant on components at the chain-recurrent set and otherwise is strictly decreasing on orbits. We conjecture as follows: With the C–0 topology on the space of flows, the set of points of semicontinuity of the nonwandering set function is precisely the set of points where the nonwandering set and the chain-recurrent set coincide (the chain-recurrent set function is upper semi-continuous everywhere).

The homotopy index of the Morse set M_k is the homotopy type $[B_k/B_{k-1}]$ meaning that obtained from a block B_k for A_k on collapsing to a point of a (smaller) block B_{k-1} for A_{k-1}. Fortunately, the index does not depend on the block chosen.

One can pursue these thoughts further and prove a Morse–Smale theorem for the Morse sets of a filtration. The statement and proof make use of an exact sequence relating the cohomologies of the index, the Morse set, and (sections of) the stable and unstable "manifolds" of the Morse set.

The Morse sets M_k and M_{k+1} are both contained in the kth Morse set of the smaller filtration $\{A_0, \ldots, A_k, A_{k+2}, \ldots, A_n\}$; call this set M. If $M = M_k \cup M_{k+1}$ then the index of M is the sum of those of M_k and M_{k+1}, where the sum of pointed spaces means that obtained by pasting them together at their distinguished point.

Note the following "boundary-value problem": Does there exist an orbit with its α-limit set in M_{k+1} and its ω-limit set in M_k? Suppose the index of M is not the sum as above; then the answer is yes. Namely, there must then be an orbit in M that is not in $M_{k+1} \cup M_k$. That means it lies in $B_{k+1} \backslash B_{k-1}$ but is not contained in $B_{k+1} \backslash B_k$ or $B_k \backslash B_{k-1}$ (cf. Fig. 4). It follows that its α-limit set is in $B_{k+1} \backslash B_k$ so in M_{k+1}, and its ω-limit set is in $B_k \backslash B_{k-1}$ so in M_{k-1}.

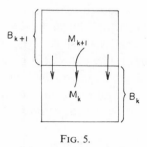

FIG. 5.

Allowing a "localization" of the above notions wherein X is replaced by a block, we give a more concrete example. Consider a flow in the plane that admits a disklike block B, as depicted in Fig. 5. Suppose there are precisely two rest points in this block, one of which is a repeller, and further suppose there is a continuous function on the block that is strictly decreasing on nonconstant orbits. Then there must be an orbit in the disk running from the repeller to the other rest point (see Fig. 6).

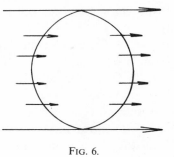

FIG. 6.

A proof could go as follows: Let x_0 be the repeller and for small ε, let $B_1 = \{x \in B \,|\, g(x) \leq g(x_0) - \varepsilon\}$. Because g is strictly decreasing on nonconstant orbits, B_1 is also a block (the new boundary points are entering points). Now points in the boundary component of B_1 near x_0 must have x_0 as their α-limit set.

This boundary component separates x_0 from the boundary of the disk B; in particular, it is not homeomorphic to a subset of an arc. But the exit set from the disk B is an arc; this implies some point in this boundary component does not leave B. Thus the ω-limit set of this point is in the disk and (again because g decreases on nonconstant orbits) therefore must be the other rest point. This orbit is the solution of the boundary-value problem.

More in line with the previous discussion of index, B_1 and $B_2 = B$ are blocks corresponding to a "local" filtration (of B). The (two) Morse sets are the rest points. The repeller must have nonzero index (the index is, in fact, a pointed two-sphere). The index of B itself (more accurately, of the maximal invariant set in B) is 0; namely, the homotopy type of the pointed point, as follows, since the set of exit points of B is a strong deformation retract of B. Since 0 cannot be a sum one of whose summands is not 0, the result follows.

Such examples arise in the problem of existence of structure for shock wave solutions of conservation laws. In another example, one finds a (local) filtration where the four Morse sets have indices Σ^3, Σ^4, Σ^5, and Σ^6 (in order increasing with the Morse sets) and where the total index is 0.* Purely algebraic considerations of a trivial sort imply the first and second as well as the third and fourth Morse sets are connected. This is because indices

---

* Here Σ^n means the pointed n-sphere.

can cancel out pairwise only in special ways—the indices of M_k, M, and M_{k+1} are related by a coexact sequence; if the index of M is 0, that of M_{k+1} is the suspension of that of M_k.

The machinery may seem a little heavy for the examples (as it is, in fact). By way of mitigation it can be pointed out that it works with no additional effort in the case where the flow in the disk (say) is weakly coupled to a system of the form $\dot{x} = Ax$, where A is a hyperbolic matrix. Here the weakly coupled flow means a small (C–0) perturbation of the product of the two flows.

Another such boundary-value problem arises in the problem of traveling wave solutions of the equation

$$u_t = u_{xx} + f(u)$$

where $f(u)$ and its integral F have graphs as depicted in Fig. 7.

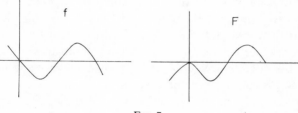

FIG. 7.

A traveling wave is a solution that depends only on $x + \theta t$, where θ is a free parameter. Thus the solution satisfies an ordinary differential equation, which, as a system, looks like

$$\dot{u} = v, \qquad \dot{v} = \theta v - f(u).$$

This system has three critical points two of which are hyperbolic and one a repelling spiral node (θ is assumed positive). The question is, Does there exist a $\theta > 0$ such that the two hyperbolic points are connected by an orbit?

The positive answer is apparent from the phase portraits for small and large θ (Fig. 8). The orbit with the arrowhead must hit the right-hand rest point for some value of θ between those pictured.

In this case the energy function $v^2/2 + F(u)$ is increasing on nonconstant orbits and provides a local filtration with three Morse sets. In order, these are M_3, repelling node; M_2, left-hand hyperbolic point; and M_1, right-hand hyperbolic point. Now, in general, if M_2 and M_1 are not connected we can choose a different filtration so that they occur in reverse order M_3,

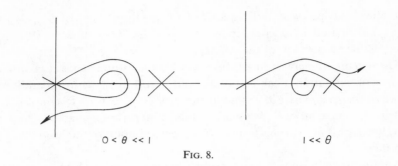

$$0 < \theta << 1 \qquad\qquad 1 << \theta$$

Fig. 8.

M_2', M_1', where $M_2' = M_1$, $M_1' = M_2$. If this were possible for all θ, then the index of M, corresponding as described in the previous example to the pair (M_2', M_3), would always be the same. For θ small, the node M_3 and the right-hand hyperbolic point M_2' are not connected, so the index of M is the sum of the indices. On the other hand for θ large there is an orbit running from the node to the right-hand hyperbolic point and one can compute that the index of M is 0 in this case (one can find a block about the connecting orbit that is a topological disk with an arc for the exit set). It follows that the interchange of M_2 and M_1 cannot be made for all θ and this implies M_2 is sometimes connected to M_1 by an orbit.

This "proof" may be too sketchy for those not familiar with the approach; however, the result is evident from Fig. 8. The point to be made might be illustrated again by the fact that the above argument will work when the equations are weakly coupled to $\dot{x} = Ax$, where A is hyperbolic; namely, the "algebraic" conditions are still satisfied.

There are several other examples of boundary-value problems where these methods prove useful, among which are the problem of existence of traveling waves of nonlinear diffusion equations of the type of Nagumo's or Hodgekin and Huxley's equation. These examples are somewhat too complicated to include here but (on paying attention to "fast" and "slow" variables) are composites of the last-mentioned example and the first one (for orbits crossing a tube).

Somewhat different qualitative features are illustrated by the classical Sturm–Liouville problem, a nonlinear version of which can be recognized in the study of the triple collision of the three-body problem.

The special aspect of Morse sets for flows on vector bundles (namely, they are subbundles for flows over chain-recurrent flows) has been studied and application made, for example, to the study of linear almost periodic differential equations. These ideas also relate to the study of hyperbolic invariant sets. For example, a compact chain-recurrent invariant set of a smooth flow is hyperbolic if and only if there are no (nontrivial) bounded orbits of the tangent flow over the invariant set. This, together with the equation for Jacobi fields, provides a proof that flows on compact surfaces of negative curvature are hyperbolic (chain recurrence is implied by volume preserving).

The notions of chain recurrence, filtrations, and Morse sets all carry over to semiflows on compact metric spaces. An example may indicate how this might be useful. Consider the equation $u_t = \Delta u + f(u)$, where the Laplacian acts on a bounded domain Ω with self-adjoint boundary conditions and $f(u)$, together with its derivative, is bounded. This equation generates a semiflow on $L^2(\Omega)$. Suppose there are three critical points [solutions of $\Delta u + f(u) = 0$] and that the linearized equations have no eigenvalues with zero real part.

Let n_1, n_2, and n_3 be the number of eigenvalues with positive real part at each of the three critical points.

The Schauder–Leray theory implies two of the n_i are even and one is odd. The Morse theory implies one of them is zero and the other two differ by 1; it also implies there are nonconstant solutions of the equation that are bounded for all time in both time directions and that connect the critical points—in fact, the dimension of the set of such orbits is at least the largest of the n_i. It is easy to conjecture that that is the exact dimension and that in general, the set of orbits that exist for all time in both time directions is contained in some finite-dimensional submanifold of $L^2(\Omega)$.

Recently, J. J. Levin has proven theorems concerning asymptotic solutions of integral equations in which the notions of filtrations and chain recurrence seem to play a role. It seems to this author that the techniques of the qualitative theory of ordinary differential equations will find an increasing number of applications in partial and integral equations.

Many authors have studied the questions briefly described here. For example, hyperbolic sets and hyperbolicity in more general situations have been studied by J. Sinai, D. Anosov, S. Smale, C. Pugh, M. Hirsch, M. Shub, R. Bowen, C. Robinson, J. Robbin, J. Moser, Z. Nitecki, J. Franks, S. Newhouse, D. Ornstein, B. Weiss, J. Palis, N. Fenechal, R. Sacker, and probably others of whom this author is not aware. At the basis of the present description lie the central ideas of T. Wazewski. The notion of filtrations has been discussed by many authors already listed above; Z. Nitecki, M. Shub, S. Smale, and J. Palis are singled out by virtue of the fact that the present author knows of their work specifically.

The notion of blocks has been discussed by R. Easton, D. Rod, R. Churchill, J. T. Montgomery, J. Selgrade, F. Wilson, J. Yorke, E. Thomas, J. Smoller, and the present author.

Chain recurrence and pseudo-orbits are discussed by J. Sinai, R. Bowen, J. Selgrade, and this author.

Almost all the things alluded to here can be found in *Mathematical Reviews* or in W. Gottschalk's, "Bibliography for Dynamical Topology" (Wesleyan University). Three items that have not yet appeared in print are R. McGehee's treatment of the triple collision in the three-body problem (*Inventiones Math.*), G. Carpenter's treatment of the Hodgkin–Huxley equation, and J. Levin's paper on Integrodifferential equations (*Advances in Mathematics*).

REFERENCES

[1] Carpenter, G., Singular perturbation solutions of nerve impulse equations, *J. Differential Equations* (to appear).

[2] Levin, J. J., On some geometric structures for integrodifferential equations, *Advances in Math.* (to appear).

[3] McGehee, R., Triple collision in the collinear three body problem, *Inventiones Math.* 27 (1974), 191–227.

The Lefschetz Fixed-Point Formula; Smoothness and Stability

MICHAEL SHUB

Department of Mathematics
Queens College, Flushing, New York

It is a great honor and pleasure for me to be speaking at the formal occasion for dedicating the Brown University Center for Dynamical Systems to the memory of Solomon Lefschetz. Lefschetz is surely one of the towering figures of mathematics in our century, and while he is mainly thought of for his contributions to algebraic geometry and topology, his contributions to dynamical systems have been great. Let me mention just two that are of special interest to me and that I shall be talking about today. First, there is the Lefschetz fixed-point formula. While in some sense this formula has become part of topology, topology used to be called analysis situs. As Dennis Sullivan put it to me one day: "Topology is the subject; analysis is the object." This, of course, is a major theme from Poincaré to the present. Lefschetz never lost sight of it. With respect to the fixed-point formula one need only read Lefschetz's discussion of it in the preface to [13] or in the introduction to [14]. In the latter, Lefschetz refers to Poincaré's interest in fixed points, which came out of dynamics. In case a vector field X has a manifold of section M, the periodic solutions of X correspond to the periodic points of the first return map on M. This is, of course, but one simple application of this very general and powerful formula. Second, I would like to mention structural stability. Structural stability was defined by Andronov and Pontryagin in 1937 [2] and they stated results on structural stability in two dimensions. While I know of no work of Lefschetz himself on structural stability we can find his influence quite clearly stated in the works of DeBaggis [6] and Peixoto [22], where some of the fundamental theorems of structural stability were first proven.

I. The Lefschetz Fixed-Point Formula and Smoothness

The Lefschetz fixed-point formula is known for quite general spaces (e.g., finite simplicial complexes) but I will restrict my discussion here to compact differentiable manifolds without boundary.

13

If 0 is an isolated fixed point of the continuous map $f\colon U \to R^m$, where U is an open subset of R^m, then the index of f at 0, $\sigma_f(0)$, is the local degree of the mapping $Id - f$ restricted to an appropriately small open set about 0. If 0 is an isolated fixed point of f^n, then $\sigma_{f^n}(0)$ is defined for all n. If $f\colon M \to M$, $f^n(p) = p$, and p is an isolated fixed point for f^n, then $\sigma_{f^n}(p)$ is defined by taking local coordinates around p. The Lefschetz numbers $L(f^n)$ are defined by the formula

$$L(f^n) \equiv \sum (-1)^i \operatorname{tr} f^n_{*i}\colon H_i(M) \to H_i(M),$$

where $H_i(M)$ is the ith homology group of M with rational coefficients and $f^n_{*i}\colon H_i(M) \to H_i(M)$ the map induced by f on these groups. $(f^n)_{*i} = (f_{*i})^n$, so it is not too difficult to compute the $L(f^n)$ once the eigenvalues of the f_{*i} are known.

The Lefschetz fixed-point formula says that the $L(f^n)$ can be computed in terms of the fixed-point indices.

(1.1) Lefschetz fixed-point theorem

$$L(f^n) = \sum_{p \in \operatorname{Fix} f^n} \sigma_{f^n}(p)$$

provided that the fixed points of f^n are isolated.

Of course, this theorem has the immediate corollary that f has a fixed point if $L(f) \neq 0$; but it says much more. Let me give an example of the type that I shall be pursuing. Let $0 < r < \infty$ and $E^r(M)$ denote the space of C^r endomorphisms $f\colon M \to M$ with the C^r-topology. $E^r(M)$ is a complete metric space, and hence the Baire category theorem applies. A subset $X \subset E^r(M)$ is said to be generic if it contains the countable intersection of open and dense sets.

By the Baire category theorem a generic set in $E^r(M)$ is dense. If a property is true for a generic subset of $E^r(M)$ we say that it is generic or that the generic f has that property. The simplest part of the Kupka–Smale theorem (see [27], for example) shows that the periodic points of period n of the generic f (i.e., the fixed points of f^n) are transversal, hence isolated, and the index of any periodic point of generic f is either $+1$ or -1. Let $N_n(f)$ be the number of period points of period n of f.

(1.2) Proposition. For the generic f in $E^r(M)$,

$$N_n(f) \geq |L(f^n)|.$$

This is an immediate consequence of the Lefschetz fixed-point formula.

Example 1. For the generic C^r, $f: S^m \to S^m$,

$$N_n(f) \geq |\deg(f)^n \pm 1|$$

where the sign is positive if m is even and negative if m is odd.

These formulas are valid for all n and thus we can try to summarize the information contained in them by a single number that measures the asymptotic exponential growth rate of the $N_n(f)$.

(1.3) Proposition. For the generic f in $E^r(M)$,

$$\lim \sup (1/n) \log N_n(f) \geq \lim \sup (1/n) \log |L(f^n)|.$$

Another way to express the number $\lim \sup (1/n) \log N_n(f)$ is that it is the reciprocal of the radius of convergence of the Artin–Mazur zeta function

$$\zeta_f(t) = \exp \left[\sum_{n=1}^{\infty} N_n(f) t^n / n \right].$$

Neither Proposition 1 nor 2 holds for all continuous, Lipschitz, or even piecewise linear mappings.

Example 2. Let $f: S^2 \to S^2$ be the map of the Riemann sphere defined by $z \to 2z^2/\|z\|$. Then $\deg(f) = 2$, but f has only two periodic points 0 and ∞. So $N_n(f) = 2$ for all n, whereas $L(f^n) = 2^n + 1$. This f is Lipschitz but not piecewise linear. There are piecewise linear homeomorphisms of manifolds even that do not satisfy this property. The Lefschetz formula is valid, of course, in the case $\sigma_{f^n}(0) + \sigma_{f^n}(\infty) = 2^n + 1$. It is not difficult to calculate that $\sigma_{f^n}(0) = 2^n$.

Let me digress for a moment to emphasize the power of formulas like Proposition (1.3). In principal if we are to know or even to estimate the growth rate of the number of periodic points of f of period n for all n, we must know f precisely and we must iterate it infinitely often. On the other hand, $\lim \sup (1/n) \log |L(f^n)|$ can be calculated. If we know the manifold M and if we know f even approximately, then by simplicial approximation the eigenvalues of the f_{*i} can be calculated, and $\lim \sup (1/n) \log |L(f^n)|$ can be determined. If, for example, the f_{*i} have a unique eigenvalue of maximal modulus, λ, then $\lim \sup (1/n) \log |L(f^n)| = \log |\lambda|$. The total effect is to replace a nonlinear problem involving infinite iterations by a finite combinatorial problem and to make it linear.

Example 2 shows that Propositions (1.2) and (1.3) fail for C^0 endomorphisms of manifolds. In fact, there are only two periodic points, while the Lefschetz numbers tend to infinity. But surprisingly enough it is not known if Proposition (1.3) is true for smooth maps.

Problem 1 [34]. Let $f: M \to M$ be smooth. Is $\limsup (1/n) \log N_n(f) \geq \limsup (1/n) \log |L(f^n)|$?

It is only recently that effects of smoothness on the Lefschetz formula have begun to be studied.

(1.4) Proposition [34]. Suppose $f: U \to R^m$ is C^1 and that 0 is an isolated fixed point of f^n for all n. Then $\sigma_{f^n}(0)$ is bounded as a function of n.

The Lefschetz trace formula now tells us:

(1.5) Corollary [34]. If $f: M \to M$ is C^1 and the Lefschetz numbers $L(f^n)$ are not bounded then the set of periodic points of f is infinite.

So the phenomenon of Example 2 cannot occur for C^1 mappings. Any C^1 mapping $f: S^2 \to S^2$ with degree 2 has infinitely many periodic points.

(1.5′) Corollary. Suppose the C^1 vector field X on N has a C^1 manifold of section M, with first return map $f: M \to M$. If the Lefschetz numbers $L(f^n)$ are unbounded, then X has infinitely many periodic solutions.

Example 3. To illustrate Corollaries (1.5) and (1.5′) we consider what has come to be called the Thom diffeomorphism of the two-dimensional torus T^2. Think of T^2 as R^2 mod the integer lattice $Z^2 \subset R^2$. $T^2 = R^2/Z^2$. The matrix $A = \binom{2\,1}{1\,1}$ has integer entries and determinant one, so A defines a map of R^2/Z^2 that is invertible, i.e., a diffeomorphism. Denote the map of T^2 it defines by A again.

$A: T^2 \to T^2$. The homology groups of T^2 with real coefficients are

$$H_0(T^2, R) = R, \qquad H_1(T^2, R) = R^2, \qquad H_2(T^2, R) = R.$$

A_{0*} and A_{1*} are the identity transformations and $A_{1*}: R^2 \to R^2$ is just A itself. Thus the Lefschetz numbers $L(A^n) = 2 - \operatorname{tr} A^n$. The eigenvalues of A are $(3 \pm \sqrt{5})/2$. So

(1) The $L(A^n)$ are unbounded; in fact,

(2) $\limsup (1/n) \log |L(A^n)| = \log [(3 + \sqrt{5})/2]$.

Consequently:

(a) Any $C^1 f$ homotopic to A has infinitely many periodic points.

(b) Any C^1 vector field X that has a C^1, T^2 manifold of section with first return map homotopic to A has infinitely many periodic solutions.

(c) The generic f in $E^r(T^2)$ that is homotopic to A has

$$\lim \sup \, (1/n) \log N_n(f) \geq \log \left[(3 + \sqrt{5})/2 \right],$$

so the number of periodic points of f of period n is growing exponentially with n.

II. Structural Stability

The theory of structural stability (and Ω-stability) for n-dimensional manifolds has made remarkable progress in the last fifteen years. We recall these concepts briefly.

(2.1) Definition. If X and Y are topological spaces, $f: X \to X$ and $g: Y \to Y$ are continuous, then f and g are topologically conjugate iff there is a surjective homeomorphism $h: X \to Y$ such that $gh = hf$.

(2.2) Definition. If X is a topological space and $f: X \to X$ is continuous, the nonwandering set of f, $\Omega(f) = \{\psi \in X \mid$ given any neighborhood U_ψ in X there is an $n > 0$ such that $f^n(U_\psi) \cap U_\psi \neq \varnothing\}$.

$\Omega(f)$ is closed, $f(\Omega(f)) \subset \Omega(f)$, and if f is a homeomorphism $f(\Omega(f)) = \Omega(f)$ and $\Omega(f) = \Omega(f^{-1})$. $\Omega(f)$ contains all the periodic points of f and all the ω-limit points of f. So as Smale has said $\Omega(f)$ is where all the action is.

(2.3) Definition. (a) $f \in \text{Diff}^r(M)$ is structurally stable if there is a neighborhood V_f of f in $\text{Diff}^r(M)$ such that any $g \in V_f$ is topologically conjugate to f.

(b) $f \in \text{Diff}^r(M)$ is Ω-stable if there is a neighborhood V_f of f in $\text{Diff}^r(M)$ such that for any $g \in V_f$, $f/\Omega(f)$ and $g/\Omega(g)$ are topologically conjugate.

Any structurally stable diffeomorphism is Ω-stable, but there are Ω-stable diffeomorphisms that are not structurally stable. This was proven by Smale, which is why he defined Ω-stability. For much more discussion of structural and Ω-stability and examples of these phenomena one should read Smale's original survey article on the subject [39], and one should consult Nitecki's book [18] as well as the further survey articles [29, 40, 42]. Lefschetz has perhaps made the strongest claim for the importance of structural stability itself [15, p. 250]:

> In a system of differential equations arising out of a practical problem the coefficients are only known approximately: various errors are inevitably involved in their determination. The phase portrait of the system must therefore be such that it is not affected by small modifications in the coefficients. In other words it must be "stable" under these conditions.

Lefschetz was, of course, talking about the definition of structural stability for differential equations, but the same may be said for diffeomorphisms. Insisting that the systems must be stable in the sense of structurally stable is perhaps extreme. Other authors, for example Thom, content themselves with less. Lefschetz was perhaps influenced by the situation for two manifolds that he was talking about, but in any case his statement gives the flavor of the importance of structurally stable systems, and much the same can be said for Ω-stable systems. Moreover, while it is to be expected that structurally stable diffeomorphisms exhibit the generic behavior of nearby systems, they will not have topological properties that are not generic. Thus structurally stable diffeomorphisms avoid pathological behavior and should be fairly simple to describe. Since $\text{Diff}^r (M)$ is separable, there are only countably many structurally stable systems up to topological conjugacy, and one might hope to find invariants that give a fairly accurate picture of the orbit structure of structurally stable systems.

To give a brief example, if $f \in \text{Diff}^r (M)$ is Ω-stable then $N_n(f) = N_n(g)$ for all n and all g in some C^r neighborhood of f. If we now apply Proposition (1.3) we have

(2.4)   **Proposition.**   If $f \in \text{Diff}^r (M)$ is Ω-stable, then

$$\limsup (1/n) \log N_n(f) \geq \limsup (1/n) \log |L(f^n)|.$$

Now we are faced with an alternative: either there is no diffeomorphism $f: T^2 \to T^2$ that is even homologous to the Thom diffeomorphism (see Example 3) and is structurally stable or there are structurally stable diffeomorphisms with an infinite number of periodic points. This was the motivation for Thom's example, and with historical hindsight the situation is clear enough. But at the time it was thought that the structurally stable diffeomorphism might be open and dense on all manifolds and would have only finitely many periodic points. The reason for this is that Peixoto had proven the openness and density of structurally stable diffeomorphisms in Diff^r (S^1) and had shown that they had only finitely many periodic points.

Actually, Smale constructed a structurally stable diffeomorphism of S^2 with infinitely many periodic points [37], and Anosov [3] later proved that the Thom diffeomorphism was structurally stable. The difference between S^1, S^2, and T^2 can be seen in the following remark. If $f: S^m \to S^m$ is a diffeomorphism, then $|L(f^n)| = 0$ or 2 for all n.

Thus Thom's example shows that infinitely many periodic points can be forced for homological reasons, because of the Lefschetz fixed-point formula. Smale's example shows that they are not pathological and that they occur in structurally stable diffeomorphisms for local as well as global topological reasons. I have dwelt on the Thom diffeomorphism because I will investigate below further homological restrictions on the growth rate of the number of periodic points of the Ω-stable diffeomorphisms that are understood. But first, I have to present the picture that evolved through the work of Smale and others describing Ω-stable and structurally stable diffeomorphisms.

(2.5) Axiom A [39]. $f \in$ Diff^r (M) satisfies Smale's Axiom A if and only if

(a) $\Omega(f)$ has a hyperbolic structure,
(b) $\Omega(f)$ is the closure of the periodic points of f.

That $\Omega(f)$ has a hyperbolic structure means that $TM|\Omega(f)$, the tangent bundle of M restricted to $\Omega(f)$, may be written as the direct sum of two Tf invariant subbundles $E^s \oplus E^u$ such that there exist constants $0 < \lambda < 1$, $0 < c$, and

$$\|Tf^n|E^s\| \le c\lambda^n \qquad \text{for} \quad n > 0,$$
$$\|Tf^n|E^u\| \le c\lambda^n \qquad \text{for} \quad n < 0.$$

When f satisfies Axiom A, Smale proved the existence of a "spectral decomposition" for $\Omega(f)$.

(2.6) Theorem [39]. If $f \in$ Diff^r (M) satisfies Axiom A, then $\Omega(f)$ is the disjoint union $\Omega(f) = \Omega_1 \cup \cdots \cup \Omega_k$, where each Ω_i is closed and invariant for f and $f|\Omega_i$ is topologically transitive.

(2.7) Corollary [39]. If $f: M \to M$ is as above, then M is the disjoint union of

(a) $M = \bigcup_{i=1}^{k} W^s(\Omega_i)$

where

$$W^s(\Omega_i) = \{\psi \in M \mid f^m(\psi) \to \Omega_i \text{ as } m \to \infty\}.$$

(b) $M = \bigcup_{i=1}^{k} W^u(\Omega_i)$

where

$$W^u(\Omega_i) = \{\psi \in M \mid f^m(\psi) \to \Omega \text{ as } m \to -\infty\}.$$

Under the circumstances we can define $\Omega_i > \Omega_j$ if $(W^u(\Omega_i) - \Omega_i) \cap (W^s(\Omega_j) - \Omega_j) \neq \varnothing$. f is said to have no cycles if $\Omega_{i_0} > \Omega_{i_1} > \cdots > \Omega_{i_j} = \Omega_{i_0}$ is impossible for any $j \geq 1$. Actually it is always impossible for $\Omega_i > \Omega_i$ for an Axiom A diffeomorphism as Smale knew, so we can say $j \geq 2$ if we want.

We can finally state Smale's Ω stability theorem:

(2.8) Theorem (Smale [41]). If f satisfies Axiom A and has no cycles then f is Ω-stable.

The converse to this theorem, which was conjectured by Smale, is one of the outstanding problems of dynamical systems.

(2.9) Conjecture (Smale [41, 42]). f is Ω-stable iff f satisfies Axiom A and the no-cycle property.

It is known that

(2.10) Theorem (Palis [20]). If f satisfies Axiom A and is Ω-stable then f has no cycles.

So the problem really is: Does Ω-stability imply Axiom A? There are some partial results here by Franks [7, 8], Guckenheimer [9], and more recently Mañé [16], but the problem is still open. Moreover, all these approaches are strictly C^1 because they rely on Pugh's general density theorem, which states:

(2.11) Theorem (Pugh [24]). For the generic f in $\text{Diff}^1 (M)$, $\Omega(f) = \overline{\text{per}\ (f)}$.

So anything that is Ω-stable in $\text{Diff}^1 (M)$ already satisfies Axiom A(b).

To know whether Pugh's closing lemma or general density theorem holds in $\text{Diff}^r (M)$, $r > 1$, is absolutely crucial for our thinking. Personally, I find the C^1 proof so difficult already that I frequently despair and try to find counterexamples.

Turning our attention to structural stability we define the strong transversality condition as in Smale [40]. If $f \in \text{Diff}^r (M)$ and $\psi \in M$,

$$W^s(\psi) = \{y \in M \,|\, d(f^n(\psi), f^n(y)) \to 0 \text{ as } n \to \infty\}$$
$$W^u(\psi) = \{y \in M \,|\, d(f^n(\psi), f^n(y)) \to 0 \text{ as } n \to -\infty\}.$$

If f satisfies Axiom A, it follows [10, 39] that $W^s(\psi)$ and $W^u(\psi)$ are $1:1$ immersed Euclidean spaces for all $\psi \in M$.

(2.12) Strong transversality condition. If f satisfies Axiom A then f satisfies the strong transversality condition iff $W^s(\psi)$ and $W^u(\psi)$ are transversal for all $\psi \in M$.

(2.13) Definition. $f \in \text{Diff}^r (M)$ is Morse–Smale iff:

(1) $\Omega(f)$ is finite,

(2) f satisfies Axiom A and the strong transversality condition.

That $\Omega(f)$ is finite means that it consists of a finite number of period points.

The first structural stability theorem for diffeomorphisms was Peixoto's.

(2.14) Theorem (Peixoto [23]). The structurally stable diffeomorphisms are open and dense in $\text{Diff}^r (S^1)$ and are the Morse–Smale diffeomorphisms.

Here the language has changed but the theorem remains the same. This theorem marked a revival in the subject. It motivated the following problems:

(1) Find open and dense or generic properties in $\text{Diff}^r (M)$.

(2) Are the Morse–Smale diffeomorphisms open and dense and structurally stable on all manifolds?

In response to (1) there are the Kupka–Smale theorem [12, 36] and Pugh's general density theorem, as well as some conjectures (see [32], for example).

In response to (2) we have seen that the Morse–Smale diffeomorphisms are not open and dense by Smale's examples and the Lefschetz formula argument. This still left:

    (2a)   Are the Morse–Smale diffeomorphisms structurally stable?

    (2b)   Are the structurally stable diffeomorphisms open and dense?

In response to (2a) there are the theorems of Palis [19] and Palis–Smale [21], which we may state as:

(2.15) Theorem (Palis–Smale). The Morse–Smale diffeomorphisms are precisely the structurally stable diffeomorphisms with a finite Ω.

And because of this theorem and the Ω-stability theorem, the conjectures of Smale [40] and Palis–Smale [21]:

(2.16) Conjecture (Palis–Smale). $f \in \text{Diff}^r(M)$ is structurally stable iff f is Axiom A and satisfies the strong transversality condition.

Robbin [25] proved half of this conjecture for f at least C^2, and Robinson [26] relaxed the hypothesis to C^1.

(2.17) Theorem (Robbin–Robinson). If f is Axiom A and satisfies the strong transversality condition, then f is structurally stable.

Once again one can prove:

(2.18) Theorem [40]. If f is structurally stable and satisfies Axiom A, then f satisfies the strong transversality condition.

So the real problem is: Does structural stability imply Axiom A?

Of course, if Ω-stability implies Axiom A so does structural stability. It is almost unthinkable that one would and the other would not.

As far as (2b) goes, Smale [38] showed in 1965 that structurally stable diffeomorphisms were not, in general, dense in $\text{Diff}^r(M)$ with the C^r topology. He invented the notion of Ω-stability and, in 1968, Abraham and Smale [1] showed that Ω-stable diffeomorphisms are also not dense in the C^r topology. There are still a few concepts of stability that might prove dense, such as topological Ω-stability [11] or the topological structural stability of attractors of Thom. Both of these concepts include diffeomorphisms that are not Ω-stable. But it is certainly conceivable that a much weaker notion of stability would be required for density.

As most of the promising approaches to the problem of whether Ω-stability implies Axiom A rely on the periodic points, it seems reasonable to pose a seemingly more general problem.

Problem 1.   Let $f \in \text{Diff}^1 (M)$. Suppose that there is a neighborhood U_f of f in $\text{Diff}^1 (M)$ such that $N_n(g) = N_n(f)$ for all $g \in U$ and all n. Does f satisfy Axiom A?

Ω-stable diffeomorphisms have this property. And the Axiom A diffeomorphisms that have this property are Ω-stable by Palis's result [20]. A diffeomorphism with this property might reasonably be called *zeta-function stable*. So we are asking: Is zeta-function stability $\Leftrightarrow$ Axiom A no cycles $\Leftrightarrow$ Ω-stability?

III. The Lefschetz Fixed-Point Formula and Stability

If f is an Axiom A diffeomorphism, Bowen [4] has proved that

$$h(f) = \lim \sup (1/n) \log N_n(f),$$

where $h(f)$ is the topological entropy of f. I will not define topological entropy here but will simply use $h(f)$ for $\lim \sup (1/n) \log N_n(f)$ when f satisfies Axiom A.

If $f: M \to M$ is continuous, we can define $s(f_*)$ to be the spectral radius of $f_*: H_*(M, R) \to H_*(M, R)$, that is, $s(f_*) = \max |\lambda|$, where the max is taken over all eigenvalues of f_{*i} and all i. It is not difficult to see that

(I) $\log s(f_*) = \lim \sup (1/n) \log |\sum \text{tr } f_{*i}^n|$.

(II)   Let $l(f) = \lim \sup (1/n) \log |\sum (-1)^i \text{tr } f_{*i}^n|$.

$\log s(f_*) \geq l(f)$ because there is no alternation of signs in (I), whereas there is in (II).

Example 4.   Let $f: M \to M$, and let $\theta: S^1 \to S^1$ be an irrational rotation. Let $g: M \times S^1 \to M \times S^1$ be $f \times \theta$. Thus by the Kunneth formula, $\text{Tr } g_{*i}^n = \text{tr } f_{*i}^n + \text{tr } f_{*i-1}^n$. Because of the alternations of signs in (II), $\sum (-1)^i \text{tr } g_{*i}^n = 0$ for all n and $l(g) = \log (0) = -\infty$. The eigenvalues of the

g_{*i} are the same as the eigenvalues of the f_{*i}, so $s(f_*) = s(g_*)$ and $\log s(f_*) = \log s(g_*)$. If $f\colon T^2 \to T^2$ were the Thom diffeomorphism then $g\colon T^3 \to T^3$, $l(g) = -\infty$, and $\log s(g_*) = \log [(3 + \sqrt{5})/2]$.

We can restate Proposition 2.4 as

(3.1) **Proposition.** If $g \in \mathrm{Diff}^r (M)$ is zeta-function stable, then

$$\lim \sup \ (1/n) \log N_n(g) \geq l(g).$$

But, in fact, one should be able to do much better:

(3.2) **Conjecture.** If $g \in \mathrm{Diff}^1 (M)$ is zeta-function stable, then

$$\lim \sup \ (1/n) \log N_n(g) \geq \log s(g_*).$$

It is difficult to see how to attack this conjecture without the tools of Axiom A and no cycles, but perhaps some sort of algebraic approximation arguments would do the trick. Bob Williams and I have recently proved this conjecture for Axiom A and no-cycle diffeomorphism by proving the following, which was conjectured in [31] and [33].

(3.3) **Theorem [35].** If $f \in \mathrm{Diff}^r (M)$ satisfies Axiom A and has no cycles, then

$$h(f) \geq \log s(f_*).$$

This theorem was previously known for Morse–Smale diffeomorphism [28, 33] and 0-dimensional Axiom A and no-cycle diffeomorphisms [5].

Let us give a corollary that includes the case of Morse–Smale diffeomorphisms.

(3.4) **Corollary.** Suppose $f \in \mathrm{Diff}^1 (M)$ is zeta-function stable and has only finitely many periodic points; then every eigenvalue of $f_*\colon H_*(M, R) \to H_*(M, R)$ is a root of unity.

To see this corollary we note that each periodic point of f must be hyperbolic. By Pugh's theorem a nearby g will have its nonwandering set equal to the set of periodic points, so it will satisfy Axiom A. By Palis's argument g will have no cycles. By the theorem $h(g) \geq \log s(g_*) = \log s(f_*)$ and $h(g) = 0$. So every eigenvalue of f_* has modulus 1. As the f_{*i} are integral matrices, every eigenvalue is a root of unity.

If we return to Example 4 for a moment we see that $g: T^3 \to T^3$ has no periodic points at all and this corresponds precisely to $L(g^n) = 0$ for all n in the Lefschetz trace formula. But any Axiom A and no-cycle f that is even homologous to g must have infinitely many periodic points; in fact, $h(f) \geq \log [(3 + \sqrt{5})/2]$. So we naturally have the question: Is g isotopic to an Axiom A and no-cycle diffeomorphism? Smale answered this question in 1971.

(3.5) Theorem (Smale [43]). Any $f \in \text{Diff}^r (M)$ is isotopic to an Axiom A and no-cycle diffeomorphism and hence an Ω-stable diffeomorphism.

This theorem was generalized in [30] with all the details carried out in [33] and [44].

(3.6) Theorem. Any $f \in \text{Diff}^r (M)$ is isotopic to an Axiom A and strong transversality diffeomorphism and hence a structurally stable diffeomorphism. Moreover, the isotopy may be chosen to be C^0 small, so the structurally stable diffeomorphisms are dense in $\text{Diff}^r (M)$ with the C^0 topology.

This theorem shows that there are no topological obstructions to structural stability and Theorem (3.3) shows that the periodic point structure of the Axiom A and strong transversality diffeomorphisms (which are the only known structurally stable diffeomorphisms) will be very rich. In many cases the growth rate of N_n will be exponential, whereas the Lefschetz formula predicts no periodic points at all. Naturally, then one is led to ask if $\log s(f_*)$ is the best lower bound one can find for $h(f)$ in terms of the homology theory of f, when f is Axiom A and no-cycles. This problem in terms of simplest diffeomorphisms is discussed at some length in [31] and [33]. Here I will restrict myself to the Morse–Smale case, because the distinction between finite and infinite periodic points is the sharpest, and the Morse–Smale diffeomorphisms are the simplest. Here there is a partial converse to Corollary (3.4).

(3.7) Theorem [33]. Let $\dim M \geq 6$ and $\pi_1(M) = 0$. If $f \in \text{Diff}^r (M)$ and every eigenvalue of f_* is a root of unity [that is, $\log s(f_*) = 0$], then there is an integer $n > 0$ such that f^n is isotopic to a Morse–Smale diffeomorphism.

A precise condition for f itself to be isotopic to a Morse–Smale diffeomorphism is given in [33], but is interesting that n cannot always be chosen equal to 1 in the theorem. There is a further obstruction related to the structure of the ideals in the ring of integers of the cyclotomic extensions of the rationals. This obstruction is still not well understood (see [33]), but it is finite.

So we are getting closer to understanding how the homological information about f gives lower bounds for $h(f)$ for any structurally stable f, and these lower bounds are sharper than the Lefschetz fixed-point formula will give us. Lower bounds are all that is possible. Smale's examples can be adapted to show that if dim $M \geq 2$, $f \in \text{Diff}^r (M)$, and $k > 0$, then there is an Axiom A and strong transversality diffeomorphism g that is isotopic to f and $h(g) > k$.

Let me close with a small discussion of a possible generalization of Theorem (3.3) in terms of topological entropy, which would give information about the orbit structure of every smooth map in terms of homological information. Here $h(f)$ will stand for the topological entropy of any continuous function; for definitions and a further discussion see [31].

(3.8) Conjecture [31]. Let $f: M \to M$ be a smooth map (or diffeomorphism); then $h(f) \geq \log s(f_*)$.

Here there is some information known. Example 2 shows that the conjecture fails for continuous maps; in fact, it fails for some piecewise linear homeomorphisms of manifolds, so smoothness is essential. We do know, however, as a special case of work of Manning that:

(3.9) Theorem (Manning [17]). Let $f: M \to M$ be continuous. Then $h(f) \geq \log s(f_{*1})$, where $f_{*1}: H_{*1}(M, R) \to H_{*1}(M, R)$.

(3.10) Corollary (Manning [17]). Let $f: M \to M$ be a homeomorphism, and let dim $M \leq 3$. Then $h(f) \geq \log s(f_*)$.

REFERENCES

[1] Abraham, R., and Smale, S., Nongenericity of Ω-stability, in "Global Analysis" (*Proc. Symp. Pure Math. 1970* **14**), p. 5. Amer. Math. Soc., Providence, Rhode Island.

[2] Andronov, A. A., and Pontryagin, L. S., Systèmes grossiers, *Dokl. Akad. Nauk.* **14** (1937), 247.

[3] Anosov, D. V., Geodesic flows on compact manifolds of negative curvature, *Trudy Mat. Inst. Steklov* **90** (1967); *Proc. Steklov Inst.*, Amer. Math. Soc. transl. (1969).

[4] Bowen, R., Topological entropy and Axiom A, see ref. [1], p. 23.

[5] Bowen, R., Entropy versus homology for certain diffeomorphisms, *Topology* **13** (1974), 61.

[6] DeBaggis, H., Dynamical systems with stable structure, *in* "Contributions to the Theory of Non-linear Oscillations" (Ann. Math. Studies, No. 29), Vol. 2, p. 37. Princeton Univ. Press, Princeton, New Jersey, 1952.

[7] Franks, J., Necessary conditions for stability of diffeomorphisms, *Trans. Amer. Math. Soc.* (1971), 158.

[8] Franks, J., Differentiably Ω-stable diffeomorphisms, *Topology* **11** (1972), 107.

[9] Guckenheimer, J., Absolute Ω-stable diffeomorphisms, *Topology* **11** (1972), 195.

[10] Hirsch, M., Palis, J., Pugh, C., and Shub, M., Neighborhoods of hyperbolic sets, *Invent. Math.* **9** (1969–70), 121.

[11] Hirsch, M., Pugh, C., and Shub, M., Invariant manifolds (to appear as Springer Lecture Notes).

[12] Kupka, I., Contribution a la theorie des champs generique, *Contribut. Differential Equations* **2** (1963), 457.

[13] Lefschetz, S., "Topology," Amer. Math. Soc. Colloq. Publ., Vol. 12. New York, 1930.

[14] Lefschetz, S., "Introduction to Topology." Princeton Univ. Press, Princeton, New Jersey, 1949.

[15] Lefschetz, S., "Differential Equations: Geometric Theory," 2nd ed. Wiley (Interscience), New York, 1963.

[16] Mañé, R., Expansive diffeomorphisms (to appear).

[17] Manning, A., Entropy and the first homology group (to appear).

[18] Nitecki, Z., "Differential Dynamics." MIT Press, Cambridge, Massachusetts, 1972.

[19] Palis, J., On Morse–Smale dynamical systems, *Topology* **8** (1969), 385.

[20] Palis, J., A note on Ω-stability, see ref. [1], p. 221.

[21] Palis, J., and Smale, S., Structural stability theorems, see ref. [1], p. 223.

[22] Peixoto, M., On structural stability, *Ann. Math.* **69** (1959), 199.

[23] Peixoto, M., Structural stability on two dimensional manifolds, *Topology* **1** (1962), 101.

[24] Pugh, C. C., An improved closing lemma and a general density theorem. *Amer. J. Math.* **89** (1967), 1010.

[25] Robbin, J., On structural stability, *Ann. Math.* **94** (1971), 447.

[26] Robinson, C., Structural stability of C^1 diffeomorphisms (to appear in *J. Differential Equations*).

[27] Shub, M., Endomorphisms of compact differentiable manifolds, *Amer. J. Math.* **XCI** (1969), 175.

[28] Shub, M., Morse–Smale diffeomorphisms are unipotent on homology, *in* "Dynamical Systems" (M. Peixoto, ed.), p. 489. Academic Press, New York, 1973.

[29] Shub, M., Stability and genericity for diffeomorphisms, see ref. [28], p. 493.

[30] Shub, M., Structurally stable diffeomorphisms are dense, *Bull. Amer. Math. Soc.* **78** (1972), 817.

[31] Shub, M., Dynamical systems, filtrations and entropy, *Bull. Amer. Math. Soc.* **80** (1974), 27.

[32] Shub, M., and Smale, S., Beyond hyperbolicity, *Ann. Math.* **96** (1972), 587.

[33] Shub, M., and Sullivan, D., Homology theory and dynamical systems, *Topology* **14** (1975), 109.

[34] Shub, M., and Sullivan, D., A remark on the Lefschetz fixed point formula for differentiable maps, *Topology* **13** (1974), 189.

[35] Shub, M., and Williams, R., Entropy and stability (to appear in *Topology*).

[36] Smale, S., Stable manifolds for differential equations and diffeomorphisms, *Ann. Scuola Norm. Sup. Pisa* **17**(3) (1963), 97.

[37] Smale, S., Diffeomorphisms with many periodic points, *in* "Differential and Combinatorial Topology," p. 63. Princeton Univ. Press, Princeton, New Jersey, 1965.

[38] Smale, S., Structurally stable systems are not dense, *Amer. J. Math.* **88** (1966), 491.

[39] Smale, S., Differentiable dynamical systems, *Bull. Amer. Math. Soc.* **73** (1967), 747.

[40] Smale, S., Notes on differentiable dynamical systems, see ref. [1], p. 277.

[41] Smale, S., The Ω-stability theorem, see ref. [1], p. 289.

[42] Smale, S., Stability and genericity in dynamical systems, "Seminaire Bourbaki," Vol. 1969–70, Lecture Notes in Math. No. 180. Springer-Verlag, Berlin, 1974.

[43] Smale, S., Stability and isotopy in discrete dynamical systems, see ref. [28], p. 527.

[44] Zeeman, E. C., C^0-density of stable diffeomorphisms and flows, *in* "Colloquium on Smooth Dynamical Systems." Dept. of Math., Univ. of Southampton.

Chapter 2: GENERAL THEORY

Nonlinear Oscillations in the Frame of Alternative Methods*

LAMBERTO CESARI

Department of Mathematics
University of Michigan, Ann Arbor, Michigan

1. Introduction

The injection of methods of functional analysis (Cesari [5, 6], 1963) in the classical bifurcation process of Poincaré [33], Lyapunov [24], and Schmidt [35] and extensive subsequent work have made this process a fine tool in nonlinear analysis, particularly in the difficult problems "at resonance" in the usual terminology. The general theory that has ensued, with all its variants and ramifications, is often referred to as bifurcation theory, or alternative methods.

A great many ideas have been brought to bear in this theory in the last few years, such as Schauder's fixed point theorem (Cesari [5], Landesman and Lazer [21], Williams [38]); Banach's fixed point theorem (Cesari [5, 6] and much subsequent work); invariance properties of topological degree (Cesari [5], Knobloch [20], Cronin [10], Williams [37], Mawhin [26, 27]); Brouwer's fixed point theorem, particularly C. Miranda's equivalent form (Cesari [6], Knobloch [20]); the theory of monotone and maximal monotone operators in Banach spaces (Gustafson and Sather [11], Cesari and Kannan [9]); Schauder's principle of invariance of domain (Kannan [17]); implicit function theorem and Newton's "polygon" method in Banach spaces (Sather [34]).

For self-adjoint problems we have presented results of the theory in a recent paper [8]. In the present paper, therefore, we shall mention recent aspects of the theory, mainly for non-self-adjoint problems. We shall present

* This research was partially supported by AFOSR Research Project 71-2122 at the University of Michigan.

first the modification of Cesari's alternative scheme proposed by Hale, Bancroft, and Sweet [15] for non-self-adjoint problems.

2. An Alternative Scheme

Let us consider the equation

$$Ex = Nx, \tag{1}$$

where E is a linear operator whose domain $\mathscr{D}(E)$ is some subspace of a Banach space X and whose range $\mathscr{R}(E)$ lies in another Banach space Y. In (1), N denotes an operator, not necessarily linear, with $\mathscr{D}(N) \subset X$, $\mathscr{R}(N) \subset Y$, $\mathscr{D}(E) \cap \mathscr{D}(N) \neq \varnothing$.

We simply assume here that there are projection operators $P: X \to X$, $Q: Y \to Y$ and a linear operator $H: \mathscr{D}(H) \to X$, $\mathscr{D}(H) \subset Y$, such that

(k_1) $H(I - Q)Ex = (I - P)x$ for all $x \in \mathscr{D}(E)$,

(k_2) $QEx = EPx$ for all $x \in \mathscr{D}(E)$,

(k_3) $EH(I - Q)Nx = (I - Q)Nx$ for all $x \in \mathscr{D}(E) \cap \mathscr{D}(N)$,

and where we assume that $\mathscr{R}(P) \subset \mathscr{D}(E)$, $\mathscr{R}(H) \subset \mathscr{D}(E)$, $(I - Q)\mathscr{R}(E) \subset \mathscr{D}(H)$.

Thus P and Q are linear bounded idempotent operators ($P^2 = P$, $Q^2 = Q$); $X_0 = PX$, $X_1 = (I - P)X$ are linear subspaces of X; $Y_0 = QY$, $Y_1 = (I - Q)Y$ are linear subspaces of Y; and $H: \mathscr{D}(H) \to X$ can be thought of, from (k_1) and (k_3), as a partial inverse of E.

Note that for any $\bar{x}$ such that $E\bar{x} = 0$, we have from (k_1) that $(I - P)\bar{x} = 0$, or $P\bar{x} = \bar{x}$, or $\bar{x} \in PX = X_0$. Thus, the null space, or kernel, of E must be contained in X_0, or briefly, ker $E \subset X_0$.

It is apparent from (k_{13}) that it is enough to define H on $(I - Q)Y = Y_1$, namely, we can well assume

$$\mathscr{D}(H) \subset Y_1 \quad \text{with} \quad \mathscr{R}(H) \subset \mathscr{D}(E) \cap X_1.$$

Moreover, if we restrict H to be a partial inverse of E, then, besides $\mathscr{R}(H) \subset \mathscr{D}(E)$, we need also $\mathscr{D}(H) \subset \mathscr{R}(E)$, and this implies that $(I - Q)\mathscr{R}(E) \subset \mathscr{R}(E) \cap Y_1$. However, since $I - Q$ acts as the identity operator on Y_1, we must also have $(I - Q)\mathscr{R}(E) \supset \mathscr{R}(E) \cap Y_1$. Thus,

$$\mathscr{D}(H) = (I - Q)\mathscr{R}(E) = \mathscr{R}(E) \cap Y_1.$$

Finally, from (k_3) we need $(I - Q)\mathscr{R}(N) \subset \mathscr{D}(H)$. This will certainly be the case if $\mathscr{R}(E) \cap Y_1 = Y_1$, and then

$$\mathscr{D}(H) = (I - Q)\mathscr{R}(E) = \mathscr{R}(E) \cap Y_1 = Y_1.$$

Finally, if Q and E are so related that $\mathscr{R}(E) = Y_1 = (I - Q)Y$, then $Y_0 = QY$ is the complementary space of Y_1 in Y, and Y_0 is then the cokernel of E, or coker $E = Y_0 = QY$. Then we have also $QEx = 0$, $EPx = 0$, for all $x \in \mathscr{D}(E)$, and $X_0 = \ker E$.

(2.i) Theorem. If (k_{123}) are satisfied, then $Ex = Nx$ for some $x \in \mathscr{D}(E) \cap \mathscr{D}(N)$ if and only if

$$x = Px + H(I - Q)Nx, \tag{2}$$

$$Q(Ex - Nx) = 0. \tag{3}$$

Proof.   First, $Ex = Nx$ implies $(I - Q)(Ex - Nx) = 0$ and

$$Q(Ex - Nx) = 0.$$

Since $Ex = Nx \in \mathscr{R}(E)$, we have $(I - Q)Ex = (I - Q)Nx \in (I - Q)\mathscr{R}(E) = \mathscr{D}(H)$. By applying H to the equation $(I - Q)(Ex - Nx) = 0$, we have $H(I - Q)Ex = H(I - Q)Nx$, and by (k_1) also $(I - P)x = H(I - Q)Nx$, and finally $x = Px + H(I - Q)Nx$, and system (2), (3) is satisfied. Conversely, if system (2), (3) is satisfied, then $(I - P)x = H(I - Q)Nx$, $(I - Q)Nx \in \mathscr{D}(H)$, $H(I - Q)Nx \in \mathscr{R}(H) \subset \mathscr{D}(E)$, thus $(I - P)x \in \mathscr{D}(E)$, and by applying E to this equation, also $E(I - P)x = EH(I - Q)Nx$. By (k_2) and (k_3) we have then $Ex - QEx = (I - Q)Nx$, and finally

$$Ex - Nx = Q(Ex - Nx) = 0.$$

Equations (2) and (3) can also be written in the form

$$x = x^* + H(I - Q)Nx, \tag{2'}$$

$$Q(Ex - Nx) = 0, \tag{3'}$$

where $x^* = Px$ is an element of $X_0 = PX$. As we shall see, for a large class of problems it is possible to choose P, Q, H in such a way that Eq. (2') is uniquely solvable for every $x^* \in X_0$. If $x = x(x^*)$, $x^* \in X_0$, is the unique solution of (2'); then (3') reduces to an equation $Mx^* = 0$ in the unknown x^* in X_0. Often, X_0 is a finite-dimensional space, and thus the original problem (1) is reduced to an *alternative problem* $Mx^* = 0$ in a finite-dimensional space X_0 (alternative methods). Equation (2') is said to be the *auxiliary* equation, and (3′) the *bifurcation,* or *determining* equation. Actually, the auxiliary equation (2′) is of the form $x - H(I - Q)Nx = x^*$, or $(I + KN)x = x^*$, that is, a Hammerstein equation.

3. Reduction to an Alternative Problem by Contraction Maps

We shall now make the following assumption:

(k_4) There are positive, not decreasing functions $\alpha(\rho)$, $\beta(\rho)$ such that for all x, x' with $\|x\|$, $\|x'\| \leq \rho$ we have

$$\|H(I - Q)Nx\| \leq \beta(\rho),$$
$$\|H(I - Q)Nx - H(I - Q)Nx'\| \leq \alpha(\rho)\|x - x'\|.$$

Thus, if $\|H(I - Q)\| \leq K$, and $\|Nx - Nx'\| \leq L\|x - x'\|$ for $\|x\|$, $\|x'\| \leq \rho$, then we can take $\alpha(\rho) = LK$.

Next, for any two positive constants c, d, we consider the sets

$$V(c) = \{x^* \in P(X) \mid \|x^*\| \leq c\},$$
$$S(x^*, c, d) = \{x \in X \mid Px = x^*, \|x\| \leq d\},$$

with $x^* \in V(c)$.

(3.i)   **Theorem.**   Under the hypotheses (k_{1234}), and positive constants $c < d$, with $\alpha(d) < 1$, $\beta(d) \leq d - c$, the map $T: X \to X$ defined by $T = P + H(I - Q)N$ maps $S(x^*, c, d)$ into itself and is a contraction.

Proof.   For x, $x' \in S(x^*, c, d)$, we have

$$Px = Px' = x^*, \qquad \|x^*\| \leq c,$$

$$\|Tx\| = \|Px + H(I - Q)Nx\| \leq \|x^*\| + \|H(I - Q)Nx\| \leq c + (d - c) = d,$$

$$\|Tx - Tx'\| = \|H(I - Q)(Nx - Nx')\| \leq \alpha(d)\|x - x'\|.$$

Thus, T is a contraction map.

Under the above hypotheses and for any $x^* \in V(c)$ there is, therefore, by Banach's theorem, a unique fixed point $x = Tx = F(x^*) \in S(x^*, c, d)$ with $Px = PTx = x^*$, which satisfies the auxiliary equation (2′). Thus, system (2′), (3′) is reduced to the single (bifurcation) equation $Mx^* = 0$, or

$$Q(E - N)F(x^*) = 0, \qquad x^* \in V(c). \tag{4}$$

Of course, in applications where $\| \ \|$ is, for instance, a square norm or an L_1-norm, it may occur that N does not satisfy the boundedness and Lipschitz requirement (k_4). Indeed, if N is the Nemitsky operator corresponding to some real function f, which may be a polynomial, then condition (k_4) and corresponding Theorem (3.i) can be shown to hold in a local sense by a number of devices, for instance, the use of an associated second norm, say a sup norm, and by restricting the search of solutions in suitable domains of the function space X, say a ball in the Sup norm (Cesari [5, 6]). Other devices are mentioned in [8] (see also Section 11).

Statement (3.i) has been repeatedly used in solving the auxiliary equation, where d is either an a priori bound for the solution, or a number to be determined at the numerical stage.

That P, Q, H can be chosen in such a way to make $k = \|H(I - Q)\|$ as small as we want, and then $\alpha(\rho) = Lk < 1$, has been shown by Cesari ([6]; see also [8]) for self-adjoint problems ($X = Y = S$ a Hilbert space), and by Hale ([13]; see Section 5) also for non-self-adjoint problems (X Hilbert, Y Banach).

These results are relevant because they show that it is "in general" possible to determine P, Q, H, and thus perform the splitting of Eq. (1) into auxiliary and bifurcation equations (2) and (3), in such a way that the auxiliary one can be solved by the Banach fixed point theorem, and thus problem (1) is theoretically reducible to the sole bifurcation equation (the alternative problem, often in a finite-dimensional space). An analogous remark holds in relation to monotone operators (see the end of Section 6).

It is also relevant to say here that the bifurcation equation can often be handled by topological degree and Leray–Schauder techniques [23].

4. Topological Degree Argument

Let us replace problem (1) by, say,

$$Ex = \lambda Nx, \qquad 0 \le \lambda \le 1, \tag{5}$$

and thus (2) and (3) are replaced by

$$x = Tx = Px + \lambda H(I - Q)Nx, \tag{6}$$

$$Q(Ex - \lambda Nx) = 0. \tag{7}$$

We assume here that the space $Y_0 = QY$ is finite-dimensional. We also assume that an a priori bound for the solution x of problem (5) can be found that is independent of λ, $0 \leq \lambda \leq 1$. Let us choose P, Q, H in such a way that T is a contraction map in the entire solid ball B with center at the origin, which is determined by the a priori bound. If $x = x(x^*, \lambda)$ is the solution of the auxiliary equation, and we write briefly the bifurcation equation (7) in the form $S_\lambda = S(x^*, \lambda) = 0$, then $S(x^*, \lambda) \neq 0$ on the boundary ∂B^* of the projection $PB = B^*$ of B. Thus, the topological degree deg $(0; S_\lambda, B^*)$ is defined and does not depend on λ, $0 \leq \lambda \leq 1$. Actually, for $\lambda = 0$ the problem is linear, and deg $(0; S_0, B^*)$ may be easy to determine. If the latter is nonzero, then deg $(0, S_1, B^*) = $ deg $(0; S_0, B^*) \neq 0$, the bifurcation equation is solvable, and so is the given problem (1).

This Leray–Schauder type argument, in connection with alternative methods, has been consistently used by Mawhin (see, e.g., [26]).

In connection with this type of argument, Williams [37] considered the self-adjoint case $X = Y$, $P = Q$, S_0 finite-dimensional, T a contraction. In this situation Williams noted that the fixed points $z = Wz$ of the map W defined by $Wz = Pz + H(I - P)Nz - P(E - N)Fz$ are exactly the solutions of the original problem (1). The map W may not be compact, but the analogous map W', defined by $W'z = WFz$, is compact, so that a Leray–Schauder degree i_{LS} for the map $I - W'$ can be defined. Williams proved that i_{LS} is then equal to the Brouwer degree of the finite-dimensional map S_1 corresponding to the bifurcation equation. In the same direction, Mawhin [27] then considered the non-self-adjoint case with $X_0 = PX = \ker E$, $Y_0 = QY = \operatorname{coker} E$, X_0 finite dimensional. Then, for any map $S: Y_0 \to X_0$ with $S^{-1}(0) = 0$, the fixed points $z = W_S z$ of the map W_S, defined by $W_S z = Pz + H(I - Q)Nz + SQNz$, are the solutions of (1). Under suitable hypotheses of boundedness on N this map W_S is compact, and a Leray–Schauder degree i_{LS} for $I - W_S$ can be defined (coincidence degree).

5. Hale's Statement Concerning the Norm of H

(a)   Let us assume that E, P, Q are given as requested, so that

$$\mathscr{R}(P) \subset \mathscr{D}(E), \qquad \ker E \subset X_0 = PX,$$

$$(I - Q)\mathscr{R}(E) = \mathscr{R}(E) \cap Y_1, \qquad QEx = EPx.$$

Let E' be the map E restricted to $D' = \mathscr{D}(E') = X_1 \cap \mathscr{D}(E)$, and let $R' = \mathscr{R}(E')$, so that $E': D' \to R'$ is one–one and onto. Then, necessarily E' has an inverse $H: R' \to D'$, which is linear and onto. As a consequence of the closed-graph theorem, we also have

(5.i)   If the linear map E has closed graph and closed range, then H is a bounded linear map.

Concerning estimates on the norm $\|H\|$ of H, the following simple remark is relevant:

(5.ii)   We have $\|H\| \le k$ if and only if $\|(I - P)x\| \le k\|Ex\|$ for all $x \in \mathscr{D}(E)$.

Indeed, for $y \in R'$ and $x \in D'$ with $y = E'x$, we have $Hy = HE'x = (I - P)x$. Then, $\|(I - P)x\| \le k\|Ex\|$ if and only if $\|Hy\| \le k\|y\|$. Thus, we need not compute H to estimate its norm $\|H\|$. The latter can be computed directly from E and P.

(b)   Let X and Y be Banach spaces, $E: \mathscr{D}(E) \to Y$ a linear operator from a linear subspace $\mathscr{D}(E)$ of X with closed graph and closed range, and $\mathscr{D}(E)$ everywhere dense in X. Let us make the assumption that there are projector operators $P_0: X \to X$, $Q_0: Y \to Y$ such that $\ker E = P_0 X = X_0$, $\operatorname{coker} E = Q_0 Y = Y_0$, and let $X = X_0 + X_1$, $Y = Y_0 + Y_1$ (direct sums), $X_1 = (I - P_0)X_0$, $Y_1 = (I - Q_0)Y_0$. Let $H_0: Y_1 \to X_1 \cap \mathscr{D}(E)$ denote the inverse operator of $E': X_1 \cap \mathscr{D}(E) \to Y_1$. Thus H_0 is a linear bounded operator.

Then, for every $x \in \mathscr{D}(E)$ we certainly have $Q_0 Ex = 0$, $EP_0 x = 0$, $E(I - P_0) = E$, $H_0 Ex = H_0 E(I - P_0)x = (I - P_0)x$, and $EH_0 y = y$ for all $y \in Y_1$, that is, (k_{123}) hold. Consequently, as in (1), we can derive from $Ex = Nx$ Eqs. (2) and (3), which now have the simpler form

$$x = P_0 x + H_0(I - Q_0)Nx, \qquad Q_0 Nx = 0. \tag{8}$$

If $S: X \to X$ is any projector operator with $SX \subset X_1 \cap \mathscr{D}(E)$, $SP_0 = 0$, then we also have $P_0 S = 0$, and $P = P_0 + S: X \to X$ is a projector operator. Moreover, $SP_0 = 0$, $(I - P)P_0 = 0$, and by applying $I - P$ and S to the first of relations (8), then the same relations (8) become

$$x = Px + (I - P)H_0(I - Q_0)Nx, \tag{9}$$

$$Sx = SH_0(I - Q_0)Nx, \qquad Q_0 Nx = 0. \tag{10}$$

Relation (9) is the auxiliary equation; relations (10) together are the bifurcation equation. It is clear that, under the same assumptions as above, relations (9) and (10) imply $Ex = Nx$. Thus, $Ex = Nx$ and system (9), (10) are equivalent.

(5.iii)   (Hale [13]).   Under the hypotheses above with X a Hilbert space and H_0 compact, given $k > 0$, it is possible to choose a projector operator

$S: X \rightarrow X$, $\|S\| = 1$, with finite-dimensional range, such that

$$SX \subset X_1 \cap \mathscr{D}(E),$$

$SP_0 = 0$, and if $P = P_0 + S$, then $\|(I - P)H_0\| < k$.

Thus, if N is Lipschitzian of constant L in the ball in X with center at the origin and radius ρ, then for all $x_1, x_2 \in X$, $\|x_1\|, \|x_2\| \le \rho$, we have

$$\|(I - P)H_0(I - Q_0)(Nx_1 - Nx_2)\| \le k\|I - Q_0\|L\|x_1 - x_2\|,$$

and $\alpha(\rho) = k\|I - Q_0\|L$ can be made as small as we want. If $\|(I - Q_0)Nx\| \le M$ for all $\|x\| \le \rho$, then $\|(I - P)H_0 Nx\| \le kM = \beta(\rho)$. Consequently, for $0 < c < d \le \rho$, $\alpha(\rho) < 1$, $\beta(\rho) \le d - c$. and for every $x^* \in PX$, $\|x^*\| \le c$, the mapping T defined by

$$Tx = x^* + (I - P) H_0(1 - Q_0)Nx$$

is a contraction on the subset $[x \in X, Px = x^*, \|x\| \le d]$ of X. Thus Eq. (1) is reducible to an alternative problem. For a proof of (5.iii), see Hale [13, p. 13].

6. Reduction to an Alternative Problem by Monotone Operators

If $Y = X^*$ is the dual space of the Banach space X, and we denote by $\langle x, y \rangle$ the pairing of x and y, or the linear operation $y(x)$, $x \in X$, $y \in X^*$, then a linear operator $T: X \rightarrow X^*$ is said to be positive provided $\langle x, Tx \rangle \ge 0$ for all $x \in X$. Let $Y = X^*$ and let us use the same notation as in (5.ii).

(6.i) We have $\langle x, -Ex \rangle \ge 0$ for all $x \in D'$ if and only if $\langle -Hy, y \rangle \ge 0$ for all $y \in R'$. If $\langle x, -Ex \rangle \ge \mu\|x\|^2$ for some $\mu > 0$, then

$$\langle -Hy, y \rangle \ge \mu\|Hy\|^2. \tag{11}$$

Indeed, as in (5.ii), we have $\langle -Hy, y \rangle = \langle x, -Ex \rangle$. Thus, again, we need not compute H to verify its positive character, which instead can be derived from E.

In our previous paper [8] we have shown that, for self-adjoint problems (in particular, $X = Y = S$, a Hilbert space), it is always possible to choose P, Q, H, in such a way that (11) is satisfied [8, formula (8)]. Also, we have shown in [8] the relevance of property (11) in proving the solvability of both the auxiliary and bifurcation equations, and therefore of the given problem (1), in self-adjoint problems (with $X = Y = S$, a Hilbert space), under monotonicity hypotheses, or analogous ones, on N. (See Cesari and Kannan [9] and also [8], where full proofs are given based on Brézis, Crandall, and Pazy's theory of monotone and maximal monotone operators.)

As we shall see in Section 7 relation (11) holds also for non-self-adjoint elliptic differential operators E, ordinary and partial, under very mild assumptions and when suitable operators P, Q, H are chosen, as Osborn and Sather have proved [32]. Here we give a statement that is analogous to another one recently proved by Osborn and Sather [32] concerning the solvability of the auxiliary equation under monotonicity conditions on N. This statement (6.ii) is an extension to non-self-adjoint problems of Cesari and Kannan's theorem (8.i) of [8].

To this effect let us assume that the operators P, Q, H have been chosen so that (k_{123}) hold and that there exists some constant $\mu > 0$ such that the following strong positivity property holds:

$$\langle x, -Ex \rangle \geq \mu \|x\|^2 \quad \text{for} \quad x \in D \cap X_1, \qquad X_1 = (I - P)X,$$

or   (12)

$$\langle -Hy, y \rangle \geq \mu \|Hy\|^2 \quad \text{for} \quad y \in Y_1, \qquad Y_1 = (I - Q)Y.$$

Concerning $N: \mathscr{D}(N) \to Y$, $X = \mathscr{D}(N)$, we shall assume that N is hemicontinuous, and that N is monotone, that is,

$$\langle Nx_1 - Nx_2, x_1 - x_2 \rangle \geq 0 \qquad \text{for all} \quad x_1, x_2 \in X. \qquad (13)$$

Also we shall assume that P and Q have been so chosen that

$$\langle x, y \rangle = 0 \qquad \text{for all} \quad x \in X_1 = (I - P)X \quad \text{and} \quad y \in Y_0 = QY. \quad (14)$$

(6.ii)   Under hypotheses (k_{123}) and (12)–(14), with X reflexive and $Y = X^*$, the auxiliary equation (2′) has a unique solution

$$x = F(x^*), \qquad x \in x^* + (X_1 \cap D) \quad \text{for every} \quad x^* \in X_0.$$

Proof.   First we note that for the linear bounded operator $-H(I - Q)$: $Y \to X_1 \cap D$ we have

$$\langle -H(I - Q)y, y \rangle = \langle -H(I - Q)y, (I - Q)y \rangle + \langle -H(I - Q)y, Qy \rangle$$
$$\geq \mu \| -H(I - Q)y \|^2 \qquad (15)$$

for all $y \in Y$, because of hypotheses (12) and (14). Actually, $-H$ is a one–one map from Y_1 to $X_1 \cap D$. If for any $x \in X_1 \cap D$ we write $x = -Hy$, $y \in Y_1$,

then $-H(I - Q)(y + Y_0) = x$ and $[-H(I - Q)]^{-1}x = y + Y_0$. Thus,
$[-H(I - Q)]^{-1}$ is a set-valued map defined for each $x \in X_1 \cap D$ with values
$y + Y_0$. Also, we see that the auxiliary equation (2′) or

$$x - x^* - H(I - Q)Nx = 0 \tag{16}$$

implies

$$[-H(I - Q)]^{-1}(x - x^*) + Nx \ni 0, \qquad x = x^* + (X_1 \cap D). \tag{17}$$

Conversely, if (17) holds, then $x - x^* - H(I - Q)x$ is in X_0, and since
$x - x^* \in X_1 \cap D$, and $X_0 \cap X_1 = 0$, we derive $x - x^* - H(I - Q)Nx = 0$.
Thus, Eqs. (16) and (17) are equivalent.

Equation (17) is of the form $(A + B)x \ni 0$, where $A: x^* + (X_1 \cap D)$,
$B: X \to X$ are the operators defined by $Ax = [-H(I - Q)]^{-1}(x - x^*)$ for
$x \in x^* + (X_1 \cap D)$, and by $Bx = Nx$ for $x \in X$, and A is a set-valued map.
Since $-H(I - Q): Y \to X_1 \cap D \subset X$ is monotone and continuous, then
$-H(I - Q)$ is maximal monotone. By hypothesis, N is monotone and hemicontinuous. Thus, $B = N$ is also maximal monotone. Finally, $\mathscr{D}(A) =$
$X_1 \cap D \subset X = \mathscr{D}(B)$, where X is the whole space. Then, $A + B$ is maximal
monotone (see, e.g., Brézis, Crandall, and Pazy [4] for a recent proof). Let
us prove that A is coercive. Indeed, for $y \in Ax = [-H(I - Q)]^{-1}x$, we
have $x = -H(I - Q)y$, $\|x\| = \|-H(I - Q)y\|$, and by (15) also

$$\|x\|^{-1}\langle x, y \rangle = \|x\|^{-1}\langle -H(I - Q)y, y \rangle \geq \|x\|^{-1}\mu\|-H(I - Q)y\|^2 = \mu\|x\|,$$

and thus $\|x\|^{-1}\langle x, Ax \rangle \to +\infty$ as $\|x\| \to +\infty$. Since A is coercive, so is
$A + B$. We conclude that $\mathscr{R}(A + B) = Y$. Hence, the equation $(A + B)x \ni 0$
is solvable, that is, (17) is solvable, and (16) is solvable.

Let us prove that, for every $x^* \in X_0$, (16) has a unique solution. Indeed,
if x_1, x_2 were two solutions, then

$$x_1 - x^* = H(I - Q)(Nx_1), \qquad x_2 - x^* = H(I - Q)(Nx_2)$$

and

$$\begin{aligned}
0 &\leq \langle Nx_1 - Nx_2, x_1 - x_2 \rangle \\
&= -\langle Nx_1 - Nx_2, -H(I - Q)Nx_1 + H(I - Q)Nx_2 \rangle \\
&\leq -\mu\|-H(I - Q)Nx_1 + H(I - Q)Nx_2\|^2 = -\mu\|x_1 - x_2\|^2,
\end{aligned}$$

where $\mu > 0$ and we have used property (12) of the operator H. Thus,
$x_1 = x_2$. We have proved that the auxiliary equation (2′), or (16), has a
unique solution $x \in x^* + (X_1 \cap D)$ for every $x^* \in X_0$.

It will be shown elsewhere that the bifurcation equation can also be discussed under monotonicity hypotheses, or analogous ones, on N, thereby extending Cesari and Kannan's results in [8, Section 8].

7. Considerations on the Positivity Property

Let $X = S_1$, $Y = S$ be real Hilbert spaces, $X \subset Y$, with inner products $(\ ,\ )_1$ and $(\ ,\ )$, and norms $\|\ \|_1$ and $\|\ \|$, respectively. Let us assume that X is everywhere dense in Y and that the identity map $j: X \to Y$ is compact. Let $E: D \to Y$ be a linear operator, with domain $D = \mathscr{D}(E) \subset X$, D dense in X, with compact resolvent $R_\lambda(E)$, satisfying

$$\text{(A)} \quad (x, -Ex) \geq a\|x\|_1^2 - b\|x\|^2 \tag{18}$$

for all $x \in D$, and some constants $a > 0$, b real.

Thus, we assume that $-E$ has a positivity property similar to the Gårding inequality for the Dirichlet problem, and we know that uniformly elliptic differential operators (ordinary and partial) with suitable boundary conditions possess this property (see, e.g., Agmon [1, 2]). Moreover, under hypotheses of denseness of the linear combinations of the generalized real and complex eigenfunctions of E^*, the dual of E, the following has been proved (Osborn and Sather [31, 32]): It is possible to extend the spaces X and Y so that relation (18) still holds in the new norms and, on the other hand, given $N > 0$ there is a projection operator P such that $\|x\|_1 \geq N\|x\|$ for all $x \in X_1 = (I - P)X$. Thus, given $\varepsilon > 0$, if we take $N^2 = b/\varepsilon$ we also have $\langle x, -Ex\rangle \geq (a - \varepsilon)\|x\|_1^2$ for all $x \in X_1$. Also, given any $\mu > 0$ if we take $N^2 = a^{-1}(\mu + b)$, we have $\langle x, -Ex\rangle \geq \mu\|x\|^2$.

Remark 1. Applications of the considerations of Sections 6 and 7 to global existence of solutions to boundary-value problems for elliptic differential equations, ordinary and partial, will be discussed elsewhere. In Sections 10 and 11 we consider boundary-value problems for similar ordinary and partial differential equations from the local viewpoint and the use of the considerations of Section 3.

Remark 2. For strong nonlinearities the requirement that N be defined in the whole space X may be too restrictive. For self-adjoint (see [8], Section 13) as well as for non-self-adjoint problems, a number of devices have been proposed, namely, suitable decompositions of the relevant operators in products of operators (Gustafson and Sather [11], Osborn and Sather [32], Kannan and Locker [18, 19]).

8. A Strongly Nonlinear Oscillation Problem

The problem of 2π-periodic solutions of the differential equation

$$x'' + x^3 = \sin t, \tag{19}$$

has been a topic of recent interest of several authors. Equation (19) was initially studied by Cesari [5] by alternative methods, namely, by the considerations of Section 3 (for self-adjoint problems) using L_2 and Sup norms, and the existence of a 2π-periodic solution was thereby proved.

9. Forced Oscillations in Liénard Systems

Let us consider the problem of existence of 2π-periodic solutions for the nonlinear differential system

$$x'' + (d/dt)[\text{grad } G(x)] + Ax = p(t), \qquad -\infty < t + \infty,$$

or   (20)

$$x_i'' + (d/dt)(\partial G/\partial x_i) + \sum_{j=1}^{n} a_{ij} x_j = p_i(t), \qquad i = 1, \ldots, n,$$

where A is a constant nonsingular $n \times n$ matrix, $p(t)$ is continuous and 2π-periodic with mean value zero, and G is a given function from R^n into R of class C^2.

Some of the results are

(a) If A is negative semidefinite, then the problem always has a solution.
(b) If $\|A\| < 1$, then the problem has at least one solution.

The proof is based on the general process of Section 6 with the selection of suitable Sobolev spaces $H^s[0, 2\pi]$ and $H^{-s}[0, 2\pi]$ for X and Y, and by taking

$$Ex = x'' \qquad \text{and} \qquad Nx = -(d/dt)[\text{grad } G(x)] - Ax + p(t).$$

Under hypothesis (a) or (b), the bifurcation equation can be chosen to be one dimensional and it is trivially solvable. The auxiliary equation is solvable by Amann's theorem [3] on the Hammerstein equation.

Results (a) and (b) coincide with those obtained by Mawhin [28], and by Kannan and Locker [19], also by alternative methods.

For $\|A\| \geq 1$ some more information on G is needed but the more general system can be taken into consideration:

$$x'' + (d/dt)(\text{grad } G(x(t))) + (d/dt)V(x(t), t) + Ax(t) = p(t), \qquad (20')$$

where $G(x)$ is a homogeneous function from R^n into R of degree $2p$, of constant sign, class C^2, satisfying $|G(x)| \geq c|x|^{2p}$ for some $p \geq 1$, constant $c > 0$. Here $V(x, t) = (V_1, \ldots, V_n)$ is continuous from R^{n+1} into R^n together with the partial derivatives $\partial V_i/\partial t$, $\partial V_i/\partial x_j$, and 2π-periodic in t, A is any $n \times n$ constant matrix, and $p(t)$ is continuous from R^1 into R^n and 2π-periodic. Then, system $(20')$ has at least one 2π-periodic solution if we know that either

(c) $p \geq 2$, $|V(x, t)| \leq Cx^p + D$ for some $C, D \geq 0$; or
(d) $p = 1$, $|V(x, t)| \leq C|x|^p + D$, with $2c > C$, and A symmetric; or
(e) V depends on x only, $V(x) = \text{grad } W(x)$, W of class C^2, satisfying

$$|V(x)| \leq C|x|^{2p-2} + D, \qquad C, D \geq 0, \quad p \geq 1,$$

and A is symmetric if $p = 1$.

Finally, we may consider the system

$$x'' = (d/dt)(\text{grad } G(x(t))) + (d/dt)V(x(t), t) + Ax(t) + g(x(t)) = p(t), \quad (20'')$$

where G, V, A, p are as before, $g: R^n \to R^n$ is any continuous function with $g(x)/|x| \to 0$ as $|x| \to \infty$, and A is nonsingular. The same statements (c), (d), (e) hold.

These and other results, which extend previous ones by Mawhin [28, 29], have been proved by Cesari and Kannan [39, 40] by alternative methods, a priori estimates, and the Borsuk–Ulam theorem.

10. Nonlinear Non-Self-Adjoint Problems for Ordinary Differential Equations. Local Analysis

We consider here, using the work of Nagle [30], boundary-value problems for systems of n first-order ordinary differential equations with homogeneous linear boundary conditions, or

$$x' - A(t)x = f(t, x, x'), \qquad a \leq t \leq b, \quad x' = dx/dt,$$
$$B_1 x(a) + B_2 x(b) = 0, \qquad\qquad\qquad\qquad\qquad (21)$$

where $x(t) = \text{col } (x_1, \ldots, x_n)$ are the n unknowns, $A(t) = [a_{ij}(t)]$ an $n \times n$ matrix with real bounded measurable entries, and B_1, B_2 constant $n \times n$ matrices. Thus, in (21) the underlying linear problem is

$$x' - A(t)x = 0, \quad a \leq t \leq b, \qquad B_1 x(a) + B_2 x(b) = 0, \qquad (22)$$

or $Ex = 0$, including the boundary conditions in the operator E defined by $Ex = x' - A(t)x$. As usual, a solution $x(t) = \text{col} (x_1, \ldots, x_n)$, $a \leq t \leq b$, is an $n \times 1$ vector with absolutely continuous (AC) entries in $[a, b]$, satisfying the differential system a.e. in $[a, b]$. The case of periodic solutions of period T is included in (21) for $a = 0$, $b = T$, $B_1 = I$, $B_2 = -I$, I the identity matrix in E^n.

Let $| \ |$ denote the Euclidean norm in E^n. In (21), $f(t, x, x') = \text{col} (f_1, \ldots, f_n)$ is an $n \times 1$ vector function defined on $[a, b] \times E^{2n}$, whose entries are measurable in t for every (x, x') and continuous in (x, x') for every t. Moreover, we assume here that for any given pair of constants R_1, R_2 there are two other constants L, M such that

$$| f(t, x, x')| \leq M, \qquad | f(t, x, x') - f(t, y, y')| \leq L[|x - y| + |x' - y'|]$$

for all $a \leq t \leq b$, $|x|$, $|y| \leq R_1$, $|x'|$, $|y'| \leq R_2$. If N denotes the operator defined in the second member of (21), then problem (21) takes the usual form $Ex = Nx$.

For any n-vector $z(t) = (z_1, \ldots, z_n)$, $a \leq t \leq b$, we denote by $\|z\|_0$ the usual Sup norm, or

$$\|z\|_0 = \sup_{a \leq t \leq b} |z(t)|,$$

and by $\|z\|_1$ the L_1-norm

$$\|z\|_1 = (b - a)^{-1} \int_a^b |z(t)| \, dt.$$

Let X denote the (Sobolev) space of all AC vector functions $x(t) = (x_1, \ldots, x_n)$, $a \leq t \leq b$, for which we may well take the norm

$$\|x\|_1' = \|x\|_0 + \|x'\|_1$$

(since this norm is equivalent to the Sobolev norm $\|x\|_1 + \|x'\|_1$). Let Y be the space of all vector functions $y(t) = (y_1, \ldots, y_n)$ with L_1-integrable entries, with norm $\|y\|_1$.

Let $\tilde{A}$ denote the transpose of any given matrix A. Then the linear problem adjoint to (15) is

$$d\tilde{y}/dt + \tilde{y}A(t) = 0, \qquad \tilde{y}(a) = \tilde{\alpha}B_1, \quad \tilde{y}(b) = -\tilde{\alpha}B_2, \tag{23}$$

or

$$dy/dt + \tilde{A}(t)y = 0, \qquad y(a) = \tilde{B}_1\alpha, \quad y(b) = -\tilde{B}_2\alpha, \tag{23'}$$

where $y(t) = \text{col} (y_1, \ldots, y_n)$, or $\tilde{y}(t) = \text{row} (y_1, \ldots, y_n)$, and where $\alpha = \text{col} (\alpha_1, \ldots, \alpha_n)$ denotes any arbitrary real $n \times 1$ vector (parametric form of the adjoint problem). Then, the linear operator E^* adjoint to E is defined by relation (23) or (23'), and as usual ker E = coker E^*, coker E = ker E^*. We denote by p and q the dimensions of ker E and coker E, respectively. Fredholm's alternative theorem now has the usual form. Moreover, projection operators $P_0: X \to X, Q_0: Y \to Y$, mapping X onto $X_0 = P_0(X)$ = ker E, and Y onto $Y_0 = Q(Y)$ = coker E, can be defined here by simple algebraic operations (see Hale [14, p. 263], Nagle [30]).

Then problem (1), or $Ex = Nx$, can be reduced to a system (2), (3) of auxiliary and bifurcation equations. By an analysis too detailed to be reported here concerning the interplay of L_1 and Sup norms, the auxiliary equation can be handled as a Lipschitz map. Thus the same equation is solvable if the Lipschitz constant is <1. For f replaced by εg in (21), where ε is a "small" parameter, this Lipschitz constant can be made <1 by taking $|\varepsilon|$ sufficiently small. Thus, we are reduced to the bifurcation equation, actually a system of q equations in p unknowns (see [30]). For $q \le p$ and g smooth, it is enough to verify that the relevant $q \times p$ Jacobian matrix for $\varepsilon = 0$, which is easy to compute, has maximum rank q. Then by the implicit function theorem, problem (21) has at least one solution.

Theorems are proved in [30] that extend to these perturbation boundary-value problems, results proved by Hale [14], Cesari [7], and Mawhin [25] for periodic solutions only, smooth solutions, and f or g independent of derivatives of maximum order. The following examples concerning periodic solutions may be of interest (for details see [30]).

For periodic solutions, as well as for other particular situations, the bifurcation equations and their Jacobian matrix at $\varepsilon = 0$ have been written explicitly.

Example 1

$$x'' + \sigma^2 x = \varepsilon(1 - x^2)x' + \varepsilon a \omega^{-1} x'' + \varepsilon b \omega(\cos \omega t + \alpha)$$

for $\varepsilon > 0$ a small parameter, σ, a, b, α constants, $\omega = T/2\pi$, has periodic solutions of the form

$$x(t, \varepsilon) = \lambda(\varepsilon)\omega^{-1} \sin [\omega(t) + \theta(\varepsilon)] + O(\varepsilon),$$
$$\lambda(\varepsilon) = \lambda_0 + O(\varepsilon), \qquad \theta(\varepsilon) = \theta_0 + O(\varepsilon),$$

where λ_0, θ_0 satisfy the equations

$$\lambda^3 - 4p\omega^3 \cos (\alpha - \theta) = 0, \qquad a\lambda + p\omega \sin (\alpha - \theta) = 0.$$

In this example $p = q = 2$.

Example 2

$$x'' = \varepsilon[-e^{x''} + x^2 \sin^2 t]$$

for $\varepsilon > 0$ a small parameter, has 2π-periodic solutions of the form $x(t) = \alpha + O(\varepsilon)$ with $\alpha = \pm 2^{1/2}$. In this example $p = q = 1$.

Hale's concept [14, p. 267] of symmetric systems is extended in [30] to boundary-value problems (21) of the form

$$x'(t) - A(t)x = f(t, x). \qquad -a \le t \le a,$$
$$B_1 x(-a) + B_2 x(a) = 0. \qquad (24)$$

We say that (24) has a symmetry property S if there is an $n \times n$ constant matrix S such that (a) $S^2 = I$; (b) $Sf(-t, Sx) = -f(t, x)$; (c) if $z(t)$, $-a \le t \le a$, satisfies $B_1 z(-a) + B_2 z(a) = 0$, so does $Sz(-t)$; (d) if $z(t)$, $-a \le t \le a$, satisfies the adjoint boundary condition $\tilde{z}(-a) = \tilde{\alpha} B_1$, $\tilde{z}(a) = -\tilde{\alpha} B_2$, so does $Sz(-t)$.

Under these hypotheses, the q components of the bifurcation equation can be proved to be linearly dependent [30], and thus we may expect a family of solutions to problem (24) depending on a suitable number of parameters.

The following two examples from [30] containing a small parameter ε both possess a family of solutions satisfying the given boundary conditions:

Example 1

$$x_1' = x_2 + \varepsilon f_1(t, x), \qquad f_1(-t, x_1, -x_2, x_3) = -f_1(t, x),$$
$$x_2' = -x_1 + \varepsilon f_2(t, x), \qquad f_2(-t, x_1, -x_2, x_3) = f_2(t, x),$$
$$x_3' = \varepsilon f_3(t, x), \qquad f_3(-t, x_1, -x_2, x_3) = -f_3(t, x)$$
$$x_1(-\pi) - x_1(\pi) = 0, \qquad x_2(-\pi) + x_2(\pi) = 0, \qquad x_3(-\pi) - x_3(\pi) = 0,$$
$$S = (1, 0, 0; 0, -1, 0; 0, 0, 1), \qquad x = (x_1, x_2, x_3).$$

The system has a two-parameter family of solutions, x_1, x_3 even, x_2 odd, of the form

$$x_1 = \lambda \cos t + O(\varepsilon), \qquad x_2 = -\lambda \sin t + O(\varepsilon), \qquad x_3 = c + O(\varepsilon),$$

λ, c arbitrary, $|\varepsilon|$ sufficiently small.

Example 2

$$x_1' = x_2 + \varepsilon f_1(t, x), \qquad x_2' = x_3 + \varepsilon f_2(t, x), \qquad x_3' = \varepsilon f_3(t, x),$$

where $\quad -a \le t \le a, \quad x_1(-a) - x_1(a) = 0, \quad x_2(-a) + x_2(a) = 0, \quad x_3(-a)$

$- x_3(a) = 0$, f_1, f_2, f_3, and S as in Example 1. The system has a two-parameter family of solutions, x_1, x_3 even, x_2 odd, of the form

$$x_1 = \alpha_1 + \alpha_2 t^2 + O(\varepsilon), \qquad x_2 = 2\alpha_2 t + O(\varepsilon), \qquad x_3 = 2\alpha_2 + O(\varepsilon),$$

α_1, α_2 arbitrary, $|\varepsilon|$ sufficiently small.

11. Nonlinear, Non-Self-Adjoint Elliptic Problems of Order m with Nontrivial Kernel—Local Analysis

Let us consider with Shaw [36] the general problem

$$
\begin{aligned}
Ex &= f(t, x, Dx, \ldots, D^m x), \qquad t \in G \subset E^v, \\
Bx &= 0, \qquad t \in \partial G, \quad t = (t_1, \ldots, t_v),
\end{aligned}
\tag{25}
$$

where E is a linear elliptic operator of order m in a region G of the v-dimensional space E^v, with linear homogeneous boundary conditions on ∂G. Of particular interest is the case where there are nonzero functions $x_0(t)$, $t \in G$, with $Ex_0 = 0$, $Bx_0 = 0$, that is, the fundamental operator pair (E, B) has nontrivial null space. A particular case is, of course, the Neumann problem with $E = \Delta$ and boundary condition $\partial x/\partial n = 0$ on ∂G, whose null space is made up of all constant functions on G. In (25) the operator E has the form

$$Ex = \sum_{|\alpha| \leq m} a_\alpha(t) D^\alpha x, \qquad D^\alpha = \partial^{|\alpha|}/\partial t_1^{\alpha_1} \cdots \partial t_v^{\alpha_v},$$

with smooth real coefficients satisfying a uniformly elliptic condition

$$c^{-1} |\xi^m| \leq \sum_{|\alpha| = m} a_\alpha(t) \xi^\alpha \leq c |\xi|^m$$

for some constant $c > 0$ and all $\xi = (\xi_1, \ldots, \xi_v)$ real.

Here G is a bounded, connected open subset of E^v with piecewise smooth boundary. In (25), $f(t, x, Dx, \ldots, D^m x)$ denotes a smooth real-valued function defined on $(G \cup \partial G) \times R^{N+v}$, where $N = 1 + v + v(v - 1)/2 + \cdots$ is the number of all possible distinct partial derivatives of orders $\leq m$ in R^v. We allow in f derivatives of order $\leq m$, that is, up to and including the order m of the operator E.

Thus, including the boundary conditions $Bx = 0$ in the definition of the operator E, and denoting by N the operator defined by the second member of (25), problem (25) takes the usual form $Ex = Nx$.

The main assumption is that the pair (E, B) of linear operators is coercive in the sense of Agmon, Douglas, Nirenberg, a condition algebraic in nature.

Then, (E, B) has important properties as a pair of differential operators.

Let us consider the Sobolev spaces $X = H^{m+r}(G)$ and $Y = H^r(G)$ for r integer, with usual norms $\| \ \|^{m+r}$, $\| \ \|^r$. Let D be the subspace of X of all $x \in X$ satisfying $Bx = 0$. Then, $E: D \to Y$ is defined on D, and a consequence of the coercivity assumption is that for every $x \in D$ we have

$$\|x\|^{m+r} \leq C(\|Ex\|^r + \|x\|^0)$$

for some constant C. Another consequence of the coercivity is that both $X_0 = \ker E$ and $Y_0 = \operatorname{coker} E$ have finite dimensions, say p and q, respectively, and there exist projection operators $P: X \to X$, $Q: Y \to Y$, with $P(X) = X_0$, $Q(Y) = Y_0$ (see [36] for proofs, details, and statements). A consequence of the assumption $r > v/2$ is that, for $x \in H^{r+m}$, by Sobolev's imbedding theorems, x and all derivatives $D^\alpha x$, $0 \leq |\alpha| \leq m$, appearing in the function f are continuous in $G \cup \partial G$, and moreover their Sup norms in $G \cup \partial G$ are not larger than $C_\alpha \|x\|^{r+m}$, $0 \leq |\alpha| \leq m$, for suitable constants C_α. Thus, the arguments in f remain bounded in the Sup norm if $\|x\|^{r+m}$ is bounded. Since f was assumed to be smooth, then f is Lipschitzian in its own arguments in each fixed ball. It can be proved that $f(t, x, \ldots, D^m x)$ is a Lipschitz function of x in the topology of $H^r(G)$ for $\|x\|^{r+m}$ below any given constant M.

By replacing f in (25) by εg, where ε is a small parameter, the auxiliary equation is solvable by contraction maps and the Banach fixed point theorem, at least for $|\varepsilon|$ sufficiently small. It turns out that the present considerations, based on Sobolev's imbedding theorems, allow remarkably good estimates of ε. For $p = q = 1$, the bifurcation equation is a real equation in one real unknown. To exhibit the power of the present approach, we present here from [36] a few examples, with $p = q = 1$, $\varepsilon = 1$, one with nonlinearity including derivatives of maximal order.

Example 1

$$\Delta x + 2x = k - \arctan x \qquad \text{on} \quad G = [(\xi, \eta), 0 \leq \xi, \eta \leq \pi],$$
$$x = 0 \qquad \text{on} \quad \partial G.$$

Here $\lambda = 2$ is an eigenvalue of the operator Δ on G with Dirichlet boundary conditions and eigenfunction $\sin \xi \sin \eta$, $(\xi, \eta) \in G$. It was shown in [36] that the problem is solvable if and only if $|k| < \pi/2$. Moreover, for $|k| < \pi/2$, the solution is then of the form

$$x(\xi, \eta) = v_d(\xi, \eta) + d \sin \xi \sin \eta,$$

with

$$\int_G v_d(\xi, \eta) \sin \xi \sin \eta \, d\xi \, d\eta = 0, \quad \|v_d\|_{L_2} \le \pi^2/3,$$

and a suitable value of the constant d.

Example 2

$$\Delta x = \sin 2\pi\xi + 2^{-1/2} x^{1/3} [x^2 (1 + x^2)^{-1}] + \cos x_{\xi\eta}, \quad (\xi, \eta) \in G,$$

$$\partial x/\partial n = 0 \quad \text{on} \quad \partial G, \quad G = [(\xi, \eta) | 0 \le \xi, \eta \le 1].$$

The problem has a solution of the form $x(\xi, \eta) = d + v(\xi, \eta)$, d a suitable constant, and $\int_G v(\xi, \eta) \, d\xi \, d\eta = 0$.

Example 3

$$\Delta x = f(\xi, \eta) \pm 10^{-1} x^3 \quad \text{on} \quad G = [(\xi, \eta) | 0 \le \xi, \eta \le 1],$$

$$\partial x/\partial n = 0 \quad \text{on} \quad \partial G,$$

where f is a given measurable bounded function on G. The problem has a solution for $|f| \le 1/20$.

12. The Theorem of Landesman and Lazer

We consider here the problem at resonance

$$
\begin{aligned}
Ex + \sigma x + g(x) &= h(t), && t \in G, \\
x &= 0, && t \in \partial G,
\end{aligned}
\tag{26}
$$

where G is a bounded domain in E^ν with smooth boundary ∂G, and E is the elliptic operator

$$Ex = \sum_{|\alpha|, |\beta| \le m} (-1)^{|\alpha|} D^\alpha (a_{\alpha\beta}(t) D^\beta x),$$

$t = (t^1, \ldots, t^\nu) \in G$, $a_{\alpha\beta} = a_{\beta\alpha}$ are bounded real-valued functions on G, with $a_{\alpha\beta}$ uniformly continuous on G for $|\alpha| = |\beta| = m$, and

$$\sum_{|\alpha| = |\beta| = m} a_{\alpha\beta} \xi_1^{\alpha_1 + \beta_1} \cdots \xi_\nu^{\alpha_\nu + \beta_\nu} \ge c(\xi_1^2 + \cdots + \xi_\nu^2)^m$$

for all $(\xi_1, \ldots, \xi_\nu) \in E^\nu$ and some $c > 0$. In (26), σ is an eigenvalue with eigenspace $W = \{w\}$ of (finite) dimension $k \ge 1$ for the underlying linear problem $Ex = 0$ on G, $x = 0$ on ∂G. Also, $h \in L_2(G)$, and $g: E^1 \to E^1$ is a given continuous function for which the limits $R = g(+\infty)$, $r = g(-\infty)$ exist and are finite, and $r \leq g(s) \leq R$ for all s real.

For any eigenfunction $w \in W$, $w \not\equiv 0$, let G^+, G^- denote the sets of points $t \in G$ where $w(t) \geq 0$, $w(t) \leq 0$, respectively, and let $w^+ = \int_{G^+} |w|\, dt$, $w^- = \int_{G^-} |w|\, dt$. Then

$$rw^+ - Rw^- \leq (h, w) \leq Rw^+ - rw^-$$

is a necessary condition for (26) to have a weak solution $x \in H^m(G)$. Moreover,

$$rw^+ - Rw^- < (h, w) < Rw^+ - rw^-$$

is a sufficient condition for (26) to have a weak solution $x \in H^m(G)$.

This theorem was proved by Landesman and Lazer [21] for $m = 2$, $k = 1$, and extended by Williams [38] as stated. Both proofs are based on the alternative method and the use of the Schauder fixed-point theorem applied to the map $S_1 \times S_0 \to S_1 \times S_0$ represented by the system of auxiliary and bifurcation equations. [Other extensions have been given by Nečas again by Schauder's theorem applied to the same map, and by Franchetta by the use of Leray–Schauder degree associated to the same map (coincidence degree).] Shaw [36] has recently extended in various ways the sufficiency condition for problems of the form $Ex = f(t, x, Dx, \ldots, D^{m-1}x)$ on G, $Bx = 0$ on ∂G, E and B as in Section 11. Under hypotheses Shaw has shown that the auxiliary equation can be solved by Schauder's fixed-point theorem yielding a set-valued function $F(x^*)$, and then the bifurcation equation is also solvable, as Shaw has proved by algebraic topology argument and exact sequences of cohomology groups. (Also, De Figuereido, Fucik, and Hess, and Fucik, Kucera, and Nečas have given independent proofs of the Landesman and Lazer theorem, and parallel results.)

REFERENCES

[1] S. Agmon, "Lectures on Elliptic Boundary Value Problems." Van Nostrand-Reinhold, Princeton, New Jersey, 1965.

[2] S. Agmon, On the eigenfunctions and on the eigenvalues of general elliptic boundary value problems. *Comm. Pure Appl. Math.* **15** (1962), 119–147.

[3] H. Amann, Existence theorems for equations of Hammerstein type. *Applicable Anal.* **2** (1973), 385–397.

[4] H. Brézis, M. Crandall, and A. Pazy, Perturbation of nonlinear maximal monotone sets. *Comm. Pure Appl. Math.* **23** (1970), 123–144.

[5] L. Cesari, Functional analysis and periodic solutions of nonlinear differential equations. *In* "Contributions to Differential Equations," Vol. 1, pp. 149–187. Wiley, New York, 1963.

[6] L. Cesari, Functional analysis and Galerkin's method. *Michigan Math. J.* **11** (1964), 385–414.

[7] L. Cesari, Existence theorems for periodic solutions of nonlinear Lipschitzian differential systems and fixed point theorems. *In* "Contributions to the Theory of Nonlinear Oscillations," Vol. 5, pp. 115–172. (Ann. Math. Study, No. 45), 1960.

[8] L. Cesari, Alternative methods in nonlinear analysis. *Internat. Conf. Differential Equations, Los Angeles, Sept. 1974,* pp. 95–198. Academic Press, New York.

[9] L. Cesari and R. Kannan, Functional analysis and nonlinear differential equations. *Bull. Amer. Math. Soc.* **79** (1973), 1216–1219.

[10] J. Cronin, Equations with bounded nonlinearities. *J. Differential Equations* **14** (1973), 581–596.

[11] K. Gustafson and D. Sather, Large nonlinearities and monotonicity. *Arch. Rational Mech. Anal.* **48** (1972), 109–122.

[12] K. Gustafson and D. Sather, Large nonlinearities and closed linear operators. *Arch. Rational Mech. Anal.* **52** (1973), 10–19.

[13] J. K. Hale, Applications of alternative problems. Lecture Notes, Brown University, Providence, Rhode Island, 1971.

[14] J. K. Hale, "Ordinary Differential Equations." Wiley (Interscience), New York, 1969.

[15] J. K. Hale, S. Bancroft, and D. Sweet, Alternative problems for nonlinear equations. *J. Differential Equations* **4** (1968), 40–56.

[16] R. Kannan, Periodically disturbed conservative systems. *J. Differential Equations* **16** (1974), 506–514.

[17] R. Kannan, Existence of periodic solutions of differential equations. *Trans. Amer. Math. Soc.* (to appear).

[18] R. Kannan and J. Locker, Operators *J*J* and nonlinear Hammerstein equations (to appear).

[19] R. Kannan and J. Locker, Nonlinear boundary value problems and operators *TT**. To appear.

[20] H. W. Knobloch, Eine neue Methode zur Approximation von periodischen Lösungen nicht linear Differentialgleichungen zweiter Ordnung. *Math. Z.* **82** (1963), 177–197.

[21] E. M. Landesman and A. Lazer, Nonlinear perturbations of linear elliptic boundary value problems at resonance. *J. Math. Mech.* **19** (1970), 609–623.

[22] P. D. Lax and A. N. Milgram, Parabolic equations. *Ann. of Math.* **33** (1954), 167–190.

[23] J. Leray and J. Schauder, Topologie et équations fonctionnelles, *Ann. Sci. École Norm. Sup.* **51** (1934), 45–78.

[24] A. M. Lyapunov, Sur les figures d'equilibre peu différentes des ellipsoids d'une masse liquide homogène douée d'un mouvement de rotation. *Zap. Akad. Nauk. St. Petersburg* **1** (1906), 1–225.

[25] J. Mawhin, Le problème des solutions périodiques en mécanique non linéaire. Thèse, Univ. Liège, 1969.

[26] J. Mawhin, Degré topologique et solutions périodiques des systèmes différentiels non linéaires. *Bull. Roy. Soc. Liège* **38** (1969), 308–398.

[27] J. Mawhin, Equivalence theorems for nonlinear operator equations and coincidence degree theory for some mappings in locally convex topological vector spaces. *J. Differential Equations* **12** (1972), 610–636.

[28] J. Mawhin, An extension of a theorem of Lazer on forced nonlinear oscillations. *J. Math. Anal. Appl.* **40** (1972), 20–29.

[29] J. Mawhin, Periodic solutions of strongly nonlinear differential systems. *Internat. Conf. Nonlinear Oscillations 5th, Kiev 1969,* Vol. 1, pp. 380–399.

[30] R. K. Nagle, Boundary value problems for nonlinear ordinary differential equations. Thesis, Univ. of Michigan, Ann Arbor, Michigan, 1975.

[31] J. E. Osborn and D. Sather, Alternative problems for nonlinear equations. *J. Differential Equations* **17** (1975), 12–31.

[32] J. E. Osborn and D. Sather, Alternative problems and monotonicity. *J. Differential Equations* **18** (1975), 393–410.

[33] H. Poincaré, Les méthodes nouvelles de la mécanique céleste. Gauthier-Villars, Paris, 1892–1899; Dover, New York, 1957.

[34] D. Sather, Branching of solutions of nonlinear equations, *Rocky Mountain J. Math.* **3** (1973), 203–250.

[35] E. Schmidt, Zur Theorie des linearen und nichtlinearen Integralgleichungen und der Verzweigung ihrer Lösungen. *Math. Ann.* **65** (1908), 370–399.

[36] H. C. Shaw, Elliptic problems of order *m* with nonzero null space. Thesis, Univ. of Michigan, Ann Arbor, Michigan, 1975.

[37] S. A. Williams, A connection between the Cesari and Leray–Schauder methods. *Michigan Math. J.* **15** (1968), 441–448.

[38] S. A. Williams, A sharp sufficient condition for solution of a nonlinear elliptic boundary value problem. *J. Differential Equations* **8** (1970), 580–586.

[39] L. Cesari and R. Kannan, Periodic solutions in the large of nonlinear ordinary differential equations. *Riv. Mat. Univ. Roma* **8** (2) (1975), 633–654.

[40] L. Cesari and R. Kannan, Solutions in the large of Liénard systems with forcing terms. *Ann. Mat. Pura Appl.* (to appear).

Topology and Nonlinear Boundary Value Problems

JEAN MAWHIN

Université de Louvain, Institut Mathématique
Louvain-la-Neuve, Belgium

1. Introduction

Topological arguments, and in particular *degree theory*, are now quite classical in existence problems for nonlinear equations and, during these last decades, progress in this field has followed very closely, and often suggested, the development of some parts of topology. Let us recall that in 1883, Poincaré [59], motivated by problems of celestial mechanics, had already used Kronecker's index [32] to give a formulation of the so-called Miranda theorem [54] and that Bohl [2] in 1904 anticipated the Brouwer fixed-point theorem [4] in his work on differential equations of mechanical systems. The applications of Poincaré and Bohl were local in nature and we had to wait for Lefschetz [38] and Levinson [41] in 1943 to see the Brouwer fixed-point theorem applied to periodic solutions problems for strongly nonlinear forced Liénard equations. From this moment, *Poincaré's method* of reducing the search of periodic solutions to that of fixed points for the operator of translation over one period and its study by topological theorems were used successfully by many authors (see, e.g., the books of Sansone and Conti [66], Lefschetz [39], Reissig *et al.* [62], and Krasnosel'skii [31]).

In the meantime, however, another approach for boundary value problems in nonlinear equations had emerged from the extension by Birkhoff and Kellog [1], Schauder [67], and Caccioppoli [9] of the Brouwer fixed-point theorem to some mappings in infinite-dimensional spaces. This current culminated in the famous paper of Leray and Schauder [40] on degree theory for compact perturbations of identity in Banach spaces, whose birth was closely related to nonlinear boundary value problems for elliptic partial differential equations (see, e.g., Cronin [16], Ladyzenskaja and Ural'ceva [34], Browder [5]). Curiously, the machinery of Leray–Schauder degree was applied somewhat later to ordinary differential equations and, in particular, only in 1952, by Stoppelli [71], to a problem of periodic solutions (see, e.g., Rouche and Mawhin [63] for a bibliography of more recent work). Up

51

to that time, all the global existence theorems proved in various fields by Leray–Schauder theory could be reduced to the common abstract form

$$Lx = N(x, 1),$$

where L has an inverse, $L^{-1}N(\cdot, \cdot)$ has compactness properties, $N(\cdot, 0) = 0$, and all possible solutions of each equation

$$Lx = N(x, \lambda), \qquad \lambda \in {]0, 1[}$$

are a priori bounded independently of λ. This abstract situation is usually known as the *Leray–Schauder continuation method*.

Problems corresponding to equations

$$Lx = Nx \qquad (1.1)$$

with L noninvertible but ker L finite dimensional had, however, received attention in the first decade of the century in the work of Lyapunov [43] and Schmidt [68] on nonlinear integral equations. When N is "small" in some sense, they reduced problems of this type to a finite-dimensional equation in ker L, the *bifurcation equation*, a process referred to today as an *alternative problem*. In 1936, Caccioppoli [10] was the first to use topological methods, essentially Brouwer degree, to study bifurcation equations, and this approach was substantially developed in 1950 by Cronin [12] and applied to some local problems for elliptic partial differential equations [13]. In his basic paper extending alternative problems to some equations with large nonlinearities, Cesari [11] also solved the generalized bifurcation equation using Brouwer degree. Later, the author [44] used simultaneously Cesari's approach, a priori estimates, and Brouwer degree to give, in the frame of periodic solutions of differential equations, a *continuation theorem for an equation of type* (1.1) *with noninvertible L*. Various proofs of this result, using only Leray–Schauder degree, were then given by Strygin [72] and the author [45], who extended it in [46] to operator equations in Banach spaces. Earlier however, Lazer [36] had used the Schauder fixed-point theorem in a subtle way to study the periodic solutions of a particular second-order differential equation with a noninvertible linear part. With Landesman [35], he applied the same approach to a Dirichlet problem for a semilinear elliptic equation, and the result was later generalized in various ways (see Section 5). In particular, Nirenberg [56] showed that topological degree was more adapted than Schauder theorem for this type of problem and was independently led, for elliptic equations, to an existence theorem close to the spirit of Mawhin's. With *coincidence degree theory* [47], extending

Leray–Schauder degree to some mappings of the form $L - N$, with L non-invertible, it appeared that the author's continuation theorem cited above was just the natural extension in this new frame of the Leray–Schauder continuation method. Therefore, it was possible to unify many proofs of old results and to solve new problems, as will be shown later.

It will be the aim of this work to give a survey of recent existence results, for operator or ordinary, functional, and partial differential equations, which fall into the scope of coincidence degree theory. As far as the applications are concerned, we have restricted ourselves here to equations with some growth restriction on the nonlinearities. For other applications to ordinary and functional differential equations, as well as for the use of coincidence degree in bifurcation theory, the reader is referred to the complementary survey papers [52] and [53]. We also note that the topological approach in classical alternative problems can be incorporated, as shown in Section 6, into the coincidence degree in the frame of α-contractions recently developed by Hetzer [29], a subject that seems promising in a number of applications. In this chapter, proofs are only given when they have not been published elsewhere or when, because of the attempt at unification, they differ substantially from the original ones.

We hope that this chapter will underline one of the basic ideas of this volume in showing how techniques in ordinary, functional, and partial differential equations can mutually influence each other when integrated in a common abstract scheme.

2. Coincidence Degree and a Generalized Continuation Theorem

The study of various existence problems for differential equations with noninvertible linear part has led the author [47] to extend Leray–Schauder degree theory [40] to some couples (L, N) of mappings between some vector spaces X and Z that we shall suppose here normed for simplicity, both norms being denoted by $|\cdot|$. More precisely, let

(i) $L: \operatorname{dom} L \subset X \to Z$ be a *linear Fredholm mapping of index zero* (i.e., such that Im L is closed in Z and

$$\dim \ker L = \operatorname{codim} \operatorname{Im} L < \infty).$$

(ii) N: cl $\Omega \subset X \to Z$, with Ω open and bounded, a mapping that is
L-compact on cl Ω, i.e., such that

(a) ΠN: cl $\Omega \to$ coker L is continuous and $\Pi N(\text{cl } \Omega)$ bounded;
(b) $K_{P, Q} N$: cl $\Omega \to X$ is compact;

where $\Pi: Z \to$ coker L is the canonical surjection and $K_{P, Q} = K_P(I - Q)$,
with $K_P = (L \mid \text{dom } L \cap \text{ker } P)^{-1}$ and $P: X \to X$, $Q: Z \to Z$ are continuous
projectors such that

$$\text{Im } P = \text{ker } L, \qquad \text{Im } L = \text{ker } Q.$$

It can be easily proved that condition (b) does not depend on the (nonunique)
choice of the projectors P and Q. Moreover, if $X = Z$ and $L = I$, the
L-compactness of N is nothing but its compactness on cl Ω.

(iii)   When $0 \notin (L - N)(\text{dom } L \cap \text{bdry } \Omega)$, an integer $d[(L, N), \Omega]$, the
coincidence degree of L and N in Ω, has been introduced in [47], which
depends only on L, N, Ω, and orientations chosen on ker L and coker L. It
is defined explicitly by the relation

$$\begin{aligned} d[(L, N), \Omega] &= d_{\text{LS}}[I - P - (\Lambda\Pi + K_{P, Q})N, \Omega, 0] \\ &= d_{\text{LS}}[I - (L + F_{P, Q, \Lambda})^{-1}(F_{P, Q, \Lambda} + N), \Omega, 0] \end{aligned}$$

where d_{LS} denotes the Leray–Schauder degree, Λ: coker $L \to$ ker L is an
orientation-preserving isomorphism, and

$$F_{P, Q, \Lambda} = \Lambda_Q^{-1} P, \qquad \Lambda_Q = \Lambda\Pi \mid \text{Im } Q.$$

It follows from conditions (i)–(iii) and from the easily proved relation

$$I - P - (\Lambda\Pi + K_{P, Q})N = (\Lambda\Pi + K_{P, Q})(L - N)$$

where $\Lambda\Pi + K_{P, Q}$ is checked to be an isomorphism between Z and dom L,
that this Leray–Schauder degree is well defined, and it is shown in [47] that
it does not depend on the choice of P, Q, and Λ. Moreover, it conserves
most of the basic properties of Leray–Schauder degree, as for example the
additivity, excision, and invariance with respect to L-compact homotopies
preserving (iii).

One can also generalize in this frame the Leray–Schauder continuation
theorem.

Theorem 2.1.   Let L and N satisfy conditions (i) and (ii) above and
suppose that the following assumptions are verified:

1. For each $x \in \operatorname{dom} L \cap \operatorname{bdry} \Omega$ and each $\lambda \in {]0, 1[}$,

$$Lx \neq \lambda Nx.$$

2. For each $x \in \ker L \cap \operatorname{bdry} \Omega$, $\Pi Nx \neq 0$ (or equivalently $QNx \neq 0$).
3. $d_B[\Lambda \Pi N | \ker L, \Omega \cap \ker L, 0] \neq 0$ (or equivalently

$$d_B[JQN | \ker L, \Omega \cap \ker L, 0] \neq 0),$$

where d_B denotes the Brouwer degree and $J: \operatorname{Im} Q \to \ker L$ is any isomorphism.

Then for each $\lambda \in [0, 1[$, the equation

$$Lx = \lambda Nx$$

has at least one solution in Ω and

$$Lx = Nx$$

has at least one solution in $\operatorname{cl} \Omega$.

For a proof of this theorem, see [47] or [51].

Remark. When $\ker L = \{0\}$, conditions (2) and (3) reduce to

$2'$: $0 \in \Omega$,

as shown in [51] by introducing a Brouwer degree in 0-dimensional spaces, and Theorem 2.1 reduces then to the usual Leray–Schauder continuation method.

3. Mapping Theorems for Quasi-Bounded Perturbations of Linear Fredholm Operators

In this section, we shall use Theorem 2.1 to give various mapping theorems relying on the concept of quasi-boundedness for a nonlinear mapping.

With X and Z still real normed spaces, let $F: X \to Z$. The following concept is due to Granas [25].

Definition 3.1. F is said to be *quasi-bounded* if the number $\|F\|$ defined by

$$\|F\| = \limsup_{|x| \to \infty} (|x|^{-1} |Fx|)$$

is finite, in which case $\|F\|$ is called the *quasi-norm* of F.

Throughout this section we shall assume that $L: \operatorname{dom} L \subset X \to Z$ is a linear Fredholm mapping of index zero and that $N: X \to Z$ is L-compact on bounded sets of X. Also, P and Q will be continuous projectors such that $\operatorname{Im} P = \ker L$, $\operatorname{Im} L = \ker Q$, and $K_{P,Q}$ is the corresponding mapping defined in Section 2.

Theorem 3.1. Suppose that the following conditions are satisfied:

(a) $K_{P,Q} N: X \to X$ is quasi-bounded.
(b) There exist $\alpha \geq 0$, $r > 0$ such that each possible solution of

$$\Pi N x = 0$$

satisfies the relation

$$|Px| < \alpha |(I - P)x| + r.$$

(c) $\|K_{P,Q} N\| < (1 + \alpha)^{-1}$.
(d) $d_B[JQN|\ker L, B(r) \cap \ker L, 0] \neq 0$.

Then

$$(L - N)(\operatorname{dom} L) \supset \operatorname{Im} L.$$

For a proof of this theorem, see [48] or [51].

When $X = Z$ and $L = I$, conditions (b) and (d) are satisfied with $\alpha = 0$ and any $r > 0$, and assumptions reduce to

$$\|N\| < 1.$$

This is a result due to Granas [25].

To obtain a useful special case of Theorem 3.1, let us suppose that X is a (possibly proper) subspace of the space $B(S, R^n)$ of bounded mappings from some set S into R^n, with a norm such that

$$|x| \geq \sup_{s \in S} |x(s)|.$$

In this case, the following result is proved in [51].

Theorem 3.2. Suppose that the following conditions are satisfied:

(a') $K_{P,Q} N: X \to X$ is quasi-bounded.
(b') There exists $\alpha > 0$ such that for each $u \in \ker L$ and each $s \in S$,

$$|u| \leq \alpha |u(s)|.$$

TOPOLOGY AND NONLINEAR BOUNDARY VALUE PROBLEMS

(b″) There exists $r_1 > 0$ such that for every $x \in \text{dom } L$ for which $|x(s)| \geq r_1$, $s \in S$, one has

$$\Pi N x \neq 0.$$

(c′) $\|K_{P,Q} N\| < (1 + \alpha)^{-1}.$

(d′) $d_B[JQN | \ker L, B(r_1) \cap \ker L, 0] \neq 0.$

Then $(L - N)(\text{dom } L) \supset \text{Im } L.$

The following special case is proved in [48] and [51].

Corollary 3.1. Suppose that X is as in Theorem 3.2 with moreover

$$|x| = \sup_{s \in S} |x(s)|$$

when x is a constant mapping and that

$$\ker L = \{x \in \text{dom } L : x \text{ is a constant mapping}\}.$$

Suppose that assumptions (a′), (b″), (d′) of Theorem 3.2 hold and that

$$\|K_{P,Q} N\| < \tfrac{1}{2}.$$

Then $(L - N)(\text{dom } L) \supset \text{Im } L.$

Let us now come back to a general normed space X.

Theorem 3.3. Suppose that the following conditions hold:

1. There exist $\delta \in [0, 1[$, $\mu \geq 0$, $\nu \geq 0$ such that for each $x \in X$,

$$|K_{P,Q} Nx| \leq \mu |x|^\delta + \nu.$$

2. For every bounded $V \subset \text{Im } L$, there exists $t_0 > 0$ such that for every $t \geq t_0$, every $z \in V$, and every $w \in \ker L \cap \text{bdry } B(1)$, one has

$$QN(tw + t^\delta z) \neq 0.$$

3. $d_B[JQN \ker L, B(t_0) \cap \ker L, 0] \neq 0.$

Then $(L - N)(\text{dom } L) \supset \text{Im } L.$

Proof. Let $y \in \text{Im } L$. We shall apply Theorem 2.1 with $Nx + y$ instead of Nx. Each equation

$$Lx = \lambda(Nx + y), \qquad \lambda \in \,]0, 1], \tag{3.1}$$

is equivalent to

$$v = \lambda K_{P, Q}(N(u + v) + y) \tag{3.2}$$

$$0 = QN(u + v) \tag{3.3}$$

with

$$u = Px, \qquad v = (I - P)x.$$

Then, for each possible solution $x = u + v$ of (3.2), we have

$$|v| \leq \mu|u + v|^\delta + v' \tag{3.4}$$

with

$$v' = v + |K_{P, Q} y|,$$

and hence, if $|u| \neq 0$,

$$|v| |u|^{-\delta} \leq \mu + \mu\delta |u|^{\delta - 1}(|v| |u|^{-\delta}) + v' |u|^{-\delta}.$$

Thus, there exists $t_1 > 0$ such that, if $|u| \geq t_1$,

$$\mu\delta |u|^{\delta - 1} \leq \tfrac{1}{2},$$

and hence

$$|v| |u|^{-\delta} \leq 2(\mu + v' t_1^{-\delta}).$$

If

$$V = \{z \in \operatorname{Im} L \colon |z| \leq 2(\mu + v' t_1^{-\delta}),$$

then, by condition (2), there exists $t_0 > 0$ such that, if $t \geq t_0$, $w \in \ker L \cap$ bdry $B(1)$, and $z \in V$,

$$QN(tw + t^\delta z) \neq 0.$$

Therefore, if $|u| \geq \max (t_0, t_1)$, $|u|^{-\delta} v \in V$, and

$$QN(u + v) = QN(|u|(u/|u|) + |u|^\delta(v/|u|^\delta)) \neq 0$$

which implies by (3.3) that $u + v$ cannot be a solution of (3.1). Thus, each possible solution $x = u + v$ of (3.1) is such that

$$|u| < t_2 = \max (t_0, t_1)$$

and therefore, by (3.4),

$$|v| \leq \mu(t_2 + |v|)^\delta + v',$$

which clearly implies that $|v| \leq t_3$, with t_3 the unique positive solution of

$$\alpha - \mu(t_2 + \alpha)^\delta - v' = 0.$$

It is then easily checked that all the conditions of Theorem 3.2 hold with $\Omega = B(t_2 + t_3)$, and the proof is complete.

When $\delta = 0$, it is clear from the proof that if assumption (2) is replaced by

2'.   There exists $t_0 > 0$ such that for every $t \geq t_0$, every $z \in \mathrm{cl}\ B(\mu + v)$, and every $w \in \ker L \cap \mathrm{bdry}\ B(1)$, one has $QN(tw + z) \neq 0$.

Then $0 \in (L - N)(\mathrm{dom}\ L)$. The corresponding theorem is due to Cronin [15].

Let us now assume that $X = Z = H$, a Hilbert space with inner product (,), that L is self-adjoint (hence dom L is necessarily dense in H), and let us give some consequences of Theorem 3.3 that were initially proved by Nečas [55] and Fučik et al. [24].

Corollary 3.2.   Suppose that assumption (1) of Theorem 3.3 and the following condition

2''.   For every bounded $V \subset \mathrm{Im}\ L$, there exists $t_0 > 0$ such that for every $t \geq t_0$, every $z \in V$, and every $w \in \ker L \cap \mathrm{bdry}\ B(1)$, one has

$$(N(tw + t^\delta z), w) > 0,$$

hold. Then $(L - N)(\mathrm{dom}\ L) \supset \mathrm{Im}\ L$.

Proof.   Using the self-adjoint character of L, one can take

$$Px = Qx = \sum_{i=1}^{n} (x, w_i)w_i,$$

where $(w_1, \ldots, w_n)$ is an orthonormed basis of ker L. P and Q are then orthogonal projectors and, using (2''), one gets

$$(QN(tw + t^\delta z), w) = (N(tw + t^\delta z), w) > 0,$$

which shows that condition (2) of Theorem 3.3 is satisfied. Also, if t_0 corresponds to $V = \{0\}$, then for $t \geq t_0$ and $w \in \ker L \cap \mathrm{bdry}\ B(1)$

$$(QN(tw), w) > 0,$$

which, by the Poincaré–Bohl theorem, implies that

$$d_B[QN|\ker L, B(t_0) \cap \ker L, 0] = 1,$$

and the proof is complete.

Corollary 3.3. Let $F: H \to H$ be L-compact on closed bounded sets of H such that assumption (1) of Theorem 3.3 holds with N replaced by F. Suppose that there exists S: ker $L \cap$ bdry $B(1) \to R$ such that for each bounded $V \subset \operatorname{Im} L$, there exists $t_0 > 0$ such that for every $t \geq t_0$, every $z \in V$, and every $w \in$ ker $L \cap$ bdry $B(1)$, one has

$$(F(tw + t^\delta z), w) \geq S(w)$$

(S is then said to be a weak δ-subasymptote to F with respect to ker L). Then for each $y \in H$ such that

$$(y, w) < S(w), \qquad w \in \text{ker } L \cap \text{bdry } B(1),$$

the equation

$$Lx = Fx - y$$

has at least one solution.

Corollary 3.4. Let F be as in Corollary 3.3 and suppose that for every $k > 0$ and every bounded $V \subset \operatorname{Im} L$ there exists $t_0 > 0$ such that for each $t \geq t_0$, each $w \in$ ker $L \cap$ bdry $B(1)$, and each $z \in V$, one has

$$(F(tw + t^\delta z), w) \geq k,$$

then $L - F$ is onto.

Corollary 3.5. Let F be as in Corollary 3.3 and suppose that there exists S^*: ker $L \cap$ bdry $B(1) \to R$ such that

$$(F(x), w) < S^*(w), \qquad x \in H, \quad w \in \text{ker } L \cap \text{bdry } B(1),$$

and such that for each $\varepsilon > 0$ and each bounded $V \subset H$, there exists $t_0 > 0$ such that for every $t \geq t_0$, every $z \in V$, and every $w \in$ ker $L \cap$ bdry $B(1)$, one has

$$\left| (F(tw + t^\delta z), w) - S^*(w) \right| < \varepsilon.$$

Then

$$Lx = Fx - y, \qquad y \in H$$

has a solution if and only if

$$(y, w) < S^*(w), \qquad w \in \text{ker } L \cap \text{bdry } B(1).$$

Remark.   The conditions in Corollaries 3.4 and 3.5 being independent of the sign of L, the same results hold for

$$Lx + Fx = y.$$

Let us come back now to normed spaces X and Z but assume that there exists an inner product space U with inner product (,) and a normed space V with norm $\|\cdot\|$, such that $X \subset U$, $Z \subset U$, and $X \subset V$ topologically. Let L: dom $L \subset X \to Z$ be a linear Fredholm mapping of index zero and $F: X \to Z$ be L-compact on closed bounded sets of X. Assume also that there exists a continuous projector $Q: Z \to Z$ such that Im $L = \ker Q$ and which is orthogonal for the inner product in U. The following result has been given by Fabry and Franchetti [20] as an application of Theorem 2.1. We give here a proof using Theorem 3.3.

Theorem 3.4.   Let us assume that the following conditions hold:

(i)   There exist $\delta \in [0, 1[$, μ, μ', v, $v' \geq 0$ such that for each $x \in X$,

$$|K_{P,Q}F| \leq \mu|x|^{\delta} + v, \qquad |Fx| \leq \mu'|x|^{\delta} + v'.$$

(ii)   There exist linear continuous mappings $H: X \to Z$, $G: X \to X$, a constant $\beta \in [0, \delta[$, with $2\delta < 1 + \beta$, and constants $a > 0$, $b \geq 0$ such that

(a) $H|\ker L$ is an isomorphism onto Im Q;
(b) $G|\ker L$ is one-to-one;
(c) $(Hx, Nx) \geq a\|Gx\|^{1+\beta} - b$ for each $x \in X$.

Then if $\beta > 0$, $L - F$ is onto and if $\beta = 0$ and $y \in Z$ is such that

$$\sup_{w \in \ker L \cap \text{bdry } B(1)} \|Gw\|^{-1}(Hw, Qy) < a \tag{3.5}$$

the equation

$$Lx - Fx = y$$

has at least one solution.

Proof.   We shall apply Theorem 3.3 with $Nx = Fx + y$ and show that condition (ii) implies conditions (2) and (3) of this theorem. If (2) is not satisfied, then there exists a bounded $\tilde{V} \subset \text{Im } L$, a sequence (t_n) with $t_n > 0$, $n \in N^*$, and $t_n \to \infty$ if $n \to \infty$, a sequence (z_n) with $z_n \in \tilde{V}$, $n \in N^*$, and a sequence (w_n) with $w_n \in \ker L \cap \text{bdry } B(1)$ such that

$$QF(t_n w_n + t_n^{\delta} z_n) + Qy = 0. \tag{3.6}$$

If necessary, by taking a subsequence we can assume that $w_n \to w$ in X, with $w \in \ker L \cap \text{bdry } B(1)$, which implies in turn that $w_n + t_n^{\delta-1} z_n \to w$ in X when $n \to \infty$. Now, using (3.6), we have, for each $n \in N^*$,

$$
\begin{aligned}
0 &= t_n^{-(1+\beta)}(H(t_n w_n), QF(t_n w_n + t_n^{\delta} z_n) + Qy) \\
&= t_n^{-(1+\beta)}(H(t_n w_n), F(t_n w_n + t_n^{\delta} z_n) + Qy) \qquad \text{(by the orthogonality of } Q) \\
&\geq t_n^{-(1+\beta)} a \| G(t_n w_n + t_n^{\delta} z_n) \|^{1+\beta} - bt_n^{-(1+\beta)} \\
&\quad - t_n^{-(1+\beta)}(H(t_n^{\delta} z_n), F(t_n w_n + t_n^{\delta} z_n)) \\
&\quad + t_n^{-\beta}(Hw_n, Qy) \qquad \text{[by the use of (ii-c)]} \\
&\geq a \| G(w_n + t_n^{\delta-1} z_n) \|^{1+\beta} - bt_n^{-(1+\beta)} - k|H| t_n^{2\delta-1-\beta} \\
&\quad \times (\mu' |w_n + t_n^{\delta-1} z_n|^{\delta} + v' t_n^{-\delta}) + t_n^{-\beta}(Hw_n, Qy) \qquad \text{[use of the}
\end{aligned}
$$

topological embedding of X and Z in U, of condition (i), and of the fact that (z_n) is a bounded sequence]

where k is a positive constant related to the above embedding and $\tilde{V}$.

Therefore, if $\beta > 0$ we obtain, if $n \to \infty$, using the fact that $X \subset V$, topologically,

$$0 \geq a \| Gw \|^{1+\beta},$$

which contradicts condition (ii-b). If $\beta = 0$, we have similarly

$$0 \geq a \| Gw \| + (Hw, Qy),$$

a contradiction with (3.5). Thus, condition (2) of Theorem 3.3 is verified.

Now we deduce from (ii-c), if $x \in \ker L$, $J = H^*$, the adjoint of $H | \ker L$ [i.e., $(Hx, \tilde{y}) = (x, H^* \tilde{y})$ for $x \in \ker L$, $\tilde{y} \in \text{Im } Q$],

$$(Hx, Fx + y) = (Hx, QFx + Qy) - (x, J(QFx + Qy))$$

and, from (ii),

$$(Hx, Fx + y) \geq a \| Gx \|^{1+\beta} + b + (Hx, Qy) > 0$$

if $|x| \geq r$, with r sufficiently large. Then, using the Poincaré–Bohl theorem,

$$d_B[J(QF(\cdot) + Qy), B(r) \cap \ker L, 0] = 1$$

and the proof is complete.

4. Periodic Solutions of Ordinary and Functional Differential Equations

Corollary 3.1 can be used to prove the existence of T-periodic solutions for a class of retarded functional differential equations. If $l \geq 0$ is an integer we shall denote by P_T^l the (Banach) space of mappings $x: R \to R^n$ that are continuous and T-periodic together with their first l derivatives, with the norm

$$|x|_l = \sum_{j=0}^{l} \sup_{t \in R} |x^{(j)}(t)|$$

with $x^{(j)} = d^j x/dt^j$, and $|\cdot|$ some norm in R^n. For some $r \geq 0$, let C_r be the (Banach) space of continuous mappings $\varphi: [-r, 0] \to R^n$ with the norm

$$\|\varphi\| = \sup_{\theta \in [-r, 0]} |\varphi(\theta)|.$$

When $r = 0$, C_r will be naturally identified with R^n. Now, if $x \in P_T^l$ and $t \in R$, and if j is an integer between 0 and l, we shall denote as usual by $x_t^{(j)}$ the element of C_r defined by

$$x_t^{(j)}: [-r, 0] \to R^n, \qquad \theta \mapsto x^{(j)}(t + \theta),$$

and when $r = 0$, $x_t^{(j)}$ will be naturally identified with $x^{(j)}(t)$.

If $k \geq 0$ is an integer, $A_1, \ldots, A_k$ are $(n \times n)$-real matrices and

$$f: R \times C_r \times \cdots \times C_r \to R^n, \qquad (t, \varphi_1, \ldots, \varphi_{k+1}) \mapsto f(t, \varphi_1, \ldots, \varphi_{k+1})$$

is T-periodic with respect to t, continuous, and takes bounded sets into bounded sets, let us consider the retarded functional differential equation

$$x^{(k+1)} + A_1 x^{(k)} + \cdots + A_k x' = f(t, x_t, x_t', \ldots, x_t^{(k)}). \tag{4.1}$$

The following existence theorem has been proved in [49] as a consequence of Corollary 3.1 with $X = P_T^k$, $Z = P_T^0$, and L and N defined in an obvious way.

Theorem 4.1. Suppose that the following conditions are satisfied:

1. The equation

$$\det (\lambda^k I + \lambda^{k-1} A_1 + \cdots + A_k) = 0$$

has no root of the form 2π im T^{-1} with m a nonzero integer.

2. $\displaystyle \limsup_{\|\varphi_1\| + \cdots + \|\varphi_{k+1}\| \to \infty} (\|\varphi_1\| + \cdots + \|\varphi_{k+1}\|)^{-1} |f(t, \varphi_1, \ldots, \varphi_{k+1})| = 0$,

uniformly in $t \in R$.

3. There exists $R > 0$ such that

$$T^{-1} \int_0^T f(t, y_t, \ldots, y_t^{(k)}) \, dt \neq 0$$

for every $y \in P_T^{k+1}$ that verifies $|y(t)| \geq R$, $t \in R$.

4. $d_B[F, B(R), 0] \neq 0$, with

$$F: R^n \to R^n, \qquad a \mapsto T^{-1} \int_0^T f(t, a, 0, \ldots, 0)\, dt.$$

Then, Eq. (4.1) has at least one T-periodic solution.

It is shown in [49] that Theorem 4.1 contains as special cases or easy consequences earlier results of Lazer [36], Villari [76], Sedsiwy [69], Reissig [61] for ordinary differential equations, and Fennell [21] for retarded functional differential equations.

For the neutral equation

$$(d/dt)D(t)x_t = f(t, x_t), \tag{4.2}$$

where f is as above and $D: R \times C_r \to R^n$, $(t, \varphi) \mapsto D(t)\varphi$, is T-periodic and continuous with respect to t, linear and continuous with respect to φ for each $t \in R$, and stable [which means that the zero solution of equation $D(t)y_t = 0$ is uniformly asymptotically stable], Hale and the author have used Theorem 3.1 (or equivalently 3.2) to prove the following result [27].

Theorem 4.2.   Suppose that the following conditions are satisfied:

1. $\lim \sup_{\|\varphi\| \to \infty} \|\varphi\|^{-1} |f(t, \varphi)| = 0$ uniformly in $t \in R$.
2. There exists $\mu > 0$ such that $|(Mc)(t)| \geq \mu|c|$ for every $t \in R$ and every $c \in R^n$, where Mc is the unique T-periodic solution of the equation $D(t)y_t = c$.
3. There exists $R_1 > 0$ such that

$$\int_0^T f(s, y_s)\, ds \neq 0$$

for every $y \in P_T^0$ verifying $\inf_{t \in R} |y(t)| \geq R_1$.
4. $d_B[F, \Omega_R, 0] \neq 0$, with

$$F: R^n \to R^n, \qquad a \mapsto T^{-1} \int_0^T f(t, (Ma)_t)\, dt$$

and

$$\Omega_R = \{a \in R^n: Ma \in B(R)\}, \qquad R = \mu^{-1}|M|R_1.$$

Then, Eq. (4.2) has at least one T-periodic solution.

Other results concerning equations of type (4.2), and especially one arising from a transmission line problem, can be found in [27].

Let us now consider the ordinary scalar differential equation

$$x'' + m^2 x = h(t) + g(t, x), \tag{4.3}$$

where $m \geq 1$ is an integer, $h: R \to R$ is 2π-periodic and continuous, and $g: R \times R \to R$ is 2π-periodic with respect to t and continuous. Using Theorem 3.4, Fabry and Franchetti [20] have proved the following theorem.

Theorem 4.3. Suppose that the following conditions are satisfied:

1. There exists $\delta \in [0, \frac{1}{2}[$ such that

$$\lim_{|x| \to \infty} |x|^\delta |g(t, x)| = 0$$

uniformly in $t \in R$.

2. There exists $\rho \geq 0$ such that for $i = 1$ or -1, for each $t \in R$, and for each $x \in R$ verifying $|x| \geq \rho$, one has

$$i(\operatorname{sign} x) g(t, x) > \tfrac{1}{4}\pi A,$$

where

$$A = (a^2 + b^2)^{1/2}, \qquad a = \pi^{-1} \int_0^{2\pi} h(t) \cos nt \, dt, \qquad b = \pi^{-1} \int_0^{2\pi} h(t) \sin nt \, dt.$$

Then Eq. (4.3) has at least one 2π-periodic solution.

This theorem generalized an earlier result of Lazer and Leach [37]. Other existence theorems for second-order vector differential equations have also been deduced from Theorem 3.4 by Fabry and Franchetti in [20].

5. Boundary Value Problems
for Some Semilinear Elliptic Partial Differential Equations

Let D be a bounded domain in R^n and $a_{\alpha\beta}$, $0 \leq |\alpha|, |\beta| \leq m$ real-valued $L^\infty(D)$-functions with $a_{\alpha\beta}$, for $|\alpha| = |\beta| = m$ uniformly continuous when $m \geq 2$. As usual, $\alpha = (\alpha_1, \ldots, \alpha_n)$, $\alpha_i \in N$, and $|\alpha| = \sum_{i=1}^n \alpha_i$. Let us assume that $a_{\alpha\beta} = a_{\beta\alpha}$ and that there exists a constant $c > 0$ such that

$$\sum_{|\alpha| = m, |\beta| = m} a_{\alpha\beta}(t) \xi^\alpha \xi^\beta \geq c |\xi|^{2m}$$

for all $\xi \in R^n$ and all $t \in D$. Let $H_0^m(D)$ and $H^m(D)$ be, respectively, the completions of the space $\mathscr{C}_0^\infty(D)$ and $C^1(D)$ for the Sobolev norm

$$|\varphi|_m = \left[\sum_{|\alpha| \le m} \int_D |D^\alpha \varphi|^2 \right]^{1/2},$$

and let us define the bilinear form

$$a(u, v) = \sum_{|\alpha| \le m, |\beta| \le m} \int_D a_{\alpha\beta}(t) D^\alpha u(t) D^\beta v(t) \, dt. \tag{5.1}$$

If, with $V = H_0^m(D)$ or $H^m(D)$,

   dom $\tilde{L} = \{u \in V : v \mapsto a(u, v)$ is continuous in V with the $L^2(D)$-norm$\}$,

then using the fact that V is dense in $L^2(D)$ and the theorem of structure of functionals in a Hilbert space, there exists a linear (but not continuous) operator $\tilde{L} \colon \text{dom } \tilde{L} \subset L^2(D) \to L^2(D)$ such that for $u \in \text{dom } \tilde{L}$ and $v \in V$,

$$a(u, v) = (\tilde{L}u, v), \tag{5.2}$$

where $(\ ,\ )$ denotes the inner product in $L^2(D)$. Hence, if $h \in L^2(D)$, each equation in V of the form

$$a(u, v) = (h, v), \qquad \forall v \in V, \tag{5.3}$$

is equivalent to the equation

$$\tilde{L}u = h. \tag{5.4}$$

It follows also from the L^2-theory of linear elliptic boundary value problems [42] that dom $\tilde{L}$ is dense in $L^2(D)$ and that $\tilde{L}$ is a Fredholm mapping of index zero that is self-adjoint and has compact right inverses on Im $\tilde{L}$.

Now if $m = 1$, $V = H^1(D)$, and D has a Lipschitzian boundary with sufficiently small Lipschitz constants, it follows from regularization results of Stampacchia [70] and Fiorenza [22] that if $h \in L^p(D)$ with $p > n$, each solution $u \in H^1(D)$ of (5.3) or (5.4) is Hölder-continuous with some coefficient $\alpha \in {]}0, 1{[}$ and that if L denotes the restriction of $\tilde{L}$ to the space $\tilde{L}^{-1}(C^0(\text{cl } D))$, with $C^0(\text{cl } D)$ the Banach space of continuous real functions on cl D, then $L \colon \text{dom } L \subset C^0(\text{cl } D) \to C^0(\text{cl } D)$ is also Fredholm of index zero and has compact right inverses on Im L (for more details, see [50] or [51]). We are now ready to study some semilinear boundary value problems.

Theorem 5.1. Let $D \subset R^n$ be a bounded open domain whose boundary is Lipschitzian with sufficiently small Lipschitz constants, $a_{ij} \colon D \to R$ as above, $i, j = 1, \ldots, n$, and f: cl $D \times R \to R$ continuous. Suppose that the following conditions hold:

1. There exists $\beta \geq 0$, $s \geq 0$ such that for each $t \in$ cl D and each $x \in R$,

$$|f(t, x)| \leq \beta|x| + s.$$

2. There exists $R > 0$ such that for each $y \in C^0($cl $D)$ such that

$$\inf_{t \in \text{cl } D} |y(t)| \geq R$$

one has

$$\int_D f(t, y(t)) \, dt \neq 0.$$

3. $\left[\int_D f(t, -R) \, dt \right] \left[\int_D f(t, R) \, dt \right] < 0.$

Then if β is sufficiently small, the variational Neumann semilinear boundary value problem

$$\int_D \left[\sum_{i, j=1}^{n} a_{ij}(t) D_i x(t) D_j v(t) \right] dt = \int_D f(t, x(t)) v(t) \, dt, \qquad (5.5)$$

for every $v \in H^1(D)$, has at least one solution x that is Hölder-continuous.

The proof, which is given in [50] or [51], consists in writing the problem in the abstract form

$$Lx = Nx$$

in $C^0($cl $D)$ with L defined above and $Nx(t) = f(t, x(t))$, and verifying that ker L is the subset of dom L of constant functions on cl D and

$$\text{Im } L = \left\{ x \in C^0(\text{cl } D): \int_D x(t) \, dt = 0 \right\}.$$

The theorem is then a consequence of Corollary 3.1.

The following result is also proved in [51].

Theorem 5.2. Suppose that f satisfies condition 1 of Theorem 5.1 and is of the form

$$f(t, x) = h(t) - g(x),$$

where h: cl $D \to R$ and $g: R \to R$ are continuous. Then if

$$g_{\pm} = \lim_{x \to \pm \infty} g(x)$$

exist (possibly infinite) and if either

$$g_- < g(x) < g_+ \qquad \text{or} \qquad g_+ < g(x) < g_-$$

for $x \in \text{cl } D$, a necessary and sufficient condition for the existence of one solution for (5.5), when β is sufficiently small, is that either

$$g_- < h_0 < g_+ \qquad \text{or} \qquad g_+ < h_0 < g_-$$

with

$$h_0 = (\text{meas } D)^{-1} \int_D h.$$

Let us now consider a semilinear Dirichlet boundary value problem for an arbitrary m. Let $f: D \times R \to R$ be a function satisfying Caratheodory conditions, i.e., such that

(i)   For each fixed $x \in R$, the function $t \mapsto f(t, x)$ is measurable in D.
(ii) For fixed $t \in D$ (a.e.), the function $x \mapsto f(t, x)$ is continuous in R.

Suppose also that there exist constants $\mu \geq 0$, $\delta \in [0, 1[$ and a function $d \in L^2(D)$ such that for $t \in D$ (a.e.),

$$|f(t, x)| \leq \mu |x|^\delta + d(t).$$

This implies in particular that the mapping $\tilde{N}$ defined by

$$\tilde{N}x(t) = f(t, x(t))$$

is a continuous mapping from $L^2(D)$ into itself and that

$$|\tilde{N}x| \leq \mu |x|^\delta + |d|,$$

where $|\cdot|$ denotes the $L^2(D)$-norm. We shall use Corollary 3.2 to give a different and shorter proof of a slight improvement of a theorem due to De Figueiredo [17].

Theorem 5.3.   Suppose that the assumptions above hold for f. If there exist functions $h_+ \in L^{2/(1-\delta)}(D)$, $h_- \in L^{2/(1-\delta)}(D)$ such that

$$\lim_{x \to \pm \infty} |x|^{-\delta} f(t, x) = h_\pm(t), \tag{5.6}$$

and if for all $v \in \ker \tilde{L} \cap \text{bdry } B(1)$, one has

$$\int_{D_+} h_+ |v|^{1+\delta} - \int_{D_-} h_- |v|^{1+\delta} > 0 \tag{5.7}$$

with $D_{\pm} = \{t \in D : v(t) \gtrless 0\}$, then the semilinear variational Dirichlet problem

$$a(x, v) = \int_D f(t, x(t))v(t)\, dt, \qquad (5.8)$$

for each $v \in H_0{}^m(D)$, with $a(x, v)$ defined in (5.1), has at least one solution.

Proof. It follows from the above discussion that (5.8) is equivalent to the abstract equation in $L^2(D)$

$$\tilde{L}x = \tilde{N}x$$

with $\tilde{L}$ and $\tilde{N}$ defined above, and assumption (1) of Corollary 3.2 holds. Let us show now that condition $(2'')$ of the same corollary is satisfied. If not, there will exist a bounded $V \subset \operatorname{Im} L$, a sequence (t_n) with $t_n > 0$ and $t_n \to \infty$ when $n \to \infty$, a sequence (z_n) with $z_n \in V$, $n \in N^*$, and a sequence (w_n) with $w_n \in \ker L \cap \operatorname{bdry} B(1)$, such that

$$\int_D f(t, t_n w_n(t) + t_n z_n(t))w_n(t)\, dt \le 0, \qquad n \in N^*. \qquad (5.9)$$

By going if necessary to subsequences, we can assume that $w_n \to w$ and $w_n + t_n^{\delta - 1} z_n \to w$ in $L^2(D)$ and $w_n(t) \to w(t)$, $w_n(t) + t_n^{\delta - 1} z_n(t) \to w(t)$, a.e. in D, when $n \to \infty$. Hence, for almost each $t \in D_+$ (resp. D_-), there exists an integer $n_0(t) > 0$ such that, if $n \ge n_0(t)$,

$$w_n(t) + t_n^{\delta - 1} z_n(t) > 0 \quad (\text{resp.} < 0),$$

which implies that a.e. in D_+ (resp. D_-),

$$t_n w_n(t) + t_n^{\delta} z_n(t) = t_n(w_n(t) + t_n^{\delta - 1} z_n(t)) \to \infty \quad (\text{resp.} -\infty)$$

if $n \to \infty$. Now, for each $n \in N^*$,

$$0 \ge \int_D t_n^{-\delta} f(t, t_n w_n(t) + t_n^{\delta} z_n(t))w_n(t)\, dt$$

$$= \int_D t_n^{-\delta} f(\cdot)w(t)\, dt + \int_D t_n^{-\delta} f(\cdot)(w_n(t) - w(t))\, dt$$

$$= \int_{D_+} t_n^{-\delta} f(\cdot)w(t)\, dt + \int_{D_-} t_n^{-\delta} f(\cdot)w(t)\, dt + \int_D t_n^{-\delta} f(\cdot)(w_n(t) - w(t))\, dt$$

$$= I_n{}^+ + I_n{}^- + I_n'. \qquad (5.10)$$

Using the Schwarz inequality and the convergence in $L^2(D)$ of $(w_n + t_n^{\delta-1} z_n)$ to w, we obtain

$$|I_n'| \le \left(\int_D t_n^{-2\delta} f^2(\cdot)\, dt \right)^{1/2} |w_n - w|$$

$$\le (\mu |w_n + t_n^{\delta-1} z_n|^\delta + t_n^{-\delta} |d|)|w_n - w| \le M|w_n - w|,$$

for some constant $M > 0$, and hence $I_n' \to 0$ if $n \to \infty$. On the other hand, the sequence $(t_n^{-\delta}(f(\cdot, t_n w_n(\cdot) + t_n^\delta z_n(\cdot))))$, bounded in $L^2(D)$, and hence in $L^2(D_\pm)$, converges weakly in $L^2(D_\pm)$ to its pointwise limit, which is, using (5.6), equal a.e. to $h_\pm |w|^\delta$. Hence,

$$\int_{D_\pm} t_n^{-\delta} f(\cdot)w(t)\, dt \to \pm \int_{D_\pm} h_\pm(t)|w(t)|^{1+\delta}\, dt.$$

Going to the limit in (5.10) we finally obtain, using (5.7),

$$0 \ge \int_{D_+} h_+(t)|w(t)|^{1+\delta}\, dt - \int_{D_-} h_-(t)|w(t)|^{1+\delta}\, dt > 0,$$

a contradiction. The proof is now complete.

An easy consequence of Theorem 5.3 is the following.

Corollary 5.1. Suppose that conditions of Theorem 5.3 hold with $\delta = 0$ and that a.e. in D and for each $x \in R$,

$$h_-(t) \le f(t, x) \le h_+(t). \tag{5.11}$$

Then condition (5.7) with nonstrict inequality and $\delta = 0$ is necessary for the existence of one solution for (5.8), the same condition with the strict inequality being sufficient.

Proof. Sufficiency has been proved. Now, if x is a solution of (5.8), we get, taking v in ker L and using the symmetry of the bilinear form a,

$$\int_D f(t, x(t))v(t)\, dt = 0,$$

i.e.,

$$\int_{D_+} f(t, x(t))|v(t)|\, dt - \int_{D_-} f(t, x(t))|v(t)|\, dt = 0,$$

which together with (5.11) implies

$$\int_{D_+} h_+(t)|v(t)|\, dt - \int_{D_-} h_-(t)|v(t)|\, dt \ge 0.$$

Corollary 5.1 with $m = 1$ is essentially the result initially proved by Landesman and Lazer [35], using in an ingenious but fairly complicated way the Schauder fixed-point theorem, and improved, using the same type of argument, by Williams [78] and Nečas [55]. Another proof, using a perturbation argument, has been given by Hess [28] and developed by De Figueiredo [17]. For various results of the same type, see also Fučik *et al.* [24], Fučik [23], and De Figueiredo [18].

Let us suppose now that D is a bounded domain of R^n with smooth boundary and that $\mathscr{L}$ is a linear elliptic partial differential operator of order m with smooth coefficients acting on scalar functions satisfying "coercive" (Lopatinsky–Shapiro) smooth boundary conditions

$$\mathscr{B}x = 0$$

on bdry D, expressed in terms of $m/2$ differential operators of order strictly smaller than m [42]. Then it is known that the operator $\mathscr{L}$ acting on such functions is of Fredholm type and we shall suppose that its index is zero. We shall consider the boundary value problem

$$\mathscr{L}x = f(t, x), \quad t \in D, \qquad \mathscr{B}x = 0, \quad t \in \text{bdry } D \qquad (5.12)$$

with $f: D \times R \to R$ continuous and having limits as $x \to \pm \infty$; for simplicity, we shall assume that

$$\lim_{x \to \pm \infty} f(t, x) = h_\pm(t)$$

uniformly for $t \in D$. The solutions of (5.1) are to be understood as functions belonging to $H^{m, \, p}(D)$, i.e., having generalized derivatives up to the other m, which belong to $L^p(D)$, for every $p < \infty$. Let $w_1, \ldots, w_d$ (resp. $w_1', \ldots, w_d'$) be smooth functions spanning ker $\mathscr{L}$ [resp. $(\text{Im } \mathscr{L})^\perp$]. If $a = (a_1, \ldots, a_d)$ is any vector in R^d, denote

$$\sum_{i=1}^d a_i w_i = a \cdot w$$

and define $\varphi: R^d \to R^d$ by

$$\varphi_i(a) = \int_{a \cdot w > 0} h_+(t)w_i'(t) \, dt + \int_{a \cdot w < 0} h_-(t)w_i'(t) \, dt, \qquad i = 1, \ldots, d.$$

We are now ready to formulate some results for (5.12), which are due to Nirenberg [56].

Lemma 5.1. Suppose that the following condition holds:

(UC) The only solution of

$$\mathscr{L}w = 0, \qquad \mathscr{B}w = 0 \qquad \text{on} \quad \text{bdry } D$$

that vanishes on a set of positive measure in D is $w = 0$.

Then, if S^{d-1} is the unit sphere in R^d, the mapping $\varphi\colon S^{d-1} \to R^d$ is continuous.

The proof of this lemma is lengthy and can be found in [56]. It depends on the following.

Proposition 5.1. Under the assumptions of Lemma 5.1, one has

$$\lim_{r \to \infty} |\Phi_i(u, ra) - \varphi_i(a)| = 0 \qquad (r > 0)$$

uniformly for u bounded in $L^1(D)$ and $a \in S^{d-1}$, where

$$\Phi_i(u, ra) = \int_D f(t, ra \cdot w(t) + u(t))w_i'(t)\, dt, \qquad i = 1, \ldots, d.$$

We now have Nirenberg's theorem, which we prove in a different way, using Theorem 3.3.

Theorem 5.4. Assume that the assumptions above and condition (UC) hold. Then if $0 \notin \varphi(S^{d-1})$ and if $d_B[\tilde{\varphi}, B(1), 0] \neq 0$, where $\tilde{\varphi}$ is any continuous extension of $\varphi|S^{d-1}$ to cl $B(1)$, then problem (5.12) has at least one solution.

Proof. To apply Theorem 3.3 we first formulate problem (5.12) in an abstract way by taking $X = C^0(\text{cl } D)$ with the uniform norm and, for some fixed $p > n$, dom $\tilde{L} = \{x \in H^{m,\,p}(D) : \mathscr{B}x = 0\}$. From the theory of linear elliptic problems [42], it is known that dom $\tilde{L} \subset C^0(\text{cl } D)$, with the canonical injection compact. Let

$$\tilde{L}\colon \text{dom } \tilde{L} \to L^p(D), \quad x \mapsto \mathscr{L}x, \qquad N\colon C^0(\text{cl } D) \to C^0(\text{cl } D), \quad x \mapsto f(\cdot, x(\cdot)),$$

$$\text{dom } L = \tilde{L}^{-1}(C^0(\text{cl } D)), \qquad L\colon \text{dom } L \to C^0(\text{cl } D), \quad x \mapsto \mathscr{L}z.$$

It follows from the regularity theory of linear elliptic problems [42] that (L, N) satisfies the required Fredholm and compactness assumptions and that each solution in dom L of

$$Lx = Nx \tag{5.13}$$

is a solution of (5.12). Now, condition (1) of Theorem 3.3 clearly holds with $\delta = 0$. Now, taking Q to be the projector such that $\text{Im } Q = \text{span}(w_1', \ldots, w_d')$ and $\ker Q = \text{Im } L$, and using the fact that $0 \notin \varphi(S^{d-1})$ and Proposition 5.1, we see that for every bounded V in $\text{Im } L$, there exists $r_0 > 0$ such that for each $r \geq r_0$, each $z \in V$, and each $a \in S^{d-1}$, one has

$$QN(ra \cdot w + z) \neq 0.$$

which is essentially condition (2) of Theorem 3.3. Now, if

$$0 < \mu \leq \inf_{a \in S^{d-1}} |\varphi(a)| = \inf_{a \in S^{d-1}} |\tilde{\varphi}(a)|$$

and if $r_1 > 0$ is so large that

$$\sup_{a \in S^{d-1}} |\Phi(0, r_1 a) - \varphi(a)| < \mu,$$

we have, by Rouché's theorem that

$$d_{\text{B}}[JQN|\ker L, B(r_1), 0] = d_{\text{B}}[\Phi(0, r_1 \cdot), B(1), 0] = d_{\text{B}}[\tilde{\varphi}, B(1), 0] \neq 0,$$

where $J: \text{Im } Q \to \ker L$, $\sum_{i=1}^{d} b_i w_i' \mapsto \sum_{i=1}^{d} b_i w_i$, and the proof is complete.

For extensions to elliptic systems of differential equations, see [56].

6. Coincidence Degree and Alternative Problems

Very recently, Hetzer [29] has shown that when X and Z are Banach spaces, the coincidence degree theory sketched in Section 1 could be developed in a similar way as in [47] by using, instead of the Leray–Schauder theory, the more general degree for α-contractions due to Nussbaum [57], Borisovich and Sapronov [3], Sadovskii [65], and others. We shall exhibit in this section the main aspects of Hetzer's article and then develop its relations with alternative problems.

Let us first recall that if Y is a metric space and B a subset of Y, the (Kuratowski) *measure $\alpha(B)$ of noncompactness of B* [33] is defined by

$$\alpha(B) = \inf \{d > 0 : B \text{ has a finite cover by sets}$$
$$\text{having a diameter smaller than } d\}.$$

If Y_1 and Y_2 are metric spaces, a continuous mapping $f: Y_1 \to Y_2$ will be said to be an *α-contraction* (or a *k-set contraction*) if there is a $k \in [0, 1[$ such that, for each bounded set $B \subset Y_1$, one has

$$\alpha(f(B)) \leq k\alpha(B).$$

When Y_2 is a Banach space and Y_1 a subset of a Banach space, an example of α-contraction $f\colon Y_1 \to Y_2$ is given by $f = f_1 + f_2$, with f_1 Lipschitzian of constant $k \in [0, 1[$ and f_2 compact on Y_1.

Now let X and Z be real Banach spaces and $L\colon \operatorname{dom} L \subset X \to Z$ a linear Fredholm mapping of index zero for which notations of Section 2 will be conserved. If $\Omega \subset X$ is open and bounded, we shall say that the mapping $N\colon \operatorname{cl} \Omega \to Z$ is an L-α-contraction if:

(a) $\Pi N\colon \operatorname{cl} \Omega \to \operatorname{coker} L$ is continuous and $\Pi N(\operatorname{cl} \Omega)$ is bounded.

(b) $K_{P,Q} N\colon \operatorname{cl} \Omega \to X$ is an α-contraction.

By noting that, if $\tilde{P}, \tilde{Q}$ are other continuous projectors such that

$$\operatorname{Im} \tilde{P} = \ker L, \qquad \operatorname{Im} L = \ker \tilde{Q},$$

we have (cf. [47] or [51])

$$K_{\tilde{P}, \tilde{Q}} N = (I - \tilde{P}) K_{P,Q} N + (I - \tilde{P}) K_P (Q - \tilde{Q}) N,$$

It is not hard, using some of the basic properties of the measure of non-compactness, to check that $K_{P,Q} N$ is an α-contraction with the same constant. Thus, the concept is independent of the choice of projectors P and Q.

Moreover, if we now assume that

$$0 \notin (L - N)(\operatorname{dom} L \cap \operatorname{bdry} \Omega), \tag{6.1}$$

then we can still define the *coincidence degree of L and N in Ω* by

$$d[(L, N), \Omega] = d_\alpha[I - P - (\Lambda\Pi + K_{P,Q})N, \Omega, 0], \tag{6.2}$$

where d_α denotes the degree for perturbations of identity by α-contractions. Hetzer has shown in [29] that the argument of [47] for proving the invariance of the right-hand member of (6.2) with respect to P, Q and orientation-preserving isomorphisms $\Lambda\colon \operatorname{coker} L \to \ker L$ still holds in this new frame. Moreover, Hetzer [29] has also shown that the basic properties of coincidence degree are preserved in this extension, as well as the generalized continuation theorem of [47] recalled in Section 2 and the generalized Granas theorem [48] recalled in Section 3. An interesting application of those existence theorems would be the existence of T-periodic solutions (or of other boundary value problems) for neutral functional differential equations of the form

$$x'(t) = f(t, x_t, x_t')$$

with f T-periodic in t, continuous, taking bounded sets into bounded sets, and Lipschitzian of constant $k \in [0, 1[$ with respect to the third variable (see Sadovskii [64]).

It is clear that very useful sufficient conditions for the L-compactness of a mapping N are either that K_P is continuous and N compact or K_P compact and N continuous and taking bounded sets into bounded sets. An interesting sufficient condition for the L-α-contractive character of N has been introduced by Hetzer in [29], when L is a closed Fredholm mapping. He introduces the (finite) number.

$$l(L) = \sup \{r \in R_+ : \text{for each bounded } B \subset \text{dom } L, r\alpha(B) \leq \alpha(L(B)).$$

He then proves the following:

Proposition 6.1. A sufficient condition for the α-contraction N: cl $\Omega \to Z$, with constant k, to be an L-α-contraction on cl Ω is that $k \in [0, l(L)[$.

We shall also note that the number $l(L)$, which is defined for closed Φ_+ and Φ_- operators, has been used by Hetzer in [29] to strengthen and unify some known results on the perturbations of semi-Fredholm operators.

Let us now develop the relations of this coincidence degree with the classical alternative problems [26, Chapter IX] when condition (b) above in the definition of L-α-contraction is replaced by the stronger condition

(b′)   There exists $k \in [0, 1[$ such that for each $x, y \in$ cl Ω,

$$|K_{P,Q}(Nx - Ny)| \leq k|x - y|.$$

Then, as shown by Browder in [7], the mapping H: cl $\Omega \to X$ defined by

$$H = I - K_{P,Q}N$$

belongs to a convex class of *permissible homeomorphisms* [6], i.e., of homeomorphisms on cl Ω such that H is a homeomorphism of Ω onto an open set $H(\Omega)$ of X that maps cl Ω homeomorphically onto cl $(H(\Omega))$. Also, as easily checked, H preserves the fibers of P, i.e.,

$$PHx = Px, \qquad x \in \text{cl } \Omega,$$

which implies at once that H^{-1}: cl $H(\Omega) \to X$ also preserves the fibers of P, i.e.,

$$PH^{-1}(x) = Px, \qquad x \in \text{cl } \Omega. \tag{6.3}$$

One has also, if $\Sigma = \Omega$ or cl Ω,

$$PH(\Sigma) = P(\Sigma), \qquad H(\Sigma) \cap \operatorname{Im} P \subset P(\Sigma). \tag{6.4}$$

We can now prove the following basic

Theorem 6.1. If conditions above hold, then

$$d[(L, N), \Omega] = d_{\mathrm{B}}[-\Lambda\Pi N H^{-1}|\ker L, H(\Omega) \cap \ker L, 0]. \tag{6.5}$$

Proof. If

$$F = I - P - (\Lambda\Pi + K_{P,Q})N, \qquad C = -\Lambda\Pi N - P,$$

$$V(x, y) = K_{P,Q} N x + \Lambda\Pi N y + P y, \qquad S(x, y) = x - V(x, y),$$

then V satisfies the conditions of Section 1 in Nussbaum [57], and hence

$$d[(L, N), \Omega] = d_{\alpha}[I - V(\cdot, \cdot), \Omega, 0] = \deg_M[F, \Omega, 0], \tag{6.6}$$

where M is the convex class of strict-contractive perturbations of identity and $\deg_M [F, \Omega, 0]$ is the Browder–Nussbaum degree [6, 8] defined here by

$$\deg_M [F, \Omega, 0] = \deg_M [[F, S], \Omega, 0] = d_{\mathrm{LS}}[I + H^{-1}C, -C^{-1}(H(\Omega)), 0].$$

On the other hand, we are also in a position to apply Theorem 3 of Browder [6] and hence

$$\deg_M [F, \Omega, 0] = \deg_1 [[F, H], \Omega, 0], \tag{6.7}$$

where the right-hand member is the Browder degree defined in [6] by

$$\deg_1 [[F, H], \Omega, 0] = d_{\mathrm{LS}}[I + CH^{-1}, H(\Omega), 0]. \tag{6.8}$$

Now, using (6.3) and the definition of Leray–Schauder degree,

$$d_{\mathrm{LS}}[I + CH^{-1}, H(\Omega), 0] = d_{\mathrm{LS}}[I - P - \Lambda\Pi N H^{-1}, H(\Omega), 0]$$
$$= d_{\mathrm{B}}[-\Lambda\Pi N H^{-1}|\ker L, H(\Omega) \cap \ker L, 0], \tag{6.9}$$

and the result follows from Eqs. (6.6)–(6.9).

Theorem 6.1 can be improved if one makes a supplementary assumption, which is classical in alternative methods.

Theorem 6.2. If conditions of Theorem 6.1 hold and if for each $a \in P(\mathrm{cl}\,\Omega)$, the mapping $T_a = x \mapsto a + K_{P,Q} N x$ is a strict contraction of cl Ω into itself such that, if $a \in P(\Omega)$, $T_a(\Omega) \subset \Omega$, then

$$d[(L, N), \Omega] = d_{\mathrm{B}}[-\Lambda\Pi N R|\ker L, P(\Omega), 0]$$

where $R(a)$ is the (unique) solution of the equation

$$y = a + K_{P,Q} N y.$$

Proof. By the assumption and the Banach fixed-point theorem, $R(a)$ exists and is unique for each $a \in P(\text{cl } \Omega)$, $R : P(\text{cl } \Omega) \to X$ is continuous, and $R(a) \in \Omega$ if $a \in P(\Omega)$. Also, necessarily, $R(a) = H^{-1}(a)$ for $a \in P(\text{cl } \Omega)$, which implies that $H(\Omega) \supset P(\Omega)$ and therefore, using (6.4), that $P(\Omega) = H(\Omega) \cap \ker L$. The result then follows from Theorem 6.1.

If we note that $\Lambda \Pi N$ can be written JQN with $J : \text{Im } Q \to \ker L$ some isomorphism, we note that $-\Lambda \Pi N R$ is nothing but the mapping defining the *bifurcation equations* of the corresponding alternative method. Hence, Theorem 6.2 explicitly relates the coincidence degree of L and N in Ω with the Brouwer degree of the bifurcation mapping at zero in $P(\Omega)$. It contains, as special cases, results of O'Neil and Thomas [58] and Thomas [73, 74] relating Browder–Nussbaum–Petryshyn degrees to Cronin's multiplicity [12, 14], which is nothing but the Brouwer degree of the bifurcation mapping of an alternative problem with $L = I - A$, A compact. See also Williams [77] for a corresponding result in the line of Cesari's approach.

7. Nonlinear Perturbations of Fredholm Mappings of Nonzero Index

All the results exposed up to now concern nonlinear perturbations of linear Fredholm mappings with zero index. When Ind $L > 0$, one can still find linear one-to-one mappings $\Lambda : \text{coker } L \to \ker L$, but the following fact, noted in [47] and [56], makes necessary the use of topological tools more sophisticated than the degree.

Proposition 7.1. If $L : \text{dom } L \subset X \to Z$, with X, Z normed spaces, is a Fredholm mapping with Ind $L > 0$, if $N : \text{cl } \Omega \subset X \to Z$ is L-compact with $\Omega \subset X$ open and bounded, and if $0 \notin (L - N)(\text{dom } L \cap \text{bdry } \Omega)$, then for each Λ as above,

$$d_{\text{LS}}[I - P - (\Lambda \Pi + K_{P,Q})N, \Omega, 0] = 0.$$

This "negative" property has, however, been used by Rabinowitz [60] to prove an existence theorem for some semilinear elliptic boundary value problems that we shall just translate here in abstract form, referring to the original paper for applications and further comments.

Theorem 7.1. If $L: \operatorname{dom} L \subset X \to Z$ is a Fredholm linear mapping with Ind $L > 0$, and $N: X \to Z$ is L-compact on closed bounded sets of X and odd, then for each open, bounded, symmetric neighborhood Ω of 0, there exists at least one $x \in \operatorname{dom} L \cap \operatorname{bdry} \Omega$ such that $Lx = Nx$.

Proof. If the theorem is not true, there exists an open, bounded, symmetric neighborhood Ω of 0 such that $0 \notin (L - N)(\operatorname{dom} L \cap \operatorname{bdry} \Omega)$ and hence, using Borsuk's theorem for degree of odd mappings, $d_{\mathrm{LS}}[I - P - (\Lambda\Pi + K_{P,Q})N, \Omega, 0]$ is odd, a contradiction with Proposition 7.1.

Rabinowitz [60] has also proved that, under conditions of Theorem 7.1, the set of zeroes of $L - N$ has a symmetric unbounded component containing 0 and, using Krasnosel'skii's concept of genus [30], has more information on this set.

We shall now state without proof an interesting topological result due to Nirenberg [56]. Let X be a Banach space, $P: X \to X$ a continuous projector with range of dimension $m > 0$, $\Phi: \operatorname{Im} P \cap \operatorname{cl} B(r) \to Y \subset \operatorname{Im} P$ a continuous mapping such that $0 \notin \Phi(\operatorname{Im} P \cap \operatorname{bdry} B(r))$, with Y a subspace of $\operatorname{Im} P$ of dimension $p < m$. Let $\Sigma: S^{m-1} \to \operatorname{Im} P \cap \operatorname{bdry} B(1)$, $\Sigma': Y \cap \operatorname{bdry} B(1) \to S^{p-1}$ be isomorphisms, and

$$\Psi: S^{m-1} \to S^{p-1}, \qquad u \mapsto \Sigma'\Phi(r\Sigma u)/|\Phi(r\Sigma u)|$$

with S^{k-1} the unit sphere in R^k.

Proposition 7.2. If Ψ has nontrivial stable homotopy, then every mapping $F: \operatorname{cl} B(r) \to X$ such that $F = \tilde{F}(\cdot, 1)$ for some compact mapping $\tilde{F}: \operatorname{cl} B(r) \times [0, 1] \to X$ verifying

$$x \neq \tilde{F}(x, \lambda), \qquad x \in \operatorname{bdry} B(r), \quad \lambda \in [0, 1], \qquad F(\cdot, 0) = P + \Phi P$$

has at least one fixed point in $B(r)$.

This result is used in [56] to extend Theorem 5.4 to elliptic operators $\mathcal{L}$ with Ind $\mathcal{L} > 0$. In this assertion, the nonvanishing of the Brouwer degree of $\tilde{\varphi}$ is replaced by the assumption of nontriviality for the stable homotopy of the mapping $\Psi: S^{d-1} \to S^{d'-1}$ defined by $\Psi = \varphi/|\varphi|$, with $d' = \dim \operatorname{coker} \mathcal{L}$. Some further results are given by Cronin in [15]. More generally, Proposition 7.2 can be used to give the following extension of Theorem 2.1, with a quite similar proof.

Theorem 7.2. Let X, Z be Banach spaces; let L: dom $L \subset X \to Z$ be a linear Fredholm mapping with Ind $L > 0$; let N: cl $B(r) \subset X \to Z$ be L-compact; and suppose that conditions (1) and (2) of Theorem 2.1 hold with $\Omega = B(r)$. Then, if the mapping

$$\Psi: S^{m-1} \to S^{p-1}, \qquad u \mapsto \Sigma' \Lambda \Pi N(r\Sigma u)/|\Lambda \Pi N(r\Sigma u)|$$

with $m = \dim \ker L$, $p = \dim \operatorname{coker} L$, $\Sigma: S^{m-1} \to \ker L \cap \operatorname{bdry} B(1)$, Σ': Im $P \cap \operatorname{bdry} B(1) \to S^{p-1}$ some isomorphisms, has nontrivial stable homotopy, the conclusions of Theorem 2.1 hold.

Nirenberg's results for elliptic equations have been somewhat extended by Tromba [75] when f is smooth, by the use of the Elworthy–Tromba degree [19] for nonlinear proper Fredholm mappings on Banach manifolds based on the Pontryagin–Thom theory of framed cobordism.

When Ind $L < 0$, no linear one-to-one Λ: coker $L \to \ker L$ exists [47], which makes the problem very difficult.

REFERENCES

[1] G. D. Birkhoff and O. D. Kellog, Invariant points in function space, *Trans. Amer. Math. Soc.* **23** (1922), 96–115.

[2] P. Bohl, Über die Bewegung eines mechanischen Systems in der Nähe einer Gleichgewichtslage, *J. Reine Angew. Math.* **127** (1904), 179–276.

[3] Ju. G. Borisovic and Yu. I. Sapronov, A contribution to the topological theory of condensing operators, *Soviet Math. Dokl.* **9** (1968), 1304–1307.

[4] L. E. J. Brouwer, Über Abbildung von Mannigfaltigkeiten, *Math. Ann.* **71** (1912), 97–115.

[5] F. E. Browder, Topological methods for nonlinear elliptic equations of arbitrary order, *Pacific J. Math.* **17** (1966), 17–31.

[6] F. E. Browder, Topology and nonlinear functional equations, *Studia Math.* **31** (1968), 189–204.

[7] F. E. Browder, Nonlinear equations of evolution and nonlinear operators in Banach spaces, *Proc. Symp. Nonlinear Functional Anal.* **18**, Part II, Amer. Math. Soc., Providence, Rhode Island, 1975.

[8] F. E. Browder and R. D. Nussbaum, The topological degree for noncompact nonlinear mappings in Banach spaces, *Bull. Amer. Math. Soc.* **74** (1968), 671–676.

[9] R. Caccioppoli, Problemi non lineari in analisi funzionale, *Rend. Sem. Math. Roma* **1**(3) (1931–1932), 13–22.

[10] R. Caccioppoli, Sulie corrispondenze funzionali inverse diramate: teoria generale e applicazioni ad alcune equazioni funzionali non lineari e al problema di Plateau, *Rend. Acc. Naz. Lincei Cl. Sci. Mat. Fis. Natur.* **24** (1936), 258–263, 416–421.

[11] L. Cesari, Functional analysis and Galerkin's method, *Michigan Math. J.* **11** (1964), 385–414.

[12] J. Cronin, Branch points of solutions of equations in Banach space, *Trans. Amer. Math. Soc.* **69** (1950), 208–231; **76** (1954), 207–222.

[13] J. Cronin, The existence of multiple solutions of elliptic differential equations, *Trans. Amer. Math. Soc.* **68** (1950), 105–131.

[14] J. Cronin, A definition of degree for certain mappings in Hilbert space, *Amer. J. Math.* **73** (1951), 763–772.

[15] J. Cronin, Equations with bounded nonlinearities, *J. Differential Equations* **14** (1973), 581–596.

[16] J. Cronin, "Fixed Points and Topological Degree in Nonlinear Analysis." Amer. Math. Soc., Providence, Rhode Island, 1964.

[17] D. G. de Figueiredo, Some remarks on the Dirichlet problem for semilinear elliptic equations, *Univ. Brasilia Trabalho Mat.* **57** (March 1974).

[18] D. G. de Figueiredo, On the range of nonlinear operators with linear asymptotes which are not invertible, *Univ. Brasilia Trabalho Mat.* **59** (April 1974).

[19] K. D. Elworthy and A. J. Tromba, Differential structures and Fredholm maps on Banach manifolds, *in* "Global Analysis," pp. 45–94 (*Proc. Symp. Pure Math.* **15**). Amer. Math. Soc., Providence, Rhode Island, 1969.

[20] C. Fabry and C. Franchetti, Nonlinear equations with growth restrictions on the nonlinear term, *J. Differential Equations* (to appear).

[21] R. E. Fennell, Periodic solutions of functional differential equations, *J. Math. Anal. Appl.* **39** (1972), 198–201.

[22] R. Fiorenza, Sulla hölderianità delle soluzioni dei problemi di derivata obliqua regolare del secondo ordine, *Ricerce Mat.* **14** (1965), 102–123.

[23] S. Fucik, Further remarks on a theorem by E. M. Landesman and A. C. Lazer, *Comm. Math. Univ. Carolinae* **15** (1974), 259–271.

[24] S. Fucik, M. Kucera, and J. Necas, Ranges of nonlinear asymptotically linear operators, *J. Differential Equations* **17** (1975), 375–394.

[25] A. Granas, The theory of compact vector fields and some of its applications to topology of functional spaces (I), *Rozprawy Mat.* **30** (1962), 1–91.

[26] J. K. Hale, "Ordinary Differential Equations." Wiley (Interscience), New York, 1969.

[27] J. K. Hale and J. Mawhin, Coincidence degree and periodic solutions of neutral equations, *J. Differential Equations* **15** (1974), 295–307.

[28] P. Hess, On a theorem by Landesman and Lazer, *Indiana Univ. Math. J.* **23** (1974), 827–830.

[29] G. Hetzer, Some remarks on Φ_+-operators and on the coincidence degree for a Fredholm equation with noncompact nonlinear perturbations, *Ann. Soc. Sci. Bruxelles* (in press).

[30] M. A. Krasnosel'skii, "Topological Methods in the Theory of Nonlinear Integral Equations." Pergamon, Oxford, 1963.

[31] M. A. Krasnosel'skii, "The Operator of Translation along Trajectories of Differential Equations." Amer. Math. Soc., Providence, Rhode Island, 1968.

[32] L. Kronecker, Über Systeme von Funktionen mehrerer Variabeln, *Monatsb. Berlin Akad.* (1869), 159–193, 688–698.

[33] C. Kuratowski, Sur les espaces complets, *Fund. Math.* **15** (1930), 301–309.

[34] O. A. Ladyzenskaya and N. N. Ural'ceva, "Equations aux dérivées partielles de type elliptique." Dunod, Paris, 1968.

[35] E. M. Landesman and A. C. Lazer, Nonlinear perturbations of linear elliptic boundary value problems at resonance, *J. Math. Mech.* **19** (1970), 609–623.

[36] A. C. Lazer, On Schauder's fixed point theorem and forced second-order nonlinear oscillations, *J. Math. Anal. Appl.* **21** (1968), 421–425.

[37] A. C. Lazer and D. E. Leach, Bounded perturbations of forced harmonic oscillators at resonance, *Ann. Mat. Pura Appl.* **82**(4) (1969), 49–68.

[38] S. Lefschetz, Existence of periodic solutions for certain differential equations, *Proc. Nat. Acad. Sci. U.S.A.* **29** (1943), 29–32.

[39] S. Lefschetz, "Differential Equations: Geometric Theory," 2nd ed. Wiley (Interscience), New York, 1963.

[40] J. Leray and J. Schauder, Topologie et équations fonctionnelles, *Ann. Ecole Norm. Sup.* **51**(3) (1934), 45–78.

[41] N. Levinson, On the existence of periodic solutions for second order differential equations with a forcing term, *J. Math. Phys.* **22** (1943), 41–48.

[42] J. L. Lions and E. Magenes, "Problèmes aux limites non homogènes et applications," Vol. 1. Dunod, Paris, 1969.

[43] A. M. Lyapunov, Sur les figures d'équilibre peu différentes des ellipsoïdes d'une masse liquide homogène dotée d'un mouvement de rotation, *Zap. Akad. Nauk. St. Petersbourg* (1906) 1–225; (1908) 1–175; (1912) 1–228; (1914) 1–112.

[44] J. Mawhin, Degré topologique et solutions périodiques des systèmes différentiels non linéaires, *Bull. Soc. Roy. Sci. Liège* **38** (1969), 308–398.

[45] J. Mawhin, Equations intégrales et solutions périodiques des systèmes différentiels non linéaires, *Acad. Roy. Belgique, Bull. Cl. Sci.* **55**(5) (1969), 934–947.

[46] J. Mawhin, Equations non linéaires dans les espaces de Banach, *Univ. Louvain, Sémin. Math. Appl. Méch.* **39**. Vander, Louvain (1971).

[47] J. Mawhin, Equivalence theorems for nonlinear operator equations and coincidence degree theory for some mappings in locally convex topological vector spaces, *J. Differential Equations* **12** (1972), 610–636.

[48] J. Mawhin, The solvability of some operator equations with a quasibounded nonlinearity in normed spaces, *J. Math. Anal. Appl.* **45** (1974), 455–467.

[49] J. Mawhin, Periodic solutions of some vector retarded functional differential equations, *J. Math. Anal. Appl.* **45** (1974), 588–603.

[50] J. Mawhin, Problèmes aux limites du type de Neumann pour certaines équations différentielles ou aux dérivées partielles non linéaires, *in* "Equations Différentielles et Fonctionnelles non Linéaires," pp. 123–134. Hermann, Paris, 1973.

[51] J. Mawhin, Nonlinear perturbations of Fredholm mappings in normed spaces and applications to differential equations, *Univ. Brasilia Trabalho Mat.* **61** (May 1974).

[52] J. Mawhin, Recent results in coincidence theory for some mappings in normed spaces, *in* "Problems in Nonlinear Functional Analysis," pp. 7–22, Bonn, 1975.

[53] J. Mawhin, Recent results on periodic solutions of differential equations *in Internat. Conf. Differential Equations*, 537–556. Academic Press, New York, 1975.

[54] C. Miranda, Un'osservazione su una teorema di Brouwer, *Boll. Un. Mat. Ital.* **3**(2) (1940), 527.

[55] J. Necas, On the range of nonlinear operators with linear asymptotes which are not invertible, *Comm. Math. Univ. Carolinae* **14** (1973), 63–72.

[56] L. Nirenberg, An application of generalized degree to a class of nonlinear problems, *in* "Troisième Coll. C.B.R.M. d'Analyse Fonctionnelle," pp. 57–74. Vander, Louvain, 1971.

[57] R. D. Nussbaum, Degree theory for local condensing maps, *J. Math. Anal. Appl.* **37** (1972), 741–766.

[58] T. O'Neil and J. W. Thomas, On the equivalence of multiplicity and the generalized topological degree, *Trans. Amer. Math. Soc.* **167** (1972), 333–345.

[59] H. Poincaré, Sur certaines solutions particulières du problème des trois corps, *Bull. Astronom.* **1** (1884), 65–74.

[60] P. H. Rabinowitz, A note on a nonlinear elliptic equation, *Indiana Univ. Math. J.* **22** (1972), 43–49.

[61] R. Reissig, Periodic solutions of a nonlinear n-th order vector differential equation, *Ann. Mat. Pure Appl.* **87**(4) (1970), 111–124.

[62] R. Reissig, G. Sansone, and R. Conti, "Qualitative Theorie Nichtlinearer Differential-gleichungen." Cremonese, Roma, 1963.

[63] N. Rouche and J. Mawhin, "Equations Différentielles Ordinaires," 2 volumes. Masson, Paris, 1973.

[64] B. N. Sadovskii, Application of topological methods in the theory of periodic solutions of nonlinear differential-operator equations of neutral type, Soviet Math. Dokl. 12 (1971), 1543–1547.

[65] B. N. Sadovskii, Limit-compact and condensing operators, Russian Math. Surveys 27 (1972), 85–156.

[66] G. Sansone and R. Conti, "Nonlinear Differential Equations." Pergamon, Oxford, 1964.

[67] J. Schauder, Der Fixpunktsatz in Funktionalräumen, Studia Math. 2 (1930), 171–180.

[68] E. Schmidt, Zur Theorie der linearen und nichtlinearen Integralgleichungen. 3 Teil. Über die Auflösung der nichtlinearen Integralgleichungen und die Verzweigung ihrer Lösungen, Math. Ann. 65 (1908), 370–399.

[69] G. Sedsiwy, Asymptotic properties of solutions of a certain n-th order vector differential equation, Rend. Acc. Naz. Lincei, Cl. Sci. Mat. Fis. Natur. 47(8) (1969), 472–475.

[70] G. Stampacchia, Problemi al contorno ellittici, con dati discontinui, dotati di soluzioni hölderiane, Ann. Mat. Pura Appl. 51(4) (1960), 1–37.

[71] F. Stoppelli, Su un'equazione differenziale della meccanica dei fili, Rend. Accad. Sci. Fis. Mat. Napoli 19(4) (1952), 109–114.

[72] V. V. Strygin, A theorem concerning the existence of periodic solutions of systems of differential equations with delayed arguments, Math. Notes USSR 8 (1970), 600–602.

[73] J. W. Thomas, The multiplicity of an operator is a special case of the topological degree for k-set contractions, Duke Math. J. 40 (1973), 233–240.

[74] J. W. Thomas, On a lower bound to the number of solutions to a nonlinear operator equation, Scripta Math. (to appear).

[75] A. J. Tromba, Stable homotopy groups of spheres and nonlinear P.D.E. (to appear).

[76] G. Villari, Soluzioni periodiche di una classe di equazioni differenziali, Ann. Mat. Pura Appl. 73(4) (1966), 103–110.

[77] S. A. Williams, A connection between the Cesari and Leray–Schauder methods, Michigan Math. J. 15 (1968), 441–448.

[78] S. A. Williams, A sharp sufficient condition for solution of a nonlinear elliptic boundary value problem, J. Differential Equations 8 (1970), 580–586.

A Survey of Bifurcation Theory*

PAUL H. RABINOWITZ

Department of Mathematics
University of Wisconsin, Madison, Wisconsin

Bifurcation is a term used in several parts of mathematics. It generally refers to a qualitative change in the objects being studied due to a change in the parameters on which they depend. We are interested in the set of solutions of functional equations, and for the examples we have in mind the following more precise definition suffices. Let X and Y be Banach spaces, $U \subset X$, and $F: U \to Y$. Suppose there is a one-to-one curve $\mathcal{I} = \{x(t) | t \in (0, 1)\} \subset U$ such that $F(z) = 0$ for $z \in \mathcal{I}$. A point $w \in \mathcal{I}$ is called a *bifurcation point for F with respect to $\mathcal{I}$* (or more simply a *bifurcation point*) if every neighborhood of w contains zeros of F not on $\mathcal{I}$. In applications, after possibly making a change of variables, one usually has $X = \mathbb{R} \times E$ with E a real Banach space, $F = F(\lambda, u)$, and $\mathcal{I} = \{(\lambda, 0) | \lambda \in (a, b) \subset \mathbb{R}\}$. We restrict our attention to this case. The members of $\mathcal{I}$ will be called trivial solutions of $F(\lambda, u) = 0$. Thus we are interested in nontrivial zeros of F.

Much of the motivation for studying bifurcation is provided by varied phenomena in the physical sciences, which can be formulated in these terms. We illustrate with some of the standard model cases.

I. Thermal Convection—The Bénard Problem

An infinite horizontal layer of a viscous incompressible fluid lies between a pair of perfectly conducting plates. A temperature gradient T is maintained between the plates, the lower plate being warmer. If T is appropriately small, the fluid remains at rest, the temperature is a linear function of the vertical coordinate, and heat is transported through the fluid solely by conduction. However, if T exceeds a certain value, the fluid undergoes time-independent motions called *convection currents* and heat is transported

---

*This paper was supported in part by the Office of Naval Research under Contract N00014-67-A-0128-0024. Reproduction in whole or in part is permitted for any purpose of the U.S. Government.

through the fluid by convection as well as conduction. In actual experiments the fluid breaks up into cells whose shape depends in part on the shape of the container. Mathematically the equilibrium configuration of the fluid is described by a system of nonlinear partial differential equations. For each value of T, there is a conduction solution satisfying these equations. Formulated in the general Banach space framework, these solutions correspond to the trivial solutions, while the value of T at which convection begins corresponds to a bifurcation point. For more on the physical and mathematical aspects of this problem, see [1, 2].

II. Rotating Fluids—The Taylor Problem

A viscous incompressible fluid lies between a pair of concentric cylinders (whose axis of rotation is vertical). The inner cylinder rotates at a constant angular velocity ω while the outer cylinder remains at rest. If ω is sufficiently small, the fluid particles move in circular orbits with velocity depending on their distance from the axis of rotation. Equilibrium states of the fluid are solutions of the time-independent Navier–Stokes equations. The above solution is called *Couette flow* and exists as a mathematical solution of the governing equations for all values of ω. When ω exceeds a certain critical value, the fluid breaks up into horizontal bands called *Taylor vortices* and a new fluid motion periodic in the vertical direction is superimposed on the Couette flow. For this example, Couette flow corresponds to the trivial solutions in the general framework and the value of ω producing the onset of Taylor vortices corresponds to a bifurcation point [1, 2].

III. Buckling Phenomena in Elasticity—The Flat Plate

A thin, planar, clamped elastic plate is subjected to a compressive force along its edges. If the magnitude λ of this force is small enough, the plate remains motionless and in equilibrium. However, if λ exceeds a certain value, the plate deflects out of its plane and assumes a new nonplanar equilibrium position called a *buckled state*. Equilibrium configurations of the plate satisfy a system of nonlinear partial differential equations called the von Kármán equations. The unbuckled state is a solution of these equations for all values of λ and corresponds to the trivial solutions, while the value of λ at which buckling occurs corresponds to a bifurcation point [3–5].

These examples have all dealt with bifurcation of time-independent solutions of a functional equation to new time-independent solutions. Of course, more complicated kinds of bifurcation may occur, such as bifurcation of time-independent to time-periodic solutions. This occurs in particular for the Taylor problem. There one finds experimentally that if ω is in the Taylor vortex flow regime and is increased still further, another bifurcation point is reached where the Taylor vortices are replaced by a new flow sometimes called wavy vortices [6], which is periodic both in the vertical direction and in time.

There is a considerable literature on bifurcation, both for particular physical problems and for the general theory. Several examples of the former are contained in [1] and [4]. As references for the latter, we suggest the survey papers of Vainberg and Trenogin [7], Prodi [8], Sather [9], Kirchgässner and Kielhöfer [2], and Stakgold [10]; the books of Krasnoselski [11] and Antman and Keller [4]; and the lecture notes of Pimbley [12], Sattinger [13], Iooss [14], and Nirenberg [15]. For a more general view of bifurcation, we refer the reader to the survey paper of Arnold [16]. Several other references will be given below but no attempt will be made to give a complete bibliography.

Returning to the general theory, there are three main questions of interest:

(i)   What are necessary and sufficient conditions for $(\mu, 0) \in \mathscr{I}$ to be a bifurcation point?

(ii)   What is the structure of the set of zeros of $F(\lambda, u)$ near $(\mu, 0)$?

(iii) In physical problems such as I–III above where there is an underlying evolution equation of which the solutions described are equilibrium solutions, determine which solutions are stable or unstable.

This survey will mainly address (i) and (ii) since (iii) involves rather different techniques. Moreover, we will focus on bifurcation of time-independent solutions of $F(\lambda, u) = 0$ to new time-independent solutions, although at the end of our survey some other cases will be discussed. Both analytical and topological methods have been employed to study (i) and (ii) and we begin with the former.

For what follows, $C^k(U, V)$ denotes k times continuously Frechet differentiable maps from U to V. The null space of a linear map is denoted by $N(L)$ and its range by $R(L)$. The dimension of a subspace Z of E or Y will be denoted by dim Z.

Suppose Ω is a neighborhood of 0 in E and $F \in C((a, b) \times \Omega, Y)$ with $F(\lambda, 0) = 0$ for $\lambda \in (a, b)$. If further F is Frechet differentiable with respect to u at $(\lambda, 0)$ and $F_u(\lambda, 0) \in C((a, b), Y)$, then a necessary condition that $(\mu, 0)$ be a bifurcation point is that $F_u(\mu, 0)$ is not an isomorphism from E to Y. Indeed, the hypotheses on F imply

$$(1) \qquad\qquad F(\lambda, u) = F_u(\lambda, 0)u + H(\lambda, u),$$

where $\|H(\lambda, u)\| = o(\|u\|)$ at $u = 0$. Therefore if $F(\lambda, u) = 0$ for (λ, u) near $(\mu, 0)$ and $F_u(\mu, 0)$ is an isomorphism,

$$(2) \qquad\qquad -u = F_u(\lambda, 0)^{-1} H(\lambda, u).$$

Since the right-hand side of (2) is $o(\|u\|)$ at $u = 0$, (2) can be satisfied near $u = 0$ only by the trivial solutions. If $F \in C^1((a, b) \times \Omega, Y)$ the result is immediate from the implicit function theorem.

This necessary condition is not sufficient as simple examples show. For example, for $E = Y = \mathbb{R}^2$,

$$(3) \qquad\qquad F(\lambda, u) = (1 - \lambda)\binom{u_1}{u_2} + \binom{u_2{}^3}{-u_1{}^3},$$

and $\mu = 1$, multiplying the first equation in (3) by u_2, the second by u_1, and subtracting shows F has no nontrivial zeros. Thus, to get bifurcation more must be assumed concerning $F_u(\mu, 0)$ or the nonlinear part of F.

If F is C^1 near $(\mu, 0)$ and $F_u(\mu, 0)$ is a Fredholm operator from E to Y, the bifurcation problem can be reduced to a finite-dimensional one via the so-called *method of Liapunov–Schmidt*. Namely, let Z be a complementary space to $N(F_u(\mu, 0))$ in E and write $u = v + w$, where $v \in N(F_u(\mu, 0))$ and $w \in Z$. If (λ, u) is near $(\mu, 0)$ and $F(\lambda, u) = 0$, by Taylor's theorem we can write

$$
\begin{aligned}
(4) \qquad 0 = F(\lambda, u) &= F(\mu, 0) + F_\lambda(\mu, 0)(\lambda - \mu) \\
&\quad + F_u(\mu, 0)w + o(\|v\| + \|w\| + |\lambda - \mu|) \\
&\equiv F_u(\mu, 0)w + H(v, w, \alpha),
\end{aligned}
$$

where $\alpha = \lambda - \mu$. Let $\tilde{Y}$ be a complementary subspace to $R(F_u(\mu, 0))$ in Y and let P (resp. $\tilde{P}$) denote the projector of Y onto $R(F_u(\mu, 0))$ (resp. $\tilde{Y}$) along $\tilde{Y}$ [resp. $R(F_u(\mu, 0))$]. Since $F_u(\mu, 0)$ is an isomorphism from Z to $R(F_u(\mu, 0))$, by the implicit function theorem there is a function $w = \varphi(\alpha, v)$ satisfying

$$(5) \qquad\qquad F_u(\mu, 0)w + PH(v, w, \alpha) = 0$$

for (α, v) near $(0, 0)$ with $\varphi(0, 0) = 0$ and φ continuously differentiable near $(0, 0)$. Thus, the bifurcation problem reduces to finding nontrivial solutions of

$$(6) \qquad\qquad \tilde{P}H(v, \varphi(\alpha, v), \alpha) = 0,$$

which represents dim $\tilde{Y}$ equations in $1 + \dim N(F_u(\mu, 0))$ unknowns. This generally is still a formidable problem and cannot be solved without more hypotheses. See, e.g., [7, 9], or [17–19] for further work on this subject.

A very useful special case and one that occurs in many applications (e.g., examples I and II above and sometimes III) arises when 0 is a "simple eigenvalue" of $F_u(\mu, 0)$. Then a fairly complete description of the solution set of $F = 0$ near $(\mu, 0)$ can be given. There are various versions of such results (see, e.g., [4, 8, 18–20]). First the notion of "simple eigenvalue" will be made more precise. Let $B(E, Y)$ denote the set of bounded linear maps from E to Y. If $L, K \in B(E, Y)$ and $r \in \mathbb{R}$, we say r *is a K-simple eigenvalue of* L if $\dim N(L - rK) = 1 = \operatorname{codim} R(L - rK)$ and

$$Kv \notin R(L - rK),$$

where v spans $N(L - rK)$. If $E = Y$, L is compact, and $K = I$, the identity map, it is easy to see that μ is an I-simple eigenvalue of L if and only if it is a simple eigenvalue of L in the usual sense (i.e., the algebraic and geometric multiplicities of μ are 1).

Theorem 7. Suppose $F \in C^2((a, b) \times \Omega, Y)$, $F(\lambda, 0) = 0$ for $\lambda \in (a, b)$, $\mu \in (a, b)$, and 0 is an $F_{\lambda u}(\mu, 0)$-simple eigenvalue of $F_u(\mu, 0)$. If Z is any complement of $N(F_u(\mu, 0)) \equiv \operatorname{span}\{v\}$ in E, then there exists a neighborhood of U of $(\mu, 0)$ in $\mathbb{R} \times E$, a $\delta > 0$, and functions $\varphi \in C^1((-\delta, \delta), \mathbb{R})$, $\psi \in C^1((-\delta, \delta), Z)$ such that $\varphi(0) = 0$, $\psi(0) = 0$, and

$$F^{-1}(0) \cap U = \{(\lambda, 0) | (\lambda, 0) \in U\} \cup \{(\varphi(s), sv + s\psi(s)) | |s| < \delta\}.$$

This theorem can be proved with the aid of the Lyapunov–Schmidt procedure but it is simpler to work directly and get the existence assertions from the implicit function theorem. The uniqueness is a consequence of the uniqueness inherent in the implicit function theorem together with an additional estimate. This version of the "simple eigenvalue theorem" is due to Crandall and Rabinowitz [21, 22]. For another interesting proof, see [15]. If F is real analytic, then by the analytic version of the implicit function theorem, (λ, u) are real analytic in s near $s = 0$ and may be obtained by a formal expansion procedure.

The smoothness assumption on F may be weakened slightly [21]. Minimal smoothness hypotheses prove useful for technical reasons when dealing with differential operators that may only be defined on a subspace of a Banach space [21].

For $E = Y = \mathbb{R}^n$, a bifurcation theorem has been obtained recently by Kopell and Howard [23] that proves to be equivalent to Theorem 7 for $E = \mathbb{R}^n$. (See also [24].) Their result is:

Theorem 8. If $F \in C^2((a, b) \times \Omega, \mathbb{R}^n)$ with $F(\lambda, 0) = 0$ for $\lambda \in (a, b)$, det $F_u(\mu, 0) = 0$, and $(d/d\lambda)(\det F_u(\lambda, 0))|_{\lambda = \mu} \neq 0$, then the conclusion of Theorem 7 is obtained.

When its hypotheses are satisfied, Theorem 7 gives a fairly satisfactory answer to questions (i) and (ii) raised above. It turns out that within its context, a partial answer can also be given to question (iii). As with ordinary differential equations, the stability of an equilibrium solution $(\bar{\lambda}, \bar{u})$ of $u_t = F(\lambda, u)$ is closely connected with the "spectrum" of $F_u(\bar{\lambda}, \bar{u})$. In fact, in the applied mathematics literature if the "spectrum" of $F_u(\bar{\lambda}, \bar{u})$ lies in an appropriate half-plane $(\bar{\lambda}, \bar{u})$ is often called stable and otherwise unstable. Likewise, the terms linearized stability and linearized instability are commonly used to describe the spectrum. The following result describes how the zero eigenvalue of $F_u(\mu, 0)$ changes along the zero set of F near $(\mu, 0)$.

Theorem 9. Suppose $E \subset Y$ with continuous injection I and 0 is an I-simple eigenvalue of $F_u(\mu, 0)$. If in addition the hypotheses of Theorem 7 are satisfied, then there exist open intervals α, $\beta \subset \mathbb{R}$ with $\mu \in \alpha$, $0 \in \beta$, and functions $\gamma \in C^1(\alpha, \mathbb{R})$, $\rho \in C^1(\beta, \mathbb{R})$, $x \in C^1(\alpha, E)$, $w \in C^1(\beta, E)$ such that

(i) $\gamma(\lambda)$ is an I-simple eigenvalue of $F_u(\lambda, 0)$ and $F_u(\lambda, 0)x(\lambda) = \gamma(\lambda)x(\lambda)$, $\lambda \in \alpha$.

(ii) $\rho(s)$ is an I-simple eigenvalue of $F_u(\lambda(s), u(s))$ and $F_u(\lambda(s), u(s))w(s) = \rho(s)w(s)$, $s \in \beta$.

(iii) $\gamma(\mu) = 0 = \rho(0)$, $x(\mu) = v = w(0)$, $x(\lambda) - v \in Z$, $w(s) - v \in Z$.

(iv) $\gamma'(\mu) \neq 0$.

(v)   Near $s = 0$, $\rho(s)$ and $-s\lambda'(s)\gamma'(\mu)$ have the same zeros and whenever $\lambda'(s) \neq 0$, the same sign.

The proof of Theorem 9 is an elementary consequence of the implicit function theorem, Theorem 7, and Taylor's theorem. Parts (i) and (ii) tell us how the zero eigenvalue of $F_u(\mu, 0)$ is perturbed along the zero set of F near $(\mu, 0)$. This leads to (v), which permits us to use qualitative knowledge of the bifurcation picture to deduce information about the spectrum of F_u. For example, for the Bénard problem with μ the smallest real number at which $F_u(\lambda, 0)$ develops a null space, it is fairly easy to show $\gamma'(\mu) < 0$ and $\lambda(s) > \mu$ for $s \neq 0$ [21, 22]. Hence $\rho(s) > 0$ for $s \neq 0$. Thus in a linearized sense the convection solutions $(\lambda(s), u(s))$ are stable since $\rho(s)$ moves into the half plane in $\mathbb{C}$ with positive real part [and the rest of the spectrum of $F_u(\lambda(s), u(s))$ remains there] while the conduction solutions $(\lambda, 0)$ are stable for $\lambda < \mu$ $[\gamma'(\mu) < 0]$ and unstable for $\gamma > \mu$. For the Taylor problem it is known experimentally that the same qualitative picture obtains but it has not yet been shown mathematically. The difficulty is in establishing the I-simplicity of 0.

The first result in the direction of Theorem 9 we know of is due to Hopf [24], who essentially treated the case $E = Y = \mathbb{R}^n$, F real analytic, and $\lambda(s) = cs + o(|s|)$ or $\lambda(s) = cs^2 + o(s^2)$ with $c \neq 0$. More recently, Sattinger [25] studied the case

$$(10) \qquad\qquad F(\lambda, u) = (L - \lambda B)u + H(\lambda, u) = 0,$$

where L and B are linear and L has a compact inverse that can be used to convert (10) to an equivalent problem

$$(11) \qquad\qquad (I - \lambda T)u + N(\lambda, u) = 0$$

with T and N compact. Under additional technical hypotheses [which implied $\gamma'(\mu) < 0$ and $\lambda'(s) \neq 0$ for $s \neq 0$] Sattinger used a topological degree argument to show $\rho(s)$ and $s\lambda'(s)$ have the same sign for $s \neq 0$. Theorem 9 is due to Crandall and Rabinowitz [22].

The next step in treating the stability question is to show how information about the spectrum of $F_u(\bar{\lambda}, \bar{u})$ with $(\bar{\lambda}, \bar{u})$ an equilibrium solution of

$$(11a) \qquad\qquad u_t = F(\lambda, u)$$

can be used to determine the behavior of solutions of (11a) for initial data near $\bar{u}$ (for λ fixed at $\bar{\lambda}$). We will not discuss this point here but refer the reader to the literature [26, 2, 13, 14], where this question is treated, often in the context of the Navier–Stokes equations.

We return again to questions (i) and (ii). In addition to analytical methods, topological degree theoretic arguments and variational methods have been successfully employed to study bifurcation problems and some of the main results obtained in this fashion will be presented next. The first such result we know of is due to Krasnoselski [11]. To state it, let $Y = E$, $L \in B(E, E)$, and μ a characteristic value of L, i.e., there exists $v \in E$, $v \neq 0$ such that $v = \mu L v$. We say μ is of multiplicity k if $\dim \operatorname{span} \bigcup_{j \in \mathbb{N}} N((I - \mu L)^j) = k$. Then Krasnoselski's result is

Theorem 12. Let $F(\lambda, u) = u - \lambda L u - H(\lambda, u)$ and Ω be a neighborhood of $(\mu, 0)$. If L is linear and compact, $H \in C(\overline{\Omega}, E)$ is compact with $H(\lambda, u) = o(\|u\|)$ at $u = 0$ uniformly on bounded λ intervals, and μ is a characteristic value of L of odd multiplicity, then $(\mu, 0)$ is a bifurcation point for F with respect to the line of trivial solutions.

The proof of Theorem 12 uses Leray–Schauder degree and exploits in particular the change in the sign of the Leray–Schauder index of $(\lambda, 0)$ as λ crosses μ. The odd multiplicity of μ is essential as the example of Eq. (3) shows.

Actually as has been shown by Rabinowitz [27], the hypotheses of Theorem 12 imply something global rather than local about the behavior of the nontrivial zeros of F. Let $\mathscr{S}$ denote the closure of the set of nontrivial solutions of $F = 0$. A component of $\mathscr{S}$ is a maximal (with respect to inclusion) closed connected subset of $\mathscr{S}$.

Theorem 13. Under the hypotheses of Theorem 12, $\mathscr{S}$ contains a component $\mathscr{C}$ that contains $(\mu, 0)$ and either intersects $\partial\Omega$ or contains $(\hat{\mu}, 0)$, where $\mu \neq \hat{\mu}$ and $\hat{\mu}$ is a characteristic value of L.

Corollary 14. If $\Omega = \mathbb{R} \times E$, the first alternative of Theorem 13 becomes $\mathscr{C}$ is unbounded.

The proofs of Theorem 13 and its corollary involve the same machinery as Theorem 12. See also Turner [28] and Ize in [15] for other proofs and extensions. Simple examples show either alternative of the theorem is possible. Several generalizations of Theorem 13 have been made and we will mention a few. The compactness requirements on L and H can be dropped at the expense of a local result. For example if $L \in B(E, E)$, μ is an isolated characteristic value of L of odd multiplicity, $I - \mu L$ has Fredholm index 0, and $H \in C^1(\Omega, E)$, then the Lyapunov–Schmidt procedure reduces the solution of $F = 0$ near $(\mu, 0)$ to a local finite-dimensional problem, so that Brouwer degree can be used to conclude bifurcation for this case. More general dependence of the linear part of F on λ can be permitted. The only property one really needs is that $(\lambda, 0)$ have different index on each side of $(\mu, 0)$. Likewise for global results, compact operators can be replaced by any more general class for which one has an appropriate notion of index, e.g., k-set contractions (see, e.g., Stuart [29]). Finally, if μ is a simple characteristic value of L, then $\mathscr{C}$ consists of two subcontinua that near $(\mu, 0)$ meet only at $(\mu, 0)$ [27]. Moreover, within the context of Corollary 14 Dancer [30] has shown each of these subcontinua has a global extension. An earlier result in this direction in [27] is not correct as stated.

Degree theoretic methods have also been successfully applied to bifurcation situations involving positive operators. The additional positivity structure permits the elimination of the odd multiplicity requirements of Corollary 14. The analog of this corollary for the positive operator case is

Theorem 15. Let K be a cone in E, $L: K \to K$ be compact and linear, and $H \in C(\mathbb{R}^+ \times K, K)$ with $H(\lambda, u) = o(\|u\|)$ at $u = 0$ uniformly on bounded λ intervals. If L has positive spectral radius and μ is the smallest positive characteristic value of L, then $F(\lambda, u) = u - \lambda(Lu + H(\lambda, u))$ has a component of nontrivial zeros $\mathscr{C}$ that meets $(\mu, 0)$ and that is unbounded in $\mathbb{R} \times K$.

The theorem was obtained independently by Dancer [31] and Turner [32]. See also Nussbaum [33] for an interesting result in this direction.

One further case in which degree theoretic arguments can be applied very effectively is when $E = Y$ is a complex Banach space and H is complex analytic. Then Schwartz [34] has shown bifurcation occurs from all characteristic values of L. Unfortunately, the complex analytic case does not seem to arise in physical applications. For further work on the complex case where analyticity is not necessarily present, see Böhme [35] and Ize [36]. See also Dancer [41] for local and global results for the real analytic case.

Next we study the class of problems arising when F is a potential operator and variational methods can be applied to the bifurcation question. Many problems in elasticity theory, in particular the example of Section III, fall into this category. To define terms more precisely, suppose E is a real Hilbert space, U is a neighborhood of 0 in E, and $f \in C^1(U, \mathbb{R})$. Then for all $u \in U, f'(u)$, the Frechet derivative of f at u, is a linear map from E to $\mathbb{R}$, i.e., $f'(u) \in E'$. Since E is self-dual, $f'(u)$ can (and will) be identified with an element of E, i.e., $f': U \to E$. We call f' a *potential operator*. The first result concerning potential operators is due to Krasnoselski [11].

Theorem 16. Suppose $f \in C^1(U, \mathbb{R})$ is weakly continuous and uniformly differentiable near 0 with $f'(u) = Lu + H(u)$, L being linear (and compact) and $H(u) = o(\|u\|)$ at $u = 0$. If μ is an eigenvalue of L, then $(\mu, 0)$ is a bifurcation point for $F(\lambda, u) = f'(u) - \lambda u$. Moreover, for all $r > 0$ and sufficiently small, F has a zero $(\lambda(r), u(r))$ with $\|u(r)\| = r$ and $\lambda(r) \to \mu$ as $r \to 0$.

The proof of Theorem 16 involves variational minimax arguments that are in part topological. A slightly weaker theorem $[f \in C^2(\Omega, \mathbb{R})]$ was proved by Marino and Prodi [37] using Morse theory. More recently, Böhme [35] and Marino [38] have, aside from their smoothness assumptions, obtained a stronger result than Theorem 16.

Theorem 17. Let $f \in C^2(\Omega, \mathbb{R})$ with $f'(u) = Lu + H(u)$, $L \in B(E, E)$, and $H(u) = o(\|u\|)$ at $u = 0$. Then every isolated eigenvalue of L of finite multiplicity is a bifurcation point for $F(\lambda, u) = f'(u) - \lambda u$. Moreover, for all $r > 0$ and sufficiently small, $F(\lambda, u)$ possesses at least two distinct zeros $(\lambda(r), u(r))$ satisfying $\|u(r)\| = r$ and $\lambda(r) \to \mu$ as $r \to 0$.

For f even, a much stronger result obtains:

Corollary 18. If f is even and dim $N(L - \mu I) = n$, then for all r sufficiently small, F possesses at least n distinct pairs of solutions as in Theorem 17.

Actually Böhme and Marino permit a slightly more general term than the u term in F. For the proof of Theorem 17, they essentially reduce the problem via the method of Lyapunov–Schmidt to considering f restricted to a finite-dimensional spherelike submanifold M of $\{\|u\| = r\}$. The maximum and minimum of f on M then provide zeros of F, λ appearing as a Lagrange multiplier. A similar theorem permitting a more general dependence of F on λ but only asserting the existence of one solution having $\|u\| = r$ has been obtained by Berger in [39] using Morse theory. Fučik *et al.* [40] also study generalizations of Theorem 17 in another direction.

It is natural to inquire whether the variational structure implies any connectivity properties for the zero set of F near $(\mu, 0)$ as was the case when μ had odd multiplicity. Böhme [35] has constructed an example with $E = \mathbb{R}^2$ and $f \in C^\infty$, where the component of $\mathscr{S}$ to which $(\mu, 0)$ belongs consists only of $(\mu, 0)$ so connectivity cannot be expected in general. However if f is real analytic a pair of curves of zeros of F bifurcate from $(\mu, 0)$ [35, 41]. Recently, in connection with a buckling problem McLeod and Turner [42] have extended Theorem 17 to the case where $f \in C^1(E, \mathbb{R})$ and $H(u)$ has a small Lipschitz constant tending to 0 as $\|u\| \to 0$.

Another approach has been taken towards the variational bifurcation question by studying the number of zeros of $F(\lambda, u)$ near $(\mu, 0)$ as a function of λ rather than of $\|u\|$ as in Theorems 16 and 17. In particular, Clark [43] treated the case where $f \in C^1(\Omega, \mathbb{R})$ is even, L is compact, $H = T + V$, where T is homogeneous of degree $k > 1$, $V(u) = o(\|u\|^k)$ at $u = 0$. Under further technical conditions on V on $N(L - \mu I)$, he gave lower bounds for the number of zeros of F for $\lambda > \mu$ (resp. $\lambda < \mu$) in terms of a topological measure of the size of the set of $u \in N(L - \mu I)$ such that $\|u\| = 1$ and $(Tu, u) \leq 0$ [resp. $(Tu, u) \geq 0$]. The method of proof here involves a Galerkin argument in conjunction with minimax techniques. Quite recently another result in this direction has been obtained by Rabinowitz [44]:

Theorem 19. Under the hypotheses of Theorem 17, either

(i) $(\mu, 0)$ is not an isolated zero of $F(\mu, \cdot)$;

(ii) There exists a one-sided neighborhood Λ of μ such that for all $\lambda \in \Lambda$, F possesses at least two distinct nontrivial zeros $(\lambda, u(\lambda))$ with $u(\lambda) \to 0$ as $\lambda \to \mu$; or

(iii) There exists a neighborhood W of μ such that for all $\lambda \in W - \{\mu\}$, F possesses at least one nontrivial zero $(\lambda, u(\lambda))$ with $u(\lambda) \to 0$ as $\lambda \to \mu$.

The proof of Theorem 19 involves variational minimax arguments. Conditions can be given for H, or more precisely its primitive h, whereby the behavior of h near $u = 0$ determines which of the alternatives of Theorem 19 occurs. There is also an analog of Corollary 18 for this case.

While Theorems 17 and 19 complement each other to some extent, it seems likely that a more general result encompassing both of them is still to be found. It is interesting to observe that the negation of alternative (i) of Theorem 19 is essentially a transversality statement, while transversality is essentially built into the framework of the proof of Theorem 17, so perhaps this is the notion required to unite the two results.

Sather [9], Kirchgässner [45], and Ize [36] have also studied bifurcation problems for $F(\lambda, u) = Lu + H(u) - \lambda u$, where H consists of a homogeneous operator plus a term of higher order at the origin, which is not a potential operator.

We conclude this survey with some results involving the bifurcation of equilibrium solutions of $u_t = F(\lambda, u)$ to time-periodic solutions and bifurcation of time-periodic to quasi-periodic solutions. The first result we know of involving the former is due to Hopf [24]:

Theorem 20. Suppose $F(\lambda, u)$ is real analytic near $(\mu, 0) \in \mathbb{R} \times \mathbb{R}^n$ with values in $\mathbb{R}^n$ and $F(\lambda, 0) = 0$ for λ near μ. Suppose further $F_u(\mu, 0)$ has exactly one pair of purely imaginary eigenvalues α_0, $\bar{\alpha}_0 \neq 0$, which are simple, and if $\alpha(\lambda)$ is the corresponding eigenvalue of $F_u(\lambda, 0)$, $\operatorname{Re} \alpha'(0) \neq 0$. Then there exist real analytic functions $(\lambda(s), u(t, s))$ and $T(s)$ for s near 0 such that $(\lambda(0), u(t, 0)) = (0, 0); u(t, s) \not\equiv 0, T(0) = 2\pi/|\alpha(0)|$, and $(\lambda(s), u(t, s))$ satisfy (11a) with period $T(s)$.

The proof of the theorem involves setting up a Poincaré map and using the implicit function theorem. Hopf also studied the stability of the periodic solutions in [24]. Hopf's work has been carried over to infinite-dimensional problems, in particular in the context of the Navier–Stokes equation, by Iudovich [46], Sattinger [13], Joseph and Sattinger [47], and Iooss [14].

Unfortunately, although the abstract machinery is now available and examples are known such as the wavy vortices for the Taylor problem, which seem to be of this type, to the best of our knowledge, nobody has succeeded in finding a fluid dynamical example where the hypotheses of the general theorems can be verified.

The Hopf bifurcation result has been carried one step further by Sacker [48]. (See also Ruelle and Takens [49] and Lanford [50].) He assumes that (11a) possesses a one-parameter family of solutions $(\lambda, u(t, \lambda))$ for λ near μ with u 2π-periodic in t. (Again u is an n-vector.) Suppose further that $n - 1$ Floquet multipliers have magnitude less than 1 for $\lambda < \mu$ and that exactly one pair of these, $\rho(\lambda) = e^{2\pi\alpha(\lambda)}$ and $\bar{\rho}(\lambda)$, leave the unit circle at $\lambda = \mu$, i.e., Re $\alpha(\mu) = 0$, Re $\alpha'(\mu) > 0$. Then if $\rho(\mu)^q \neq 1$, $q = 2, 3, 4$, F is sufficiently smooth (C^5), and another technical condition is satisfied, $(\lambda, u(t, \lambda))$ bifurcates into an asymptotically stable torus as λ increases through μ.

Sacker gives two proofs of his result: the first involves setting up a Poincaré map to find an invariant curve and the second requires the solution of a first-order quasi-linear partial differential equation that is carried out using an elliptic regularization argument. The analog for the Navier–Stokes equations of Sacker's result has been carried out recently by Iooss [51] using [49] and [50].

REFERENCES

[1] Chandrasekar, S., "Hydrodynamic and Hydromagnetic Stability." Oxford Univ. Press, London and New York, 1961.

[2] Kirchgässner, K., and H. Kielhöfer, Stability and bifurcation in fluid dynamics, *Rocky Mountain Math. J.* 3 (1973), 275–318.

[3] Berger, M. S., and P. C. Fife, Von Kármán's equations and the buckling of a thin elastic plate, II, *Comm. Pure Appl. Math.* 21 (1968), 227–241.

[4] Keller, J. B., and S. Antman (eds.), "Bifurcation Theory and Nonlinear Eigenvalue Problems." Benjamin, New York, 1969.

[5] Friedrichs, K. O., and J. Stoker, The nonlinear boundary value problem of the buckled plate, *Amer. J. Math.* 63 (1941), 839–888.

[6] Coles, D., Transition in circular Couette flow, *J. Fluid Mech.* 21 (1965), 385–425.

[7] Vainberg, M. M., and V. A. Trenogin, The methods of Lyapounov and Schmidt in the theory of nonlinear equations and their further development, *Russian Math. Surv.* 17 (1962), 1–60.

[8] Prodi, G., Problemi di diramazione per equazioni funzionali, *Boll. Un. Mat. Ital.* 22 (1967), 413–433.

[9] Sather, D., Branching of solutions of nonlinear equations in Hilbert space, *Rocky Mountain Math. J.* 3 (1973), 203–250.

[10] Stakgold, I., Branching solutions of nonlinear equations, *SIAM Rev.* **13** (1971), 289–332.

[11] Krasnoselski, M. A., "Topological Methods in the Theory of Nonlinear Integral Equations." Macmillan, New York, 1964.

[12] Pimbley, G. H., Jr., "Eigenfunction Branches of Nonlinear Operators and Their Bifurcations." Springer, New York, 1969.

[13] Sattinger, D. H., "Topics in Stability and Bifurcation Theory." School of Math., Univ. of Minnesota, Minneapolis, Minnesota, 1972.

[14] Iooss, G., "Stabilité et Bifurcation." Dept. of Math., Univ. of Paris Sud, Orsay, 1973.

[15] Nirenberg, L., "Topics in Nonlinear Functional Analysis." Courant Inst. of Math. Sci., New York Univ., New York, 1974.

[16] Arnold, V. I., Lectures on bifurcations in versal families, *Russian Math. Surveys* **28** (1973), 54–123.

[17] Vainberg, M. M., and Aizengendler, P. G., Theory and methods of investigation of branch points of solutions, *Progr. Math.* **2** (1968), 1–72.

[18] Cronin, J., Branch points of solutions of equations in Banach space I & II, *Trans. Amer. Math. Soc.* **69** (1950), 208–231; **76** (1954), 207–222.

[19] Bartle, R., Singular points in functional equations, *Trans. Amer. Math. Soc.* **75** (1953), 366–384.

[20] Krasnoselski, M. A., "Positive Solutions of Operator Equations." Noordhoff, Gronningen, 1964.

[21] Crandall, M. G., and P. H. Rabinowitz, Bifurcation from simple eigenvalues, *J. Functional Anal.* **8** (1971), 321–340.

[22] Crandall, M. G., and Rabinowitz, P. H., Bifurcation, perturbation of simple eigenvalues, and linearized stability, *Arch. Rational Mech. Anal.* **52** (1973), 161–180.

[23] Kopell, N. and L. N. Howard, Bifurcations under Nongeneric Conditions, preprint.

[24] Hopf, E., Abzweigung einer Periodischen Lösung von einer Stationäsen Lösung eines Differentialsystem, *Ber. Sächs. Akad. Wiss. Leipzig Math. Phys. Kl.* **94** (1942), 3–22.

[25] Sattinger, D. H., Stability of bifurcating solutions by Leray–Schauder degree, *Arch. Rational Mech. Anal.* **43** (1971), 154–166.

[26] Prodi, G., Teoremi di tipo locale per il sistema di Navier–Stokes e stabilita delle soluzioni stazionaire, *Rend. Sem. Mat. Univ. Padova* **32** (1962), 374–397.

[27] Rabinowitz, P. H., Some global results for nonlinear eigenvalue problems, *J. Functional Anal.* **7** (1971), 487–513.

[28] Turner, R. E. L. T., Transversality in nonlinear eigenvalue problems, *in* "Contributions to Nonlinear Functional Analysis" (E. H. Zarantonello, ed.). Academic Press, New York, 1971.

[29] Stuart, C. A., Some bifurcation theory for k-set contractions, *Proc. London Math. Soc.* **27** (1973), 531–550.

[30] Dancer, E. N., On the Structure of Solutions of Nonlinear Eigenvalue Problems, preprint.

[31] Dancer, E. N., Global solution branches for positive mappings, *Arch. Rational Mech. Anal.* **52** (1973), 181–192.

[32] Turner, R. E. L., Transversality and cone maps, *Arch. Rational Mech. Anal.* **58** (1975), 151–179.

[33] Nussbaum, R. D., A Global Bifurcation Theorem with Applications to Functional Differential Equations, preprint.

[34] Schwartz, J. T., Compact analytic mappings of B-spaces and a theorem of Jane Cronin, *Comm. Pure Appl. Math.* **16** (1963), 253–260.

[35] Böhme, R., Die Lösung der Verzweigungsgleichungen für nichtlineare Eigenwertproblem, *Math. Z.* **127** (1972), 105–126.

[36] Ize, G. A., Bifurcation Theory for Fredholm Operators, thesis, New York Univ., 1974.

[37] Marino, A., and Prodi, G., La teoria di Morse per gli spazi di Hilbert, *Rend. Sem. Mat. Univ. Padova* **41** (1968), 43–68.

[38] Marino, A., La biforcazione nel caso variazionale, *Proc. Conf. Sem. Mat. Dell'Univ. Bari* (Nov. 1972) (to appear).

[39] Berger, M. S., Bifurcation theory and the type numbers of Marston Morse, *Proc. Nat. Acad. Sci. U.S.A.* **69** (1972), 1737–1738.

[40] Fučik, S., Nečas, J., Souček, J., and Souček, V., Spectral analysis of nonlinear operators, *Springer Lect. Notes Math.* **343**, 1973.

[41] Dancer, E. N., Global structure of the solutions of real analytic eigenvalue problems, *Proc. London Math. Soc.* **27** (1973), 747–765.

[42] McLeod, B. and R. E. L. Turner, Bifurcation for nondifferentiable operators (in progress).

[43] Clark, D. C., Eigenvalue Bifurcation for Odd Gradient Operators, preprint

[44] Rabinowitz, P. H., A bifurcation theorem for potential operators (in progress).

[45] Kirchgässner, K., Multiple eigenvalue bifurcation for holomorphic mappings *in* "Contributions to Nonlinear Functional Analysis" (E. H. Zarantonello, ed.), pp. 69–100. Academic Press, New York, 1971.

[46] Iudovich, V. I., Appearance of auto-oscillations in a fluid, *J. Appl. Math. Mech.* **35** (1971), 638–655.

[47] Joseph, D., and Sattinger, D. H., Bifurcating time periodic solutions and their stability, *Arch. Rational Mech. Anal.* **45** (1972), 75–109.

[48] Sacker, R. J., On invariant surfaces and bifurcation of periodic solutions of ordinary differential equations, Tech. Rep. IMM-NYU 344, Courant Inst. Math. Sci., New York Univ. (October 1964).

[49] Ruelle, D., and Takens, F., On the nature of turbulence, *Comm. Math. Phys.* **20** (1971), 167–192.

[50] Lanford, O. E., III, Bifurcation of periodic solutions with invariant tori: The work of Ruelle and Takens, *Springer Lect. Notes Math.* **322** (1973), 159–192.

[51] Iooss, G., Bifurcation of a Periodic Solution into an Invariant Torus for Navier-Stokes Equations and their Respective Stabilities, preprint.

Generalized Linear Differential Systems and Associated Boundary Problems*

WILLIAM T. REID

Department of Mathematics
The University of Oklahoma, Norman, Oklahoma

1. Introduction

The considered generalized differential system is

$$\Delta[y, z](t) \equiv -dz(t) + [C(t)u_y(t) - D(t)z(t)] \, dt + [dM(t)]u_y(t) = 0,$$
$$L[y, z](t) \equiv A_1(t)u_y'(t) + A_0(t)y(t) - B(t)z(t) = 0, \qquad u_y(t) = A_2(t)y(t), \tag{S}$$

where on a given real interval $[a, b]$ the matrix functions A_0, A_1, A_2, B, C, D, and M are, respectively, of dimensions $n \times n$, $n \times n$, $n \times n$, $n \times m$, $m \times n$, $m \times m$, and $m \times n$, with A_1, A_2 nonsingular. The notation employed is consistent with that of references [6–8] of the author. It is assumed that $A_0, A_1, A_2, A_1^{-1}, A_2^{-1}, B, C$, and D are of class $\mathfrak{L}^\infty$ on $[a, b]$, while $M \in \mathfrak{B}\mathfrak{V}_{mn}$. For $\hat{\mathfrak{D}} = \{y : y \in \mathfrak{L}_n, \ y = A_2^{-1}u_y \text{ with } u_y \in \mathfrak{A}_n\}$, a solution of (S) is a pair $(y, z) \in \hat{\mathfrak{D}} \times \mathfrak{B}\mathfrak{V}_m$ such that $L[y, z] = 0$ a.e., while $\Delta[y, z] = 0$ in the sense of the Riemann–Stieltjes integral equation

$$z(t) = z(\tau) + \int_\tau^t [Cu_y - Dz] \, ds + \int_\tau^t [dM]u_y = 0, \qquad (t, \tau) \in [a, b] \times [a, b]. \tag{1}$$

For a system of ordinary differential equations to which (S) reduces when M is constant, some of the merits of introducing matrix coefficients A_1, A_2 are discussed in [8, pp. 109, 110; also Problem 8, p. 116]. For (S) in case $A_1 \equiv A_2 \equiv E_n$, see [5, 6]; in case $A_2 \equiv E_n$, see [7]. The real scalar generalized second-order differential equations treated by Sz-Nagy [9], Feller [1], and Kac and Krein [2] are particular cases of (S). The comprehensive significance of systems (S) is herein illustrated by applications to systems with interface conditions, to integrodifferential systems with integral boundary conditions, and to systems of difference equations.

* Partially supported by the National Science Foundation under Grant GP-36120.

For brevity, an $(r + h) \times (s + k)$ matrix M is written $(M^{11}, M^{12}; M^{21}, M^{22})$ in terms of the component matrices

$$M^{11} = [M_{\alpha\beta}], \quad M^{12} = [M_{\alpha, s+j}], \quad M^{21} = [M_{r+i, \beta}], \quad M^{22} = [M_{r+i, s+j}],$$

where $\alpha = 1, \ldots, r, \beta = 1, \ldots, s, i = 1, \ldots, h, j = 1, \ldots, k$. Correspondingly, an $(r + h)$-dimensional vector v is written $(v^1; v^2)$, where $v^1 = (v_\alpha), v^2 = (v_{r+i})$.

2. Basic Properties

The symbol $\mathfrak{D}$ will denote the class of $y \in \hat{\mathfrak{D}}$ for which there is a $z \in \mathfrak{L}_m^\infty$ such that $L[y, z](t) = 0$ a.e. on $[a, b]$. Also, we set

$$\mathfrak{D}^0 = \{y : y \in \mathfrak{D}, u_y(a) = 0 = u_y(b)\}.$$

Correspondingly,

$$\hat{\mathfrak{D}}_* = \{\zeta : \zeta \in \mathfrak{L}_n^\infty, \zeta = A_1^{*-1} v_\zeta \text{ with } v_\zeta \in \mathfrak{B}\mathfrak{B}_n\},$$

and a solution of the system

$$\begin{aligned}
\Delta_*[\eta, \zeta](t) &\equiv -dv_\zeta(t) + [C^*(t)\eta(t) + A_2^{*-1}(t)A_0^*(t)\zeta(t)]\, dt \\
&\quad + [dM^*(t)]\eta(t) = 0, \\
L_*[\eta, \zeta](t) &\equiv \eta'(t) - D^*(t)\eta(t) - B^*(t)\zeta(t) = 0,
\end{aligned} \tag{S_*}$$

is a pair $(\eta, \zeta) \in \mathfrak{A}_m \times \hat{\mathfrak{D}}_*$ such that $L_*[\eta, \zeta] = 0$ a.e. on $[a, b]$, and $\Delta_*[\eta, \zeta] = 0$ in the sense of a Riemann–Stieltjes integral equation related to Δ_* as (1) is related to Δ. Correspondingly, $\mathfrak{D}_*$ signifies the class of $\eta \in \mathfrak{A}_m$ for which there is a $\zeta \in \mathfrak{L}_n^\infty$ such that $L_*[\eta, \zeta] = 0$ a.e. and $\mathfrak{D}_*^0 = \{\eta : \eta \in \mathfrak{D}_*, \eta(a) = 0 = \eta(b)\}$.

One may readily establish results corresponding to Theorems 2.1–2.3 of [6] and Lemmas 2.2 and 2.3 of [7], relating (1) to the functional

$$J[y, z; \eta, \zeta] = \int_a^b [\zeta^* Bz + \eta^* Cu_y]\, dt + \int_a^b \eta^*[dM]u_y.$$

In particular, as an extension of Theorem 2.3 of [6], we have that if $(f, \psi) \in \mathfrak{L}_n \times \mathfrak{B}\mathfrak{B}_m$ then (y, z) is a solution of

$$\Delta[y, z] = d\psi, \qquad L[y, z] = f, \tag{S'}$$

iff $(\hat{y}, \hat{z}) = (y, z - Mu_y) \in \hat{\mathfrak{D}} \times \mathfrak{B}\mathfrak{B}_m$ and is a solution of the system

$$\hat{\Delta}[\hat{y}, \hat{z}] = d\psi + MA_1^{-1} f\, dt, \qquad \hat{L}[\hat{y}, \hat{z}] = f, \tag{$\hat{S}'$}$$

where $\hat{\Delta}[\hat{y},\ \hat{z}] = -d\hat{z} + [\hat{C}\hat{A}_2\,\hat{y} - \hat{D}\hat{z}]\,dt$, $\hat{L}[\hat{y},\ \hat{z}] = \hat{A}_1[\hat{A}_2\,\hat{y}]' + \hat{A}_0\,\hat{y} - \hat{B}\hat{z}$, while $\hat{A}_1 = A_1$, $\hat{A}_2 = A_2$, $\hat{B} = B$, $\hat{A}_0 = A_0 - BMA_2$, $\hat{D} = D + MA_1^{-1}B$, and $\hat{C} = C - DM + MA_1^{-1}A_0 A_2^{-1} - MA_1^{-1}BM$. In particular, if ψ is absolutely continuous then $(\hat{S}')$ is an ordinary differential system. For $p = (u_y\,;\,z)$ and $q = (\eta;\,v_\zeta)$, let $\partial p = (p(a);\,p(b))$ and $\partial q = (q(a);\,q(b))$. If P is an $r \times 2(n + m)$ matrix of rank r, then corresponding to the classical solvability theorem for ordinary differential equations [8, Theorem III.6.2], for $(f, \psi) \in \mathfrak{L}_n \times \mathfrak{B}\mathfrak{B}_m$ the system

$$\Delta[y, z](t) = d\psi(t), \qquad L[y, z](t) = f(t), \qquad P(\partial p) = 0 \qquad (\text{S}^\text{B})$$

has a solution iff $\int_a^b (\eta^* \, d\psi + \zeta^* f \, dt) = 0$ for all solutions $(\eta,\ \zeta)$ of the homogeneous adjoint problem

$$\Delta_*[\eta, \zeta](t) = 0, \qquad L_*[\eta, \zeta](t) = 0, \qquad Q(\partial q) = 0, \qquad (\text{S}^\text{B}_*)$$

where Q is a $[2(n + m) - r] \times 2(n + m)$ matrix of rank $2(n + m) - r$ satisfying $PK^*Q^* = 0$, with $K = (-J, 0;\, 0,\, J)$ and J the $(n + m) \times (n + m)$ matrix $(0,\, -E_m;\, E_n,\, 0)$.

3. Applications

Three types of applications will be presented.

A. SYSTEMS WITH INTERFACE CONDITIONS

Consider a system

$$\Delta[y, z] \equiv -dz + [Cy - Dz]\,dt = d\psi, \qquad L^0[y, z] \equiv A_1{}^0 y' + A^0 y - B^0 z = 0,$$
$$(\text{S}^0)$$

with coefficient matrix functions satisfying conditions analogous to those specified in the introduction, $\psi \in \mathfrak{B}\mathfrak{B}_m$, and interface values a_α $(\alpha = 1, \ldots, \sigma)$ satisfying $a = a_0 < a_1 < \cdots < a_\sigma < a_{\sigma+1} = b$, with the condition that (y, z) is a bounded solution of (S^0) on each subinterval $(a_{\beta-1}, a_\beta)$, $\beta = 1, \ldots, \sigma + 1$, having unilateral limits at each division point, and subject to end and interface conditions $P^0(\partial p) = 0$, $y(a_\alpha{}^+) = H_\alpha y(a_\alpha{}^-)$ $(\alpha = 1, \ldots, \sigma)$ where P^0 is an $r \times 2(n + m)$ matrix of rank r, and H_α $(\alpha = 1, \ldots, \sigma)$ is a nonsingular $n \times n$ matrix. If $u = u_y$ is defined as

$$u(t) = y(t) \qquad \text{on} \quad (a_\sigma, b),$$
$$u(t) = H_\sigma \cdots H_\alpha\, y(t) \qquad \text{on} \quad (a_{\alpha-1}, a_\alpha) \quad (\alpha = 1, \ldots, \sigma),$$

then for

$$A_2(t) \equiv E \qquad \text{on} \quad (a_\sigma, b),$$

$$A_2(t) \equiv H_\sigma \cdots H_\alpha \qquad \text{on} \quad (a_{\alpha-1}, a_\alpha) \quad (\alpha = 1, \ldots, \sigma),$$

$$A_1(t) = A_1{}^0(t)A_2^{-1}(t) \qquad \text{on} \quad \bigcup_{\beta=1}^{\sigma+1} (a_{\beta-1}, a_\beta),$$

the system $(\mathbf{S^{0B}})$ consisting of $(\mathbf{S^0})$, $P^0(\partial p) = 0$, and interface conditions, is equivalent to a system $(\mathbf{S^B})$ with $M \equiv 0$, $P = P^0(R, 0; 0, E_{n+m})$, and R the $(n+m) \times (n+m)$ matrix $(H_1^{-1} \cdots H_\sigma^{-1}, 0; 0, E_m)$.

B. BOUNDARY PROBLEMS INVOLVING INTEGRODIFFERENTIAL SYSTEMS

Many considered boundary problems may be written in the form

$$x'(t) + K(t)x(t) + \Omega(t)\int_a^b [d\Psi(s)]x(s) = \phi(t), \qquad \int_a^b [d\Theta(s)]x(s) = 0, \quad \textbf{(ID)}$$

where the vector functions x and ϕ are k-dimensional, K and Ω are, respectively, $k \times k$ and $k \times r$ matrix functions of class $\mathfrak{L}^\infty$, while $\Psi \in \mathfrak{BV}_{rk}$ and $\Theta \in \mathfrak{BV}_{lk}$. Let $S(t) = 0$ for $t \in [a, b)$, $S(b) = -E_r$, and define the $(k+r) \times (k+r)$ matrix function $\mathscr{A}_0(t)$ and the $(l+r) \times (k+r)$ matrix function $\mathscr{M}(t)$ as

$$\mathscr{A}_0(t) = (K(t), \Omega(t); 0, 0), \qquad \mathscr{M}(t) = (\Theta(t), 0; \Psi(t), S(t)).$$

Equation $(\textbf{ID})$ is satisfied by x iff y and z are vector functions of respective dimensions $k+r$ and $l+r$ with $x = (y_\alpha)$ $(\alpha = 1, \ldots, k)$ that satisfy with $f = (\phi; 0)$ the system

$$-dz + [d\mathscr{M}]y = 0, \qquad y' + \mathscr{A}_0 y = f, \qquad z(a) = 0, \qquad z(b) = 0, \quad (2)$$

which is of the form $(\mathbf{S^B})$ with $n = k+r$, $m = l+r$, $A_1 = A_2 = E_n$, B, C, and D zero matrices, while $\mathscr{A}_0$ and $\mathscr{M}$ are as defined above. Moreover, in terms of $\eta = (\eta_1; \eta_2) \in \mathfrak{A}_l \times \mathfrak{A}_r$ and $\zeta = (\zeta_1; \zeta_2) \in \mathfrak{BV}_k \times \mathfrak{BV}_r$, the system adjoint to (2) is

$$\eta' = 0, \qquad -d\zeta + \mathscr{A}_0{}^*\zeta + [d\mathscr{M}^*]\eta = 0, \qquad \zeta(a) = 0, \qquad \zeta(b) = 0. \quad (3)$$

Then $\hat\eta = \eta$, $\hat\zeta = \zeta - \mathscr{M}^*\eta$ are solutions of a system of ordinary differential equations, and hence absolutely continuous. In terms of $\hat\eta = (\eta_1; \eta_2)$, where

η_1 and η_2 are constant vectors of respective dimensions l and r, and $\hat{\zeta} = (\hat{\zeta}_1; \hat{\zeta}_2) \in \mathfrak{A}_k \times \mathfrak{A}_r$ with $\hat{\zeta}_1 = \zeta_1 - \Theta^*\eta_1 - \Psi^*\eta_2$, $\hat{\zeta}_2 = \zeta_2 - S^*(t)\eta_2$, this system may be written

$$-\hat{\zeta}_1' + K^*(t)\hat{\zeta}_1 + K^*(t)G(t) = 0, \qquad -\hat{\zeta}_2' + \Omega^*(t)\hat{\zeta}_1 + \Omega^*(t)G(t) = 0,$$

$$\hat{\zeta}_1(a) + G(a) = 0, \qquad \hat{\zeta}_2(a) = 0, \qquad \hat{\zeta}_1(b) + G(b) = 0, \qquad \hat{\zeta}_2(b) - \eta_2 = 0, \tag{4}$$

where $G(t) = \Theta^*(t)\eta_1 + \Psi^*(t)\eta_2$. For the case of

$$\int_a^b [d\Psi]x = Cx(a) + Dx(b), \qquad \int_a^b [d\Theta]x = Ax(a) + Bx(b) + \int_a^b Hx\,ds,$$

(4) reduces to the system adjoint to (ID) as determined by Krall [3]. For differential and integrodifferential equations involving integral boundary conditions the recent literature has been rather extensive, but for brevity reference is limited to Krall [3], which contains a fairly extensive bibliography. For (ID), the above formulation is a unifying one, whose simplification is due largely to the fact that the given k-dimensional x-problem is extended to a $(k + l + 2r)$-dimensional problem, wherein the boundary conditions are satisfied by all absolutely continuous vector functions $(y; z)$ that vanish at a and b. It is to be remarked that for a special case of a self-adjoint system (ID) the reduction to an equivalent ordinary differential equation problem was employed by Reid [4].

C. RELATED DIFFERENTIAL AND DIFFERENCE SYSTEMS

Let $m = n$, $A_1 = A_2 = B = E_n$, and A_0, C, and D be zero matrices, while $a = t_0 < t_1 < \cdots < t_{\rho+1} = b$ is a partition of $[a, b]$, and M is constant on each open subinterval $(t_{\alpha-1}, t_\alpha)$ $(\alpha = 1, \ldots, \rho + 1)$. Then for

$$\mu_0 = M(a^+) - M(a), \qquad \mu_\alpha = M(t_\alpha^+) - M(t_\alpha^-) \quad (\alpha = 1, \ldots, \rho),$$

$$\mu_{\rho+1} = M(b) - M(b^-),$$

$(y; z)$ is a solution of (S) iff y is the polygonal function whose graph joins the successive points $(t_j, y(t_j))$ $(j = 0, \ldots, \rho + 1)$, and the values $y(t_j)$ satisfy the linear second-order difference equation

$$(1/\delta_\alpha)[y(t_{\alpha+1}) - y(t_\alpha)] - (1/\delta_{\alpha-1})[y(t_\alpha) - y(t_{\alpha-1})] - \mu_\alpha y(t_\alpha) = 0$$
$$(\alpha = 1, \ldots, \rho),$$

where $\delta_\beta = t_{\beta+1} - t_\beta$ $(\beta = 0, 1, \ldots, \rho)$, while

$$z(t) = y'(t) = (1/\delta_\beta)[y(t_{\beta+1}) - y(t_\beta)] \quad \text{on} \quad (t_\beta, t_{\beta+1}), \qquad \beta = 0, 1, \ldots, \rho,$$
$$z(a) = (1/\delta_0)[y(t_1) - y(t_0)] - \mu_0 \, y(a),$$
$$z(t_{\beta+1}) = (1/\delta_\beta)[y(t_{\beta+1}) - y(t_\beta)] + [M(t_{\beta+1}) - M(t_\beta)]y(t_{\beta+1}),$$
$$\beta = 0, 1, \ldots, \rho.$$

The intimate relationship between scalar second-order linear homogeneous differential equations and systems of difference equations is well known. In the real self-adjoint case the existence of analogous comparison and oscillation properties of solutions dates from the work of Sturm. For the author the relationship between differential and difference equations remained in the domain of analogy, however, until his study of generalized differential systems in [5], and the realization that both were special instances of a general system, wherein for the self-adjoint case one has extensions of the classical Sturmian theory.

REFERENCES

[1] Feller, W., Generalized second order differential equations and their lateral conditions, *Illinois J. Math.* **1** (1957), 459–504.
[2] Kac, I. S., and Krein, M. G., On the spectral function of the string, *Amer. Math. Soc. Transl.* **103**(2) (1974), 19–102.
[3] Krall, A. M., Differential–boundary operators, *Trans. Amer. Math. Soc.* **154** (1971), 429–458.
[4] Reid, W. T., An integro-differential boundary value problem, *Amer. J. Math.* **60** (1938), 257–292.
[5] Reid, W. T., Generalized linear differential systems, *J. Math. Mech.* **8** (1959), 705–726.
[6] Reid, W. T., Generalized linear differential systems and related Riccati matrix integral equations, *Illinois J. Math.* **10** (1966), 701–722.
[7] Reid, W. T., Variational methods and boundary problems for ordinary differential equations, *Proc. US–Japan Sem. Differential Functional Equations, Univ. of Minnesota, Minneapolis, Minnesota, 1967* 267–299.
[8] Reid, W. T., "Ordinary Differential Equations." Wiley, New York, 1971.
[9] Sz-Nagy, B., Vibrations d'une corde non-homogène, *Bull. Soc. Math. France* **75** (1947), 193–209.

Some Stochastic Systems Depending on Small Parameters*

WENDELL H. FLEMING and C. P. TSAI
Lefschetz Center for Dynamical Systems
Division of Applied Mathematics
Brown University, Providence, Rhode Island

1. Introduction

Let us consider a finite-dimensional stochastic system whose state $x(t) = (x_1(t), \ldots, x_n(t))$ is a vector-valued random variable. We suppose that $x(t)$ evolves according to stochastic differential equations, written in vector-matrix notation as

$$dx = f(x(t))\, dt + \sigma(x(t))\, dw, \tag{1.1}$$

where $w = (w_1, \ldots, w_d)$ is a brownian motion, and (1.1) is interpreted in the Ito sense.

It is generally quite difficult to calculate the probability distribution of $x(t)$. Actually, one often wants not the distribution itself, but rather certain related quantities (for example, the moments). Suppose that f or σ depends on a small parameter ε, and that the quantities of interest can be found exactly when $\varepsilon = 0$. One wishes to compute these quantities approximately for small $\varepsilon \neq 0$. This kind of problem arises in many applications, and various approximation schemes exist, including system linearization, perturbation expansions in powers of the parameter ε, and truncation schemes.

In Section 2 we briefly review some literature dealing with nearly deterministic systems. In Section 3 we turn to systems that we call nearly linear. Some results are stated about truncating the infinite system of differential equations for moments. Proofs of these results, and extensions of them appear in [20]. In Section 4 the same truncation scheme is applied to a one-dimensional problem in population genetics theory, concerning the rate of decay of mean heterozygosity in case of nearly neutral genes. It appears that this technique may be useful for more dimensional genetics models. Finally,

---

* This research was supported by the National Science Foundation under GP-28931X2.

in Section 5 we mention some results about optimal stochastic control problems that are either nearly deterministic or nearly of linear regulator type.

2. Nearly Deterministic Systems

Now suppose that (1.1) takes the form

$$dx = f(x(t)) \, dt + (2\varepsilon)^{1/2} \sigma(x(t)) \, dw, \tag{2.1}$$

with ε a small positive parameter. A variety of questions can be asked; which of these are well posed and interesting depends to a considerable extent on the structure of the unperturbed dynamical system [$\varepsilon = 0$ in (2.1)]. One can always consider results for fixed t (finite). For small ε, the distribution of $x(t)$ will be approximately gaussian and centered about the unperturbed system trajectory. A "ray" method for finding the distribution approximately has been given by Cohen and Lewis [4]. It involves the solution of a nonlinear first-order partial differential equation.

In many problems, it suffices to compute expectations $E\Phi[x(t)]$ for various choices of function $\Phi(x)$, rather than to compute the distribution of $x(t)$. An example is the moments of $x(t)$. Now $x(t)$ is a Markov diffusion process with generator

$$\mathscr{L} = \mathscr{L}^0 + \varepsilon \mathscr{M},$$

$$\mathscr{L}^0 = f \cdot \nabla, \qquad \mathscr{M} = \sum_{i,j=1}^{n} a_{ij} \frac{\partial^2}{\partial x_i \, \partial x_j}, \qquad a_{ij} = \sum_{k=1}^{d} \sigma_{ik} \sigma_{jk}. \tag{2.2}$$

Let us suppose that $x(0) = y$, and regard the desired expectation as a function of t and y:

$$\phi(t, y) = E\Phi(x(t)).$$

Under suitable technical assumptions ϕ satisfies the backward equation

$$\partial \phi / \partial t = \mathscr{L} \phi,$$

with the initial data $\phi(0, y) = \Phi(y)$. One can seek an asymptotic expansion:

$$\phi = \phi^0 + \varepsilon \theta_1 + \varepsilon^2 \theta_2 + \cdots + \varepsilon^m \theta_m + O(\varepsilon^{m+1}). \tag{2.3}$$

The coefficients $\theta_1, \theta_2, \ldots$ satisfy linear first-order partial differential equations, obtained by formal substitution in (2.3). Under suitable assumptions, the validity of (2.3) can be shown either by methods of partial differential equations, or by a probabilistic method [6].

Another quantity of interest is $E\Phi(x(\tau))$, where τ is the exit time of $x(t)$ from some bounded open set B. If we denote this by $\phi(y)$, then $\mathscr{L}\phi = 0$ in B and $\phi = \Phi$ on the boundary ∂B. An expansion like (2.3) holds in any "regular" subregion B' of B from which characteristic curves lead to ∂B and intersect ∂B nontangentially. See [6, Section 5]. Holland [12] has given a probabilistic treatment of the boundary layer expansion, near points of ∂B from which the characteristics lead into a regular subregion of B. If all characteristics are asymptotically stable to a point inside B, then $\phi(y)$ behaves entirely differently for small ε. See Vent'sel and Freidlin [21].

In a number of physical applications $\Phi(x)$ can be regarded as "energy" associated with state x, and $\mathscr{L}^0\Phi = 0$. Energy is constant along the deterministic trajectories ($\varepsilon = 0$). Then $\mathscr{L}\Phi = \varepsilon\mathscr{M}\Phi$; it is reasonable to consider times of order ε^{-1}. In some problems (arising for instance in wave propagation in random media) there is an approximating diffusion process on a new scale $\tau = \varepsilon t$, after introducing new state variables and appropriately centering the process. See Papanicolaou [18, Section 2; 19, Sections 4, 8]. In other cases, in which $f(x)$ is nonlinear, only information about the rate of growth of mean energy is known. See Carrier [3] and Fleming [6, Section 3, Example 2].

If the process $x(t)$ is ergodic, one can ask for an asymptotic expansion in powers of ε for $\int \Phi(x)\mu(dx)$, where μ is the equilibrium distribution (the integral is over n-dimensional space). Such expansions are obtained by Holland [10].

A method of perturbing the time scale in powers of ε, following Poincaré, was applied by Amazigo [1] to a problem of buckling for columns with random imperfections.

3. Nearly Linear Systems

Let us now suppose that (1.1) has the form

$$dx = [Ax(t) + \varepsilon g(x(t))]\, dt + \left[\alpha + \sum_{i=1}^{n} \beta_i x_i(t)\right] dw, \qquad (3.1)$$

and $g(x)$ is a polynomial of degree l. We summarize a truncation method for the moments, which is analyzed in detail in [20]. For the validity of the method, some further conditions on g are needed. These conditions imply, in particular, that solutions of (3.1) do not explode in finite time, and that

$$(d/dt)E\Phi(x(t)) = E\{\mathscr{L}\Phi(x(t))\} \qquad (3.2)$$

for any polynomial $\Phi(x)$.

When $\varepsilon = 0$ we denote the solution by $x^0(t)$:

$$dx^0 = Ax^0(t)\, dt + \left[\alpha + \sum_{i=1}^{n} \beta_i x_i^0(t)\right] dw. \qquad (3.1^0)$$

We call the system (3.1^0) linear [some authors say bilinear, in view of the products $x_i^0(t)\, dw$]. If each $\beta_i = 0$, then $x^0(t)$ is gaussian. Its distribution can be found explicitly by solving linear ordinary differential equations for the means and convariances. See, for instance, [6, Section 3, Example 1]. If some $\beta_i \neq 0$, the distribution of $x^0(t)$ can no longer be found exactly. However, a system of linear ordinary differential equations for the moments of order up to any given m can still be derived. This is seen by applying (3.2) to any polynomial Φ, and observing that $\mathscr{L}\Phi$ is a polynomial of degree no more than the degree of Φ.

For $\varepsilon > 0$, $\mathscr{L}\Phi$ is a polynomial of degree $\leq m + l - 1$ if Φ is a polynomial of degree $\leq m$. By applying (3.2) to any polynomial of degree $\leq m$, we get a linear system of differential equations for the moments of $x(t)$ of order $\leq m$. This system is not closed, since the right side may involve moments of orders $m + 1, \ldots, m + l - 1$. However, an approximation for the moments of orders $\leq m$ that is correct up to order ε^{k+1} can be obtained by truncating the infinite system of differential equations for the moments. This is done by ignoring all moments of order greater than some m_k.

Let us describe the results when $x(t)$ and $w(t)$ are scalar processes. For the vector case, only notational changes are needed to describe the truncation method. However, in that case the conditions given in [20] to ensure its validity are more complicated. Consider the moments

$$\mu_j(t) = E[x(t)]^j, \qquad j = 0, 1, 2, \ldots.$$

Assume that the polynomial

$$g(x) = \sum_{i=0}^{l} g_l x^l$$

has odd degree l, and that $g_l < 0$. By (3.2) with $\Phi(x) = x^j$, we get

$$d\mu_j/dt = \tfrac{1}{2}j(j-1)[\alpha^2\mu_{j-2} + 2\alpha\beta\mu_{j-1} + \beta^2\mu_j] + jA\mu_j + \varepsilon j \sum_{i=0}^{l} g_i \mu_{j+i-1}. \qquad (3.3)$$

For $\varepsilon = 0$, this linear system of equations for μ_j, $j \leq m$, is closed. Given m and k, let

$$m_k = m + k(l - 1),$$

and let μ_j^k, $j = 0, 1, \ldots, m_k$ be the solution of the linear system obtained by omitting from (3.3) all terms μ_r with $r > m_k$. The initial data are

$$\mu_j(0) = \mu_j^k(0) = y^j, \qquad y = x(0).$$

Then it can be shown that, for fixed T, there exists a constant $C = C(T, k, m)$ such that

$$|\mu_j(t) - \mu_j^k(t)| \leq C\varepsilon^{k+1}, \qquad j = 1, \ldots, m, \quad 0 \leq y \leq T. \tag{3.4}$$

See [20, Theorem 3.1]. Actually, for the proof of this result, as well as for calculations in examples, it is more convenient to rewrite the truncated system as a system of linear differential equations for $\tilde{\mu}_j^k$, defined by

$$\tilde{\mu}_j^k(t) = \mu_j^k(t), \qquad j = 0, 1, \ldots, m,$$
$$\tilde{\mu}_j^k(t) = \varepsilon\mu_j^k(t), \qquad j = m + 1, \ldots, m_1$$
$$\vdots$$
$$\tilde{\mu}_j^k(t) = \varepsilon^k\mu_j^k(t), \qquad j = m_{k-1} + 1, \ldots, m_k.$$

Under stronger assumptions the approximation of $\mu_j(t)$ by $\mu_j^k(t)$ is uniform for all $t \geq 0$. In fact, if

$$A < 0, \qquad (m_{k+1} - 1)\beta^2 < 2|A|,$$

then the constant $C = C(k, m)$ in (3.3) can be taken independent of T. See [20, Corollary 3.1].

In formal truncation schemes proposed for various applications, moments of orders exceeding the truncation level m_k are often replaced by appropriate combinations of moments of orders $\leq m_k$, not simply omitted as we have done. See for instance, Amazigo et al. [2, p. 1352] and Kushner [14]. Such modifications should improve the estimate (3.4); but we have not attempted a rigorous treatment of this idea. For another class of truncation procedures in the theory of turbulence, see Lundgren [15].

Let us conclude this section with a two-dimensional example [20, p. 42].

Example. Let

$$dx_1 = x_2 \, dt, \qquad dx_2 = -(x_1 + bx_2 + \varepsilon x_1^3) \, dt + \sigma x_1 \, dw,$$

with b, ε, σ positive, b, σ fixed, and ε small. Let

$$\mu_{i, j-i}(t) = E[x_1(t)^i x_2(t)^{j-i}], \qquad i = 0, 1, \ldots, j.$$

The truncation method consists of omitting terms in the equations corresponding to (3.3) of order $j > m + 2k$. If $\sigma^2 < b(m + 2k - 3)^{-1}$, then the estimate corresponding to (3.4) holds uniformly for all $t \geq 0$. For instance, suppose that we wish to estimate moments of order 2 up to terms involving ε, ε^2. In this example, the equations for even- and odd-order moments are uncoupled. The truncated system consists of 15 linear differential equations for

$$\tilde{\mu}^2_{i,\, 2-i} = \mu^2_{i,\, 2-i}, \qquad \tilde{\mu}_{i,\, 4-i} = \varepsilon\mu_{i,\, 4-i}, \qquad \tilde{\mu}^2_{i,\, 6-i} = \varepsilon^2\mu^2_{i,\, 6-i}.$$

4. An Application in Population Genetics Theory

We consider a one-dimensional example of a population with two possible gene types (alleles) at a given locus on a chromosome. This example has already been analyzed in detail by classical analysis. See Crow and Kimura [5, Chap. 8] and Miller [16]. By a truncation method like that in Section 3 we can reproduce fairly easily some of the known results, for the case of nearly neutral genes. The truncation method could be applied to multi-dimensional models where classical analysis does not seem available, as we shall indicate at the end of the section.

Let $x(t)$ denote the frequency of one of the two gene types, $0 \leq x(t) \leq 1$. It is a one-dimensional diffusion; the generator takes the form

$$\mathscr{L} = x(1 - x)(d^2/dx^2) + \varepsilon x(1 - x)[h + (1 - 2h)x]\, d/dx,$$

provided time is measured in units of $4N$, where N is the population size. Moreover, $\varepsilon = 4Ns$ where s and h are certain constants appearing in coefficients of selective advantage. The disappearance of a gene type corresponds to $x = 0$ or $x = 1$. There is no mechanism such as recurrent mutation in the model for reintroducing an extinct type; the boundary conditions at 0 and 1 are absorbing. We take ε small (near neutrality, neutrality when $\varepsilon = 0$).

When $\varepsilon = 0$ the eigenvalues of $\mathscr{L}^0 = x(1 - x)\, d^2/dx^2$ are $\lambda_i^0 = -i(i + 1)$, $i = 1, 2, \ldots$. See Crow and Kimura [5, p. 383]. The quadratic $\Phi_1(x) = x(1 - x)$ is an eigenfunction corresponding to the dominant eigenvalue $\lambda_1^0 = -2$. Let

$$H(t) = E[x(t)(1 - x(t))] = E\Phi_1(x(t)).$$

Then $2H(t)$ is an important quantity called the mean heterozygosity. When $\varepsilon = 0$, we get from (3.2) and $\mathscr{L}^0\Phi_1 = -2\Phi_1$ the well-known formula

$$H^0(t) = y(1 - y)e^{-2t},$$

where $y = x(0)$ is the initial gene frequency.

We are interested in estimating λ_1 and $H(t)$ for ε small. Once λ_1 is estimated, one can also estimate such quantities of interest as the rate of absorption of $x(t)$ at 0 or 1 and the expected time to absorption. For this purpose consider the polynomials

$$\Phi_{2i-1} = [x(1 - x)]^i, \qquad i = 1, 2, \ldots,$$
$$\Phi_{2i}(x) = [x(1 - x)]^i(1 - 2x), \qquad i = 1, 2, \ldots.$$

Also let

$$H_1(t) = H(t) = E\Phi_1(x(t)),$$
$$H_j(t) = \varepsilon^i E\Phi_j(x(t)), \qquad j = 2i, 2i + 1, \quad i = 1, 2, \ldots,$$
$$\mathbb{H}^k = (H_1, \ldots, H_{2k+1}).$$

By applying (3.2) to $\Phi = \Phi_i$, $i = 1, \ldots, 2k + 1$, we get a system of linear differential equations of the form

$$d\mathbb{H}^k/dt = \Lambda^k\mathbb{H}^k + \varepsilon^{k+1}\Gamma_k, \tag{4.1}$$

where Λ^k is a $(2k + 1) \times (2k + 1)$ matrix depending on ε and each component of Γ_k is a linear combination of $E\Phi_{2k+2}(x(t))$ and $E\Phi_{2k+3}(x(t))$. When $\varepsilon = 0$, Λ^k is an upper triangular matrix, with diagonal elements λ_1^0, λ_2^0, λ_3^0, $\ldots$, λ_{2k+1}^0. Instead of (4.1) consider the truncated system

$$d\hat{\mathbb{H}}^k/dt = \Lambda^k\hat{\mathbb{H}}^k \tag{4.2}$$

with $\hat{\mathbb{H}}^k(0) = \mathbb{H}^k(0) = \mathbb{H}_0^k$. Then

$$\mathbb{H}^k(t) = \exp(\Lambda^k t)\mathbb{H}_0^k + \varepsilon^{k+1} \int_0^t \exp[\Lambda^k(t - s)]\Gamma_k(s)\,ds, \tag{4.3}$$

$$\hat{\mathbb{H}}^k(t) = \exp(\Lambda^k t)\mathbb{H}_0^k. \tag{4.4}$$

From (4.3) and (4.4) we see that $\mathbb{H}^k(t) - \hat{\mathbb{H}}^k(t)$ is of order ε^{k+1} uniformly for t in any finite interval. Therefore, in the expansion

$$H(t) = H^0(t) + \varepsilon I_1(t) + \varepsilon^2 I_2(t) + \cdots + \varepsilon^k I_k(t) + O(\varepsilon^{k+1}) \tag{4.5}$$

we have

$$I_j = \frac{1}{j!} \frac{\partial^j \hat{H}_1}{\partial \varepsilon^j}\bigg|_{\varepsilon=0}, \qquad j = 1, \ldots, k,$$

where $\hat{H}_1$ is the first component of the vector $\hat{\mathbb{H}}$. By repeated differentiation of (4.2) with respect to ε, we get linear differential equations for $I_1, \ldots, I_k$. The initial data are $I_j(0) = 0$. We solved these equations to get

$$I_1(t) = H^0(t)[c_1 - \tfrac{1}{10}(1 - 2h)t + \tfrac{1}{8}(1 - 2y)e^{-4t} - \tfrac{1}{5}(1 - 2h)c_2 e^{-10t}],$$

$$c_1 = \tfrac{1}{8}(1 - 2y) + \tfrac{1}{5}(1 - 2h)c_2, \qquad c_2 = y(1 - y) - \tfrac{1}{5}, \qquad y = x(0),$$

with a more complicated expression for $I_2(t)$.

Estimates for the dominant eigenvalue λ_1. Let $\hat{\lambda}_1$ be the dominant eigenvalue for the matrix Λ^k. We know that, when $\varepsilon = 0$, $\hat{\lambda}_1{}^0 = \lambda_1{}^0 = -2$. Let us show that, for small $\varepsilon \neq 0$,

$$|\lambda_1 - \hat{\lambda}_1| \leq C\varepsilon^{k+1} \tag{4.6}$$

for some constant C. The proof will use the fact, from eigenfunction expansion of the probability density of $x(t)$ [5, p. 408], that $\exp(-\lambda_1 t)E\Phi_j(x(t))$ is bounded for each j, and $\exp(-\lambda_1 t)E\Phi_1(x(t)) = \exp(-\lambda_1 t)H(t)$ is bounded away from 0. Let V_1 be an eigenvector for Λ_k corresponding to the eigenvalue $\hat{\lambda}_1$. For small ε, the eigenvalues of Λ^k are simple with $\hat{\lambda}_1$ near -2. We can take a basis of eigenvectors $V_1, V_2, \ldots, V_{2k+1}$ with V_1 near $V_1{}^0 = (1, 0, \ldots, 0)$. Let $[Y]_1$ denote the component of a $(2k + 1)$-dimensional vector Y with respect to V_1. Then

$$[\exp(\Lambda^k t)Y]_1 = \exp(\hat{\lambda}_1 t)[Y]_1.$$

By taking components in (4.3)

$$[\mathbb{H}^k(t)]_1 = \exp(\hat{\lambda}_1 t)[\mathbb{H}_0{}^k]_1 + \varepsilon^{k+1}\int_0^t \exp[\hat{\lambda}_1(t - s)][\Gamma_k(s)]_1 \, ds.$$

When $\varepsilon = 0$, $\mathbb{H}_0{}^k = (y(1 - y), 0, \ldots, 0)$ and thus $[\mathbb{H}_0{}^k]_1 = y(1 - y)$. Since Λ^k does not depend on the initial data $y = x(0)$, we may take $y = \tfrac{1}{2}$ in this proof. Then $[\mathbb{H}_0{}^k]_1 \geq \tfrac{1}{8}$ for small ε. Since $\Gamma_k(s)$ is a linear combination of $E\Phi_{2k+2}(x(s))$ and $E\Phi_{2k+3}(x(s))$,

$$|[\Gamma_k(s)]_1| \leq c \exp(\lambda_1 s) \qquad \text{for some} \quad c.$$

Case 1. $\lambda_1 > \hat{\lambda}_1$. For some $c_1 > 0$ we have

$$c_1 \exp(\lambda_1 t) \leq [\mathbb{H}^k(t)]_1 \leq \exp(\hat{\lambda}_1 t)[\mathbb{H}_0{}^k]_1 + \frac{\varepsilon^{k+1}c}{\lambda_1 - \hat{\lambda}_1}[\exp(\lambda_1 t) - \exp(\hat{\lambda}_1 t)].$$

We take t large and find that

$$\lambda_1 - \hat{\lambda}_1 \leq cc_1^{-1}\varepsilon^{k+1}.$$

Case 2. $\lambda_1 < \hat{\lambda}_1$.   For some c_2 we have

$$c_2 \exp(\lambda_1 t) \geq \tfrac{1}{8} \exp(\hat{\lambda}_1 t) - \frac{\varepsilon^{k+1} c}{\hat{\lambda}_1 - \lambda_1}[\exp(\hat{\lambda}_1 t) - \exp(\lambda_1 t)].$$

We again take t large and find that

$$\hat{\lambda}_1 - \lambda_1 \leq 8 c \varepsilon^{k+1}.$$

These two estimates give (4.6).

We computed λ_1 approximately using (4.6) in two cases, namely, $h = \tfrac{1}{2}$ and $h < 0$. When $h = \tfrac{1}{2}$, $g(x) = \tfrac{1}{2}x(1 - x)$ is quadratic. The number of linear differential equations needed for accuracy of order ε^{k+1} is now $k + 1$ rather than $2k + 1$. For $k = 2$, one finds by applying the implicit function theorem to the 3×3 matrix Λ^2 that

$$\hat{\lambda}_1 = -2(1 + \tfrac{1}{40}\varepsilon^2) + O(\varepsilon^4),$$

which agrees with Kimura's power series [5, p. 398] for λ_1 up to fourth-order terms. The case $h < 0$ is called overdominant. We estimated λ_1 numerically using (4.6), and obtained results agreeing well with those of Kimura and Miller [16] [5, p. 412] in the range about $0 < \varepsilon < 3$, i.e., $Ns < 0.75$.

The above is just a different treatment of a problem already thoroughly studied by classical analysis. Let us next turn to a two-dimensional example, in which classical analysis does not seem to be an available tool. In this example, the population is situated in two niches, with a certain rate m of exchange of genes between niches. Let N be the number of individuals in each niche, and

$$\alpha = 4Nm, \qquad \varepsilon = 2Ns, \qquad h = \tfrac{1}{2}.$$

Let $x_i(t) = $ gene frequency in niche, $i = 1, 2$.

The generator of the two-dimensional diffusion $(x_1(t), x_2(t))$ is

$$\begin{aligned}
\mathscr{L} = \; & x_1(1 - x_1)\frac{\partial^2}{\partial x_1{}^2} + x_2(1 - x_2)\frac{\partial^2}{\partial x_2{}^2} \\
& + \alpha\left[(x_2 - x_1)\frac{\partial}{\partial x_1} + (x_1 - x_2)\frac{\partial}{\partial x_2}\right] \\
& + \varepsilon\left[x_1(1 - x_1)\frac{\partial}{\partial x_1} + x_2(1 - x_2)\frac{\partial}{\partial x_2}\right].
\end{aligned}$$

Consider the problem of finding

$$H_1(t) = E[x_1(t)(1 - x_1(t))]$$
$$H_2(t) = E[(1 - 2x_1(t))(x_2(t) - x_1(t))]$$
$$H_3(t) = E[x_2(t)(1 - x_2(t))].$$

$2H_1(t)$, $2H_3(t)$ are the within-niche heterozygosities and $H_2(t) + 2H_1(t)$ is the between-niche heterozygosity. When $\varepsilon = 0$, the vector $\mathbb{H} = (H_1, H_2, H_3)$ satisfies

$$d\mathbb{H}/dt = \Lambda \mathbb{H}, \qquad \Lambda = \begin{bmatrix} -2 & \alpha & 0 \\ 4 - 2\alpha & -4\alpha & 4\alpha \\ 2\alpha & \alpha & -(2\alpha + 2) \end{bmatrix},$$

$$\lambda_1 = -(1 + 2\alpha) + (1 + 4\alpha^2)^{1/2}, \qquad \lambda_2 = -(1 + 2\alpha),$$

$$\lambda_3 = -(1 + 2\alpha) - (1 + 4\alpha^2)^{1/2}.$$

For $\varepsilon \neq 0$ one is dealing with a selection–migration model with finite population sizes in each colony. Near neutrality means ε near 0. The truncation method outlined above could be applied, but we have not done the calculation.

A similar method of linear differential equations (or difference equations in discrete time) has been used for other problems in neutral gene theory. We refer, in particular, to the problem of linkage disequilibrium for 2-locus models; see Hill and Robertson [8], Kimura and Ohta [13, Chapter 7], and Nei and Li [17]. Again our truncation method could be applied if near neutrality holds, at the expense of increasing substantially the number of linear differential equations that must be simultaneously treated.

5. Optimal Stochastic Control

Let us now suppose that the stochastic differential equation describing the process $x(t)$ depends on a control. For a nearly deterministic system, instead of (2.1) let us take

$$dx = f(x(t), u(t)) \, dt + (2\varepsilon)^{1/2}\sigma(x(t)) \, dw, \tag{5.1}$$

where $u(t)$ is the control applied at time t. Given a criterion of system performance, the problem is to calculate approximately the optimal expected performance and an optimal control. For closed-loop (feedback) controls this problem is treated in [7] and [9], and for open-loop controls in [11].

In [20, Chapter 3] a nonlinear perturbation of the stochastic linear regulator is treated. The system equations are now

$$dx = \{Ax(t) + Bu(t) + \varepsilon[g(x(t))]\} \, dt + \sigma(t) \, dw(t) \qquad (5.2)$$

with no constraints on the control $u(t)$ and with the usual quadratic performance criterion. Under suitable assumptions on the uncontrolled system [with $u(t) \equiv 0$], it is shown that there is an expansion of the optimal feedback control in powers of ε. If $x(t)$ is a scalar-valued process, these assumptions hold if $g(x) = g_l x^l + g_{l-1} x^{l-1} + \cdots + g_0$, with l odd and $g_l < 0$. The method is first to find an expansion of the corresponding solution of the dynamic programming equation in powers of ε. A crucial estimate is an a priori linear growth rate as $|x| \to \infty$ for the optimal feedback control as a function of state x.

REFERENCES

[1] Amazigo, J. C., Dynamic buckling of structures with random imperfections, *Proc. Symp. Probl. Mech.* (H. Leipholz, ed.). Univ. of Waterloo Press, 1974.

[2] Amazigo, J. C., Budianski, B., and Carrier, G., Asymptotic analyses of the buckling of imperfect columns on nonlinear elastic foundations, *Internat. J. Solids Structures* **6** (1970), 1341–1356.

[3] Carrier, G., Stochastically perturbed dynamical systems, *J. Fluid Mech.* **44** (1970), 249–264.

[4] Cohen, J. K., and Lewis, R. M., A ray method for the asymptotic solution of the diffusion equation, *J. Inst. Math. Appl.* **3** (1967), 266–290.

[5] Crow, J., and Kimura, M., "Introduction to Population Genetics Theory." Harper, New York, 1970.

[6] Fleming, W. H., Stochastically perturbed dynamical systems, *Rocky Mountain Math. J.* **4** (1974), 407–433.

[7] Fleming, W. H., Stochastic control for small noise intensities, *SIAM J. Contr.* **9** (1971), 473–518.

[8] Hill, W. C., and Robertson, A., Linkage disequilibrium in finite populations, *Theor. Appl. Genet.* **38** (1968), 226–231.

[9] Holland, C. J., A numerical technique for small noise stochastic control problems. *J. Optimizat. Theory Appl.* **13** (1974), 74–93.

[10] Holland, C. J., Ergodic expansions in small noise problems, *J. Differential Equations* **16** (1974), 281–288.

[11] Holland, C. J., Small noise open loop control problems, *SIAM J. Contr.* **12** (1974).

[12] Holland, C. J., Singular perturbations in elliptic boundary value problems, *J. Differential Equations* (to appear).

[13] Kimura, M., and Ohta, T., "Theoretical Aspects of Population Genetics," Monographs in Population Biology. Princeton Univ. Press, Princeton, New Jersey, 1971.

[14] Kushner, H. J., Approximations to optimal nonlinear filters, *IEEE Trans. Automat. Contr.* **AC-12** (1967), 546–556.

[15] Lundgren, T. S., A closure hypothesis for the hierarchy of equations for turbulent probability distribution functions, *in* "Statistical Models and Turbulence" (M. Rosenblatt, ed.). *Springer Lect. Notes Phys.* **12** (1972), 70–100.

WENDELL H. FLEMING AND C. P. TSAI

[16] Miller, G. F., The evaluation of eigenvalues of a differential equation arising in a problem of genetics, *Proc. Cambridge Phil. Soc.* **58** (1962), 588–593.

[17] Nei, M., and Li, W. H., Linkage disequilibrium in subdivided populations, *Genetics* **75** (1973), 213–219.

[18] Papanicolaou, G. C., Some problems and methods for the analysis of stochastic equations, *in* "Stochastic Differential Equations" (J. B. Keller and H. P. McKean, eds.), pp. 21–34 (*SIAM–AMS Proc.* **6**). Amer. Math. Soc., Providence, Rhode Island, 1973.

[19] Papanicolaou, G. C., Asymptotic analysis of transport processes, *Bull. Amer. Math. Soc.* **81** (1975), 330–392.

[20] Tsai, C. P., Perturbed Stochastic Dynamical Systems, Ph.D. Thesis, Brown Univ., 1974.

[21] Vent'sel, A. D., and Freidlin, M. I., On small random perturbations of dynamical systems, *Usp. Mat. Nauk* **25** (1970), 3–55 [*English transl.: Russian Math. Surveys.*]

Bifurcation*

KLAUS KIRCHGÄSSNER and JÜRGEN SCHEURLE
Mathematisches Institut A der Universität Stuttgart
Stuttgart, Germany

1. Introduction

It is well known that Hadamard's postulates for a well-posed problem are inadequate for those problems of mathematical physics described by quasilinear elliptic equations. Even if a solution exists it is not unique in general. The stationary Navier–Stokes equations, e.g., possess a continuum of solutions in many physically interesting cases. Among all possible solutions nature selects according to stability properties (cf. [17]).

The question arises naturally whether there are constructive methods simulating the behavior of nature and selecting the physically relevant solution. Usually, analytic and numerical procedures describe the whole set of solutions. Those methods are known for bifurcation problems near a simple eigenvalue (cf. [18]). Recently algorithms have been proposed on a combinatorial basis that determine all fixed points of a mapping [1].

We describe an iteration process that, under certain assumptions, possesses the required selection property. It can be considered a nonlinear version of the well-known Picard–Poincaré–Neumann method. The sufficient part of the criterion for local convergence is trivial, whereas the necessary part is deeper and important for the selection. We show the applicability for some semilinear problems in a partially ordered Banach space. Finally, it is shown that certain stability–exchange phenomena of bifurcation theory are simulated by this iteration process.

2. A Selective Iteration Procedure

Let X be a real Banach space and $\mathfrak{L}(X)$ the space of bounded endomorphisms on X with the uniform operator-norm. We call $L \in \mathfrak{L}(X)$ 1-separable if the spectrum $\sigma(L)$ of the natural complexity of L decomposes

---

* English-language version of an article in "Jahresbericht der Deutschen Mathematiker-Vereinigung," Volume 77, 1975, appearing through the courtesy of B. G. Teubner, Stuttgart.

into two disjoint parts σ_1 and σ_2, σ_2 being nonempty, and if

$$\sup_{\sigma_1} |\lambda| < \inf_{\sigma_2} |\lambda| = a, \qquad a > 1$$

holds. Compact linear operators L having spectral radius spr (L) greater than 1 are 1-separable.

The theorem of Picard-Poincaré-Neumann states that $L^n x$ converges to 0 for all $x \in X$ if spr $L < 1$. Necessary for $N = X$, where

$$N = \{x \in X \mid \lim L^n x = 0\}$$

is spr $L \le 1$ [25].

For a 1-separable L let P (resp. Q) denote the Dunford-projectors to σ_1 (resp. σ_2) [13, p. 178]:

$$P = -\frac{1}{2\pi i} \int_{|z|=b} (L - z\,id)^{-1}\, dz, \qquad Q = id - P$$

where $\sup_{\sigma_1} |\lambda| < b < a$ holds.

One has $X = PX \oplus QX$. The closed subspaces PX, QX are invariant with respect to L. There are constants $a > 1$, $\rho > 0$, and a norm $x = \|Px\| + \|Qx\|$ equivalent to the original norm in X such that

$$\|LQx\| \ge a\|Qx\|, \qquad \|LPx\| \le (a - \rho)\|Px\| \tag{2.1}$$

holds. Subsequently, this norm will be used.

A Fréchet-differentiable mapping $F: X \to X$ is called a Fredholm operator if $F'(x)$ is a Fredholm operator for every $x \in X$. F is called real analytic in X if, for all $x, h \in X$, there are $P_n(x; h) \in X$, $P_n(x; \cdot)$ being a continuous operator homogeneous of degree n, and $P_n(\cdot; h)$ being continuous in X, such that

$$F(x + h) = \sum_{n=0}^{\infty} \frac{1}{n!} P_n(x; h)$$

is satisfied, where the series converges normally [11, p. 769], and $\bar{B}$ denotes the closure, $\mathring{B}$ the interior of $B \subset X$, and $B_r(x) = \{y \in X \mid \|y - x\| < r\}$.

Let $F: X \to X$ with $F(0) = 0$ be continuously differentiable in some $B_r(0)$, i.e., $F': B_r(0) \to \mathfrak{L}(X)$ exists and is continuous; then $F^n(x)$ converges to 0 for all x in a neighborhood of 0, if spr $F'(0) < 1$. If, however, $F'(0)$ is 1-separable one can construct examples, even for C^∞-mappings, such that the set $N = \{x \in X \mid F^n(x) \to 0\}$ is a neighborhood of 0 [30].

Theorem 2.1. Let $F: X \to X$ be continuously differentiable in a neighborhood of 0, and let $F'(0)$ be 1-separable. Furthermore, let one of the following conditions be satisfied:

(i) $\overline{F(\overset{\circ}{B_\rho(x)})} \neq \varnothing$ for all $x \in X$, $\rho > 0$.

(ii) F is continuously differentiable in X and Fredholm, with $\dim QX = \infty$.

(iii) F is real analytic and a Fredholm operator with ind $F'(0) = 0$ and $\dim X/F'(0)^n X \leq n_0 < \infty$ for all $n \in \mathbb{N}$.

Then, the set N is of first category.

All applications yet made use condition (iii). Some will be described in the next sections. Only a sketch of the proof will be given. For a more detailed presentation cf. [30].

Lemma 2.2. Let $F: X \to X$ be continuous and continuously differentiable in a neighborhood of 0 and let $F'(0)$ be 1-separable. Then, there exists a double cone $K \subset X$ and a ball $B_s = B_s(0)$ such that $x, y \in \bar{B}_s$, $x - y \in \dot{K}$ implies for some $n \in \mathbb{N}$,

$$\|F^n(x)\| > s \qquad \text{or} \qquad \|F^n(y)\| > s$$

Proof. Let a, ρ be as in (2.1). Choose $q, \eta > 0$ such that $a \geq (1 + \eta)(1 + q)$ and define K as follows:

$$K = \{x \in X \mid \|Px\| \leq q\|Qx\|\}, \qquad \dot{K} = K - \{0\}$$

Define $\sigma = \rho q/a$, $\delta = \min(\eta, \sigma/(1 + q)^2)$, and set $L = F'(0)$. There is a ball $\bar{B}_s$ where $\|L - F'(y)\| < \delta$ holds. For $x \in K$ one obtains immediately

$$\begin{aligned}
\|PLx - PF'(y)x\| &\leq \delta(1 + q)\|QLx\| \\
\|QLx - QF'(y)x\| &\leq \delta(1 + q)\|QLx\|
\end{aligned} \tag{2.2}$$

We show first:

(1) $L\dot{K} \subset \overset{\circ}{K}$,

(2) $\|Lx\| \geq (1 + \eta)\|x\|$ for $x \in K$, and

(3) for arbitrary $y_1, \ldots, y_n \in \bar{B}_s$ and $x \in \dot{K}$,

$$v = \sum_{j=1}^{n} \alpha_j F'(y_i)x \in \overset{\circ}{K}, \qquad \|v\| \geq (1 + \eta - \delta)\|x\|$$

holds if $\alpha_j > 0$ and $\sum_{j=1}^{n} \alpha_j = 1$.

In view of (2.1) and $x \in \dot{K}$ one obtains

$$\|PLx\| = \|LPx\| \leq (a - \rho)q\|Qx\| \leq (q - \sigma)\|QLx\|$$

and hence (1). Furthermore, $x \in \dot{K}$ yields (2):

$$\|Lx\| = \|PLx\| + \|QLx\| \geq a\|Qx\| \geq (1 + \eta)\|x\|$$

For the proof of (3) one uses (1) and (2.2):

$$\|Qv\| \geq \|QLx\| - \left\| Q \sum_{j=1}^{n} \alpha_j(F'(y_j) - L)x \right\| \geq (1 - \delta(1 + q))\|QLx\|$$

$$\|Pv\| \leq \|PLx\| + \left\| P \sum_{j=1}^{n} \alpha_j(F'(y_j) - L)x \right\| \leq (q - \sigma + \delta(1 + q))\|QLx\|$$

Since $QLx \neq 0$, one obtains $\|Pv\| < q\|Qv\|$. Furthermore, (2) yields:

$$\|v\| \geq \|Lx\| - \left\| \sum_{j=1}^{n} \alpha_j(F'(y_j) - L)x \right\| \geq (1 + \eta - \delta)\|x\|$$

For $x, y \in \bar{B}_s$, $x - y \in \dot{K}$, $\lambda \in [0, 1]$, one has $x + \lambda(y - x) \in \bar{B}_s$ and

$$F(y) - F(x) = \int_0^1 F'(x + \lambda(y - x))(y - x) \, d\lambda \in K$$

In view of the closedness of K, the last relation follows already, if the corresponding Riemann sums are in K. This, however, is a consequence of (3). The same argument, together with (3), yields the inequality

$$\|F(y) - F(x)\| \geq (1 + \eta - \delta)\|y - x\|$$

hence $F(y) - F(x) \in \dot{K}$. It follows inductively that $F^n(y) - F^n(x) \in \dot{K}$ as long as $F^k(x), F^k(y) \in \bar{B}_s$ for $1 \leq k \leq n - 1$. On the other hand,

$$\|F^n(y) - F^n(x)\| \geq (1 + \eta - \delta)^n\|y - x\|$$

shows that $F^k(y), F^k(x) \in \bar{B}_s$ cannot hold for all $k \in \mathbb{N}$.

Proof of Theorem 2.1.   Let $\bar{B}_s$ be the ball of the preceding lemma and define

$$A_n = \{x \in X \,|\, \|F^k(x)\| \leq s \text{ for } k \geq n\}$$

The sets A_n are closed, and we have

$$N \subset \bigcup_{n \in \mathbb{N}} A_n$$

It remains to show that A_n is nowhere dense. Assume $\mathring{A}_n \neq \varnothing$ and $B_\rho(x) \subset \mathring{A}_n$; then

(1) $F^k B_\rho(x) \subset \bar{B}_s$ for $k \geq n$, and
(2) $(F^k)'(y)X \subset X - \mathring{K}$ for $k \geq n$, $y \in B_\rho(x)$.

Otherwise there exist $y \in B_\rho(x)$, $z \in X$, and $k \geq n$ with $0 \neq (F^k)'(y)z \in \mathring{K}$. Hence there is a $\lambda \in \mathbb{R}$ such that $y + \lambda z \in B_\rho(x)$ and

$$0 \neq F^k(y + \lambda z) - F^k(y) \in K \cap \bar{B}_s$$

which contradicts Lemma 2.2 by (1). This proves (2).

(i) The interior of $\overline{F^n(\overset{\circ}{B_\rho(x)})}$ is nonempty. Let

$$B_r(y) \subset \overline{F^n(\overset{\circ}{B_\rho(x)})} \qquad \text{and} \qquad y \in F^n(B_\rho(x))$$

Since $\mathring{K} \neq \varnothing$ there exists a $u \neq y$ with

$$u \in (y + K) \cap B_r(y) \cap F^n(B_\rho(x))$$

such that $\|F^k(y)\| \leq s$ and $\|F^k(u)\| \leq s$ for all $k \in \mathbb{N}$ and $u - y \in \dot{K}$ leads to a contradiction by Lemma 2.2.

(ii) Let $(e_1, \ldots, e_m) \subset QX$ be linearly independent and $[e_j]$ be the corresponding elements of $X/(F^k)'(y)X$, $y \in B_\rho(x)$. By (2) one obtains from

$$\sum_{j=1}^{m} \alpha_j[e_j] = [0], \qquad z = \sum_{j=1}^{m} \alpha_j e_j \in (X - \mathring{K}) \cap QX$$

Hence $z = 0$ and all $\alpha_j = 0$. Therefore, we have

$$\dim X/(F^k)'(y)X \geq \dim QX = \infty$$

contradicting the Fredholm property.

(iii) We show by induction

(3) $\dim X/(F^{mn})'(y)X \geq m$ for all $y \in X$, $m \in \mathbb{N}$

Setting

$$Z(y) = (F^{mn})'(y), \qquad V(y) = (F^{(m-1)n})'(F^n(y))$$

one obtains for $y \in B_\rho(x)$ by the chain rule, (1), (2), and the validity of (3) for $m - 1$:

(4) $Z(y)X \subset V(y)X \cap (X - \mathring{K})$

According to (3) of the proof of Lemma 2.2 one gets for any $e \in \mathring{K}$:

(5) $0 \neq b = V(y)e \in \mathring{K} \cap V(y)X$

Let $([a_1]_v, \ldots, [a_q]_v)$, $a_j \in X$ be a basis of $X/V(y)X$ and consider $[a_1]_z, \ldots,$ $[a_q]_z$, $[b]_z$ as elements of $X/Z(y)X$. From

$$\sum_{j=1}^{q} \alpha_j [a_j]_z + \beta [b]_z = [0]_z$$

it follows by (4) that

$$\sum_{j=1}^{q} \alpha_j a_j + \beta b \in V(y)X \cap (X - \mathring{K})$$

and by (5) that

$$\sum_{j=1}^{q} \alpha_j a_j \in V(y)X$$

Hence $\alpha_j = 0$. Therefore,

$$\beta b \in (X - \mathring{K}) \cap (\mathring{K} \cup \{0\})$$

implying $\beta = 0$. Hence we have

$$\dim X/Z(y)X \geq \dim X/V(y)X + 1 \geq m$$

The mapping $Z: X \to \mathfrak{L}(X)$ is analytic and $Z(y)$ is Fredholm with ind $(Z(y)) = 0$ for all $y \in X$. Assertion (3) for all of X follows by a theorem of Gohberg and Krein [9, p. 21] (concerning the necessary generalization see [30]).

3. Positone Problems

The selection property of certain iteration schemes can be demonstrated effectively via Theorem 2.1 for so-called positone problems. These are roughly described by equations leaving invariant a partial order of the given Banach space (for the nomenclature cf. [15]). Because of their practical importance they have been treated extensively (cf. [8] and also [12]). Existence and the qualitative behavior of "positive" solutions have been studied in [3, 14, 15], whereas the multiplicity question was discussed in [2, 22, 23].

We are interested in the convergence of a certain iteration scheme and the stability properties of its limits. The iteration procedure has often been used. Bandle [4] was the first to point out that a sequence converging from above to a nonminimal solution cannot exist, if certain conditions are met. We owe to Amann the hint that this can be concluded already from Theorem F in [2] and earlier results of Laetsch [23].

Somewhat related to our results are those of Fujita [7] and Sattinger [18, 29], which show the connection between minimality and stability. Stability is understood with respect to solutions of a corresponding evolution equation. One can show that the iteration procedure can be considered a discrete-time version of this evolution equation. However, the discrete case requires stronger assumptions. We shall rely on Theorem 2.1 (iii) and thus on the analytic case. A special application follows at the end of this section. It can be easily extended to those boundary value problems whose Green's function and normal derivative are positive.

Let (X_0, X_1, X_2) be a triple of real Banach spaces with $X_2 \subset X_1$. In X_1 (resp. X_0) there are given two cones K_1 (resp. K_0), i.e., closed, convex sets with $\alpha K_j \subset K_j$ for $\alpha \in \mathbb{R}^+$ and $K_j \cap (-K_j) = \{0\}$. We use the notation $\dot{K}_j = K_j - \{0\}$, $\mathring{K}_j$ for the interior of K_j and assume that $\mathring{K}_1$ is not empty. In X_j, K_j generates a partial order consistent with the algebraic and topologic structure (cf. [19]). We write $u \leq v$ if $v - u \in K_j$, $u < v$ if $v - u \in \dot{K}_j$, $u \ll v$ if $v - u \in \mathring{K}_j$.

H1. Let $A \in \mathfrak{L}(X_2, X_0)$ be an isomorphism and let A^{-1} considered a mapping from X_0 into X_1 be compact and strongly positive, i.e., $A^{-1}\dot{K}_0 \subset \mathring{K}_1$.*

H2. Let $F: X_1 \to X_0$ with $F(0) \geq 0$ be real analytic in X_1 with the properties:

$$u \leq v \quad \text{implies} \quad F(u) \leq F(v), \qquad u \ll v \quad \text{implies} \quad F(u) < F(v)$$

We look for solutions $u \in X_2$ of the equation

$$Au = \lambda F(u), \qquad \lambda \geq 0 \tag{3.1}$$

or equivalently for $u \in X_1$, satisfying

$$u = \alpha u + (1 - \alpha)\lambda A^{-1}F(u) = T(\lambda, u) \tag{3.1'}$$

for some $\alpha \in (0, 1)$. Obviously, $T(\lambda, \cdot)$ is a real analytic Fredholm map, and $T'(\lambda, u)$ has index 0 for every $u \in X_1$ (T' denotes the Fréchet derivative with respect to u). The spectrum of T'—except for the point α—consists of isolated eigenvalues of finite multiplicities. If $\lambda > 0$, $u < v$ implies $T(\lambda, u) \ll T(\lambda, v)$ and $T'(\lambda, u)K_1 \subset K_1$. spr $T'(\lambda, u)$ is less (resp. greater) than 1 if and

* By replacing sequences by subsequences in the proofs, one could weaken this assumption to $A^{-n}\dot{K}_0 \subset \mathring{K}_1$ for some $n \in \mathbb{N}$.

only if spr $\lambda A^{-1}F'(u)$ is less (resp. greater) than 1. It is independent of α. If 0 is the only spectral point of $\lambda A^{-1}F'(u)$, then for both operators the spectral radius is less than 1. Otherwise, $\lambda A^{-1}F'(u)$ possesses, according to [21, Theorem 6.1], a positive eigenvalue $1/\lambda_0$ of maximum modulus. A simple calculation verifies the assertion.

A solution $\tilde{u}$ of (3.1') is called *weakly stable* if $\sigma(T'(\lambda, \tilde{u})) \subset \overline{B_1(0)}$ holds; $\tilde{u}$ is called *stable* if $\sigma(T'(\lambda, \tilde{u})) \subset B_1(0)$ holds. A nonweakly stable solution is called *unstable*.

To justify this definition we point out that if all eigenvalues of $A^{-1}F'(\tilde{u})$ are real, $\tilde{u}$ is stable if the spectrum of $id - \lambda A^{-1}F'(\tilde{u})$ is positive. A physically relevant criterion of stability is $\sigma(A - \lambda F'(\tilde{u})) \subset \mathbb{C}^+$. It is well known for ordinary differential equations that this criterion guarantees the asymptotic stability of $\tilde{u}$ with respect to solutions of $du/dt + Au - \lambda F(u) = 0$. Moreover, $\sigma(A - \lambda F'(\tilde{u})) \subset \overline{\mathbb{C}^+}$ is a necessary condition for Lyapunov stability. Generalizations to parabolic equations are given in [16].

In general, very little can be said about the location of $\sigma(A - \lambda F'(\tilde{u}))$ if $\sigma(id - \lambda A^{-1}F'(\tilde{u}))$ is known, and vice versa. If, however, the spectrum of $A - \lambda F'(\tilde{u})$ is real and if H1 and H2 are satisfied, it can be readily seen by means of Theorem 6.3 in [21], that the numerical and physical notions of stability coincide for small $|\lambda|$.

Lemma 3.1. Assume H1 to hold, let $F: X_1 \to X_0$ be real analytic, and $\tilde{u} \in X_2$ be a solution of (3.1'). If $T^n(\lambda, u)$ converges to $\tilde{u}$ for all u in an open set of X_1, then $\tilde{u}$ is weakly stable.

For the proof observe that for fixed λ, $G(v) = T(\lambda, \tilde{u} + v) - T(\lambda, \tilde{u})$ satisfies the assumptions of Theorem 2.1(iii); $T^n(\lambda, u)$ converges to $\tilde{u}$ if $G^n(v)$ converges to 0, and vice versa.

A sequence (u_n) in X_0 is called *bounded from above* (*below*) if there is a $u \in X_0$ with $u_n \leq u$ ($u_n \geq u$) for all $n \in \mathbb{N}$. We call K_0 *weakly regular* if every monotonically increasing (decreasing) sequence that is bounded from above (below) converges weakly in X_0. $u_0 \in X_1$ satisfying $u_0 \gg T(\lambda, u_0)$ $[u_0 \ll T(\lambda, u_0)]$ is called a strong supersolution (subsolution).

Theorem 3.2. Let H1, H2, $\lambda \geq 0$, be satisfied, and let K_0 be weakly regular. If u_0 is a positive, strong supersolution then $T^n(\lambda, u_0)$ converges to a weakly stable solution $\tilde{u}$ of (3.1') with $\tilde{u} \geq 0$.

Proof. The sequence $u_{n+1} = T(\lambda, u_n)$ consists of positive, strong super-solutions and $u_n \gg u_{n+1}$ holds. Hence, $F(u_n) > 0$ decreases monotonically. In view of the weak regularity of K_0, $F(u_n)$ converges weakly in X_0. Since A^{-1} is compact, $\lambda A^{-1} F(u_n)$ converges strongly in X_1. Hence, $u_{n+1} - \alpha u_n$ converges strongly in X_1, $\alpha \in (0, 1)$, which implies the convergence of u_n to a solution $\tilde{u}$. For the weak stability of $\tilde{u}$, observe that the set of strong supersolutions is open. Let $U(u_0) \subset X_1$ be a neighborhood of strong, positive supersolutions of u_0, $[\tilde{u}, u_0] = \{u \in X_1 | \tilde{u} \leq u \leq u_0\}$ and $\Omega = U(u_0) \cap [\tilde{u}, u_0]$; then, as will be shown, $T^n(\lambda, u)$ converges to $\tilde{u}$ for every $u \in \Omega$, and $\overset{\circ}{\Omega}$ is nonempty; the assertion follows via Lemma 3.1.

For $v_0 \in \Omega$ the sequence $v_{n+1} = T(\lambda, v_n)$ converges to some $\tilde{v}$ and $\tilde{u} \leq v_n \leq u_n$ holds for all $n \in \mathbb{N}$. Therefore, $\tilde{u} = \tilde{v}$. The interior of $[\tilde{u}, u_0]$ is nonempty since u_1 belongs to it. For small positive α, $u_\alpha = (1 - \alpha) u_0 + \alpha u_1 \in U(u_0) \cap [\tilde{u}, \overset{\circ}{u}_0]$. Hence, $\overset{\circ}{\Omega}$ is nonempty. Q.E.D.

Corollary 3.3. Let H1, H2, $\lambda \geq 0$ be satisfied, and let K_0 be weakly regular. If u_0 is a strong subsolution then the sequence $T^n(\lambda, u_0)$ is either unbounded or converges to a weakly stable solution $\tilde{u}$ of (3.1′) with $u_0 \ll \tilde{u}$.

The sequence $T^n(\lambda, u_0)$ never converges to an unstable solution, if u_0 is a strong super- or subsolution. Naturally, all stable solutions are local attractors.

A solution $\check{u}$ of (3.1′) is called *minimal* if there is no solution $\tilde{u}$ with $\tilde{u} < \check{u}$. If $F(0) > 0$, hence $T(\lambda, 0) \gg 0$ for $\lambda > 0$, holds, then 0 is a strong subsolution and there is at most one positive minimal solution. In addition, this solution is weakly stable. If $F(u) > 0$ for all $u \in X_1$, then by using the implicit function theorem one concludes that a positive, minimal solution exists for λ in some interval $(0, \lambda_0)$ for which 1 is not an eigenvalue of $T'(\lambda, \check{u}(\lambda))$. There $\check{u}(\lambda)$ is stable and $T^n(\lambda, 0)$ converges to it from below.

If in addition to H2, F is strictly convex, i.e.,

$$F(\beta u + (1 - \beta) v) < \beta F(u) + (1 - \beta) F(v) \tag{3.2}$$

holds for $u \neq v$, $\beta \in (0, 1)$, then every nonminimal solution $\tilde{u}$ is unstable, whenever $1 \notin \sigma(T'(\lambda, \tilde{u}))$. Namely, let $\bar{u} < \tilde{u}$ be solutions of (3.1′); $u_0^\beta = \beta \bar{u} + (1 - \beta) \tilde{u}$ is a strong supersolution for $\beta \in (0, 1)$.

The sequence $T^n(\lambda, u_0^\beta)$ converges from above to u_*^β, $\bar{u} \leq u_*^\beta < \tilde{u}$. Thus, in every neighborhood of $\tilde{u}$ there are u_0 such that $T^n(\lambda, u_0)$ does not converge to $\tilde{u}$. Hence, it is not stable. Since $T'(\lambda, \tilde{u}): K_1 \to K_1$, there is a positive eigenvalue of maximal modulus [21, Theorem 6.1] that has to be greater than 1.

If F is strictly concave [(3.2) holds in reversed order], then every minimal stable solution $\check{u}$ is maximal even in the class of subsolutions. Since $\check{u} < \tilde{u}$, and $\tilde{u}$ a subsolution of (3.1') implies that $\beta\tilde{u} + (1 - \beta)\check{u}$ is a strong subsolution for every β. The argument proceeds as above.

Corollary 3.4. Let H1, H2 be satisfied, and let K_0 be weakly regular.

(i)   If $F(0) > 0$, there exists at most one positive, minimal solution and it is weakly stable.

(ii)   If F is strictly convex in X_0, then every nonminimal solution $\tilde{u}$ is unstable if 1 is not an eigenvalue of $T'(\lambda, \tilde{u})$.

(iii)   If F is strictly concave in X_1, then every stable, minimal solution is maximal, even among the class of subsolutions.

As an application we consider a semilinear Dirichlet problem describing, e.g., the temperature distribution in a chemically active gas [8] (for other applications see [12]). $D \subset \mathbb{R}^n$ denotes a bounded domain with a C^2-boundary and Δ the n-dimensional Laplacian. Choose $p > n$, $X_0 = L_p(D)$, $X_1 = \mathring{C}^1(D) = \{u \in C^1(\bar{D}) | u(\mathbf{x}) = 0 \text{ for } \mathbf{x} \in \partial D\}$,

$$X_2 = \mathring{W}_p^2(D) = \{u \in W_p^2(D) | u(\mathbf{x}) = 0 \text{ for } \mathbf{x} \in \partial D\}.$$

W_p^2 can be considered a compact subset of $C^1(\bar{D})$. $A: u \to -\Delta u$ is a topological isomorphism between $\mathring{W}_p^2$ and L_p. Hence, $A^{-1}: L_p \to \mathring{C}^1(D)$ is compact. A^{-1} has a representation by the Green function G that satisfies $G > 0$ in $D \times D$ and $\partial G/\partial v > 0$ in $\partial D \times D$ (cf. [26, p. 85]); v denotes the interior unit normal. Let $K_0 = \{u \in L_p(D) | u(\mathbf{x}) \geq 0 \text{ a.e. in } D\}$, $K_1 = \{u \in \mathring{C}^1(D) | u(\mathbf{x}) \geq 0, \mathbf{x} \in \bar{D}\}$. Then A^{-1} is strongly positive, since for every $u \in \mathring{K}_0$ we have $(A^{-1}u)(\mathbf{x}) > 0$ for $\mathbf{x} \in D$ and $(\partial(A^{-1}u)/\partial v)(\mathbf{x}) > 0$ for $\mathbf{x} \in \partial D$. K_0 is weakly regular in view of [10, Theorem 13.44].

If F satisfies H2 then all conclusions of this section are true; e.g., $F(u) = e^u$ implies the existence of a unique, positive, stable minimal solution $\check{u}(\lambda)$ in some interval $(0, \lambda_0)$ [3]. For u_0 sufficiently close to 0, $T^n(\lambda, u_0)$ converges to $\check{u}(\lambda)$. Every nonminimal solution is unstable if 1 is not an eigenvalue of $T'(\lambda, \tilde{u})$. There exists no strong supersolution U with $U > \tilde{u}$.

4. Bifurcation

In this section we describe selection properties for some specific bifurcation phenomena, namely, when F, in addition to H2, satisfies $F(0) = 0$ and is either concave or convex, $u = 0$ is always a solution of (3.1') (the trivial solution). Under suitable hypothesis there exists a positive, simple character-istic value λ_0 of minimal modulus of $A^{-1}F'(0)$. The nontrivial solutions of (3.1') emanating in λ_0 form an analytic branch that is stable for $\lambda > \lambda_0$ and unstable for $\lambda < \lambda_0$ (cf. [5]). Here we show that the sequence $T^n(\lambda, u_0)$, for certain u_0, always converges to the stable solution. Moreover, one easily obtains existence and uniqueness results.

For strictly concave F and $u > 0$, $\alpha \in (0, 1)$, one has $F(\alpha u) \geq \alpha F(u) > 0$. Hence, $F'(0)$: $\dot{K}_1 \to \dot{K}_0$ and $T'(\lambda, 0)$: $\dot{K}_1 \to \overset{\circ}{K}_1$ for $\lambda > 0$. Theorem 6.3 in [22] guarantees that there is a positive, simple characteristic value λ_0 of minimal modulus of $A^{-1}F'(0)$. The corresponding eigenfunction satisfies $\varphi \gg 0$. Using [5] one obtains $u = 0$, $\lambda = \lambda_0$ as a point of bifurcation; the nontrivial branch forms locally a real analytic curve.

Theorem 4.1. Let H1, H2 be satisfied, and let K_0 be weakly regular. In addition, assume $F(0) = 0$, F strictly concave. Then, $A^{-1}F'(0)$ possesses a positive, simple characteristic value λ_0 of minimal modulus. Equation (3.1') has no positive solution in $[0, \lambda_0]$. Nontrivial positive solutions exist in a right-neighborhood of λ_0; they form a real analytic branch $\tilde{u}(\lambda)$ with $\tilde{u}(\lambda_0) = 0$.

The trivial solution is stable in $(0, \lambda_0)$ and unstable for $\lambda > \lambda_0$, where $\tilde{u}(\lambda)$ is stable. The sequence $T^n(\lambda, u_0)$ converges to 0 if $u_0 > 0$ is a strong supersolution and if $\lambda \in [0, \lambda_0)$; in some interval (λ_0, λ_1) the sequence converges to $\tilde{u}(\lambda)$.

Proof. The bifurcation picture near λ_0 follows directly from [21] and [5]. For global existence of the positive branch see [6]. $u = 0$ is a stable solution of (3.1') in the interval $[0, \lambda_0)$. Assume that $U > 0$ is an additional solution for some $\lambda \in (0, \lambda_0)$. In view of the concavity of F, one obtains $T(\lambda, \beta U) \gg \beta T(\lambda, U) = \beta U \gg 0$ for $\beta \in (0, 1)$. Hence, βU is a positive, strong sub-solution, and $T^n(\lambda, \beta U)$ defines a monotonically increasing sequence of strong subsolutions. This yields a contradiction to the stability of 0 since βU can be chosen arbitrarily close to 0.

The same argument shows that there is no positive, strong subsolution in $[0, \lambda_0)$. Now, assume $U > 0$ to be a solution of (3.1') for $\lambda = \lambda_0$. Then $T(\lambda_0, \beta U) \gg \beta U$ holds for $\beta \in (0, 1)$. For fixed β, βU is a positive, strong subsolution for $\lambda < \lambda_0$, yielding a contradiction.

$u = 0$ is unstable for $\lambda > \lambda_0$, since λ/λ_0 is the eigenvalue of maximal modulus for $A^{-1}F'(0)$. According to [5], $\tilde{u}(\lambda)$ is stable for sufficiently small $\lambda - \lambda_0$. Moreover, $\tilde{u}(\lambda)$ is a minimal, positive solution. In view of its stability it is maximal as well among all positive solutions (same argument as above). The sequence $T^n(\lambda, u_0)$ converges to $\tilde{u}(\lambda)$ from above if u_0 is a positive, strong supersolution and if λ is sufficiently close to λ_0. Q.E.D.

For positive, strong subsolutions u_0, which are small in norm, $\lambda > \lambda_0$, $T^n(\lambda, u_0)$ converges to $\tilde{u}(\lambda)$ from below. Without the restriction for the norm the sequence may be unbounded.

Theorem 4.2. Let H1, H2 be satisfied, let K_0 be weakly regular, and let F be strictly convex, $F(0) = 0$, $F'(0)$: $\dot{K}_1 \rightarrow \dot{K}_0$. Then $A^{-1}F'(0)$ possesses a simple, positive characteristic value λ_0 of minimal modulus. For $\lambda \geq \lambda_0$ Eq. (3.1') has the unique solution $u = 0$ in K_1; $u = 0$ is stable for $\lambda \in (0, \lambda_0)$ and unstable for $\lambda > \lambda_0$.

The positive solutions branching off at $\lambda = \lambda_0$ exist for λ in a left-neighborhood of λ_0. They form a real analytic curve $\tilde{u}(\lambda)$. In the class of positive solutions $\tilde{u}(\lambda)$ is minimal and maximal; it is unstable.

There are no positive, strong supersolutions for $\lambda > \lambda_0$. The sequence $T^n(\lambda, u_0)$ is unbounded if u_0 is a positive, strong subsolution and $\lambda > \lambda_0$. For $\lambda < \lambda_0$ and $\lambda_0 - \lambda$ sufficiently small, there are no positive, strong subsolutions that are small in norm; the sequence $T^n(\lambda, u_0)$ converges to 0 if u_0 is a positive, strong supersolution.

Proof. The proof of the spectral properties and the bifurcation behavior proceeds as in Theorem 4.1. If $\lambda > \lambda_0$ is different from all characteristic values of $A^{-1}F'(0)$ and if $U > 0$ is a solution of (3.1') then, in view of the convexity of F, βU is a positive, strong supersolution for $\beta \in (0, 1)$. The sequence $T^n(\lambda, \beta U)$ converges to a solution U_1 with $\beta U \gg U_1 \gg 0$ (0 is unstable). Since $u = 0$ is isolated one obtains a contradiction. The same argument excludes the existence of strong, positive supersolutions.

If λ coincides with a characteristic value λ_j of $A^{-1}F'(0)$, then βU is a positive, strong supersolution for $\lambda > \lambda_j$. Hence, 0 is the unique solution of (3.1') for $\lambda \geq \lambda_0$. The stability properties of the trivial solution are obvious.

The branching solution $\tilde{u}(\lambda)$ is unstable [5] and isolated for $\lambda < \lambda_0$, λ close to λ_0. Among the positive solutions it is minimal. If $U > \tilde{u}(\lambda)$ is another solution of (3.1') then $u_0 = \beta U + (1 - \beta)\tilde{u}(\lambda)$ is a positive, strong supersolution and $T^n(\lambda, u_0)$ converges to a solution v of (3.1') with $\tilde{u}(\lambda) \leq v \ll u_0$. Since β can be chosen small, one obtains a contradiction to the instability and isolation of $\tilde{u}(\lambda)$. Hence, $\tilde{u}(\lambda)$ is maximal.

If U is a positive, strong supersolution, βU, $\beta > 0$, is also one. The sequence $T^n(\lambda, u_0)$, $\lambda > \lambda_0$, converges from above to some v, $\beta U \gg v \gg 0$. This yields a contradiction to the instability and isolation of $u = 0$ for $\lambda \neq \lambda_j$. For $\lambda = \lambda_j$ the argument proceeds as above. The unboundedness of $T^n(\lambda, u_0)$, $\lambda > \lambda_0$, follows from the uniqueness of the trivial solution, if u_0 is a positive, strong subsolution.

Choose $\tilde{\lambda} < \lambda_0$ so that $\tilde{u}(\lambda)$ is the minimal, positive solution for $\lambda \in (\tilde{\lambda}, \lambda_0)$. Since $0 \ll u_0 \ll \tilde{u}(\lambda)$ holds for positive, strong subsolutions u_0 that are small in norm, $T^n(\lambda, u_0)$ converges to $\tilde{u}(\lambda)$ from below for $\lambda \in (\tilde{\lambda}, \lambda_0)$, a contradiction to the instability of $\tilde{u}(\lambda)$. The convergence of $T^n(\lambda, u_0)$ to 0 follows, for positive, strong supersolutions u_0, from the maximality and the instability of $\tilde{u}(\lambda)$. Q.E.D.

We close this section with some remarks on local selection properties of the iteration scheme near the bifurcation points generated by a simple real, or a pair of simple, purely imaginary eigenvalues of the linearization (for proofs, see [30]). Let X be a real Banach space, V a neighborhood of 0 in $\mathbb{R}$, and $T: V \times X \to X$ a C^∞-mapping. We denote by D_λ (resp. D_u) the partial derivatives with respect to λ (resp. u). Moreover, $T(\lambda, 0) = 0$ should hold for all $\lambda \in \mathbb{R}$. We consider the equation

$$u = T(\lambda, u) \tag{4.1}$$

Let 1 be a simple eigenvalue of $D_u T(0, 0)$, φ the normed eigenfunction, Q the corresponding Dunford projector, and $P = id - Q$. Furthermore, the following notations are used:

$$Y = PX, \qquad K^+ = Y + \mathbb{R}^+ \varphi, \qquad K^- = Y + \mathbb{R}^- \varphi$$

Assume that $D_{u\lambda} T(0, 0)\varphi \in K^+$, $D_u^j T(\lambda, 0) = 0$, $j = 2, \ldots, n$, and (a) $D_u^{n+1} T(0, 0)\varphi \in K^-$, (b) $D_u^{n+1} T(0, 0)\varphi \in K^+$ holds for sufficiently small $|\lambda|$. Then $u = 0$, $\lambda = 0$ is a point of bifurcation. If n is odd, there exists exactly one nontrivial branching solution of (4.1) for every sufficiently small $|\lambda|$. If n is even, then in case (a), $\lambda > 0$, and case (b), $\lambda < 0$, there exists exactly one nontrivial branching solution $u^+(\lambda) \in K^+$ [resp. $u^-(\lambda) \in K^-$].

Analogous results with a slightly different notion of simplicity are proved in [5] (cf. [24] also). The proof of the present assertion uses the center-manifold theorem of Ruelle and Takens [27].

For the selection property of the iteration scheme one needs in addition that T is a real analytic Fredholm operator with index 0 and $\dim X/D_u T(\lambda, 0)^n X \leq n_0(\lambda) < \infty$ for all $n \in \mathbb{N}$. The trivial solution $u = 0$ is stable for $\lambda < 0$; the sequence $T^n(\lambda, u_0)$ converges to 0 for small $\|u_0\|$. If

$\lambda > 0$, the trivial solution is unstable and the set $N = \{x \in X/T^n(\lambda, x) \to 0\}$ is—according to Theorem 2.1—of first category. If a nontrivial branching solution $u^+(\lambda) \in K^+$ [resp. $u^-(\lambda) \in K^-$] exists, then for small $\|u_0\|$, $T^n(\lambda, u_0)$ converges in a set of second category to $u^+(\lambda)$ [resp. $u^-(\lambda)$] depending on whether $u_0 \in K^+$ or $u_0 \in K^-$ holds. Hence, these solutions are weakly stable.

If at $\lambda = 0$ a pair of conjugate complex, simple, nonreal eigenvalues of $D_u T(\lambda, 0)$ leaves the closed unit circle with nonvanishing "velocity," and if all other spectral points remain for small $|\lambda|$ in the open unit circle, then according to [27] there exists a two-dimensional attractive, invariant manifold M_λ of $T(\lambda, \cdot)$ in a neighborhood $V(0)$ of 0. Under certain additional assumptions M_λ contains a closed attractive curve S_λ, which contains 0 in its interior. Then $u_n = T^n(\lambda, u_0)$ converges to S_λ for almost all u_0 of a neighborhood of 0, i.e., the distance $d(u_n, S_\lambda)$ converges to 0. For $\lambda < 0$ the sequence converges to the stable solution $u = 0$.

REFERENCES

[1] Allgower, E. L., and Jeppson, M. M., The approximation of solutions of nonlinear elliptic boundary value problems having several solutions, *Lecture Notes Math.* **333** (1973), 1–20.

[2] Amann, H., On the number of solutions of nonlinear equations in ordered Banach spaces, *J. Functional Analysis* **11** (1972), 346–384.

[3] Amann, H., On the existence of positive solutions of nonlinear elliptic boundary value problems, *Indiana Univ. Math. J.* **21** (1971), 125–146.

[4] Bandle, C., "Habilitationsschrift." ETH, Zürich, 1973.

[5] Crandall, M. G., and Rabinowitz, P. H., Bifurcation from simple eigenvalues, *J. Functional Analysis* **8** (1971), 321–340.

[6] Dancer, E. N., Global solution branches for positive mappings, *Arch. Rational Mech. Anal.* **52** (1973), 181–192.

[7] Fujita, H., On the nonlinear equations $\Delta u + e^u = 0$ and $\partial v/\partial t = \Delta v + e^v$, *Bull. Amer. Math. Soc.* **75** (1969), 132–135.

[8] Gelfand, I. M., Some problems in the theory of quasilinear equations, *Amer. Math. Soc. Transl. Ser. 2* **29** (1963), 295–381.

[9] Gohberg, J. C., and Krein, M. G., "Introduction to the Theory of Linear Nonself-Adjoint Operators." *Amer. Math. Soc.*, Providence, Rhode Island, 1969.

[10] Hewitt, E., and Stromberg, K., "Real and Abstract Analysis." Springer, New York, 1969.

[11] Hille, E., and Phillips, R. S., "Functional Analysis and Semigroups." *Amer. Math. Soc.*, Providence, Rhode Island, 1969.

[12] Joseph, D. D., and Sparrow, E. M., Nonlinear diffusion induced by nonlinear sources, *Quart. Appl. Math.* **28** (1963), 327–342.

[13] Kato, T., "Perturbation Theory for Linear Operators." Springer, New York, 1966.

[14] Keller, H. B., Positive solutions of some nonlinear eigenvalue problems, *J. Math. Mech.* **19** (1969), 279–296.

[15] Keller, H., and Cohen, D., Some positone problems suggested by nonlinear heat generation, *J. Math. Mech.* **16** (1967) 1361–1376.

[16] Kielhöfer, H. J., Stability and semilinear evolution equations in Hilbert spaces, to appear in *Arch. Rational Mech. Anal.*

[17] Kirchgässner, K., and Kielhöfer, H. J., Stability and bifurcation in fluid dynamics, *Rocky Mountain J. Math.* **3** (1973), 275–318.

[18] Krasnoselskii, M. A., "Topological Methods in the Theory of Nonlinear Integral Equations." Macmillan, New York, 1964.

[19] Krasnoselskii, M. A., "Positive Solutions of Operator Equations." Noordhoff, Groningen, 1964.

[20] Krasnoselskii, M. A., "Approximate Solution of Operator Equations." Moscow, 1969.

[21] Krein, M. F., and Rutman, M. A., Linear operators leaving invariant a cone in a Banach space, *Amer. Math. Soc. Transl. Ser. 1* **10** (1962), 1–128.

[22] Laetsch, T. W., Existence and bounds for multiple solutions of nonlinear equations, *SIAM J. Appl. Math.* **18** (1970), 389–400.

[23] Laetsch, T. W., On the number of solutions of boundary value problems with convex nonlinearities, *J. Math. Anal. Appl.* **35** (1971), 389–404.

[24] Nirenberg, L., "Topics in Nonlinear Functional Analysis." Courant Inst., New York, 1973/1974.

[25] Patterson, W. M., Iterative methods for the solutions of a linear operator equation in Hilbert spaces—a survey, *Lecture Notes Math.* **394** (1974).

[26] Protter, M. H., and Weinberger, H. F., "Maximum Principles in Differential Equations." New York, 1967.

[27] Ruelle, D., and Takens, F., On the nature of turbulence, *Commun. Math. Phys.* **20** (1971), 167–192; Note concerning our paper "On the Nature of Turbulence," *ibid.* **23** (1971), 343–344.

[28] Sattinger, D. H., Topics in stability and bifurcation theory, *Lecture Notes Math.* **309** (1973).

[29] Sattinger, D. H., Monotone methods in nonlinear elliptic and parabolic boundary value problems, *Indiana Univ. Math. J.* **21** (1972), 979–1000.

[30] Scheurle, J., Selektive Iterationsverfahren, Univ. Stuttgart, 1974.

Chapter 3 : EVOLUTIONARY EQUATIONS

An Introduction to Evolution Governed by Accretive Operators

MICHAEL G. CRANDALL*
Department of Mathematics
University of California, Los Angeles, California
and
Mathematics Research Center
University of Wisconsin, Madison, Wisconsin

Introduction

One of the purposes of this symposium is to discuss the effect of techniques of ordinary differential equations and the point of view of dynamical systems on problems in partial differential equations. This is a broad topic and the author is competent to discuss only small parts of it. Perhaps the most obvious observation about the theory of ordinary differential equations vis-à-vis partial differential equations concerns a striking difference. There are a few basic theorems about ordinary differential equations concerning the existence, uniqueness, and dependence on given data of solutions of initial-value problems. Among other things these theorems guarantee a rich supply of solutions for most ordinary differential equations, and provide convenient parametrizations of the solutions of a given equation and basic information used throughout the development of more special topics and results. The theory of partial differential equations, on the other hand, does not have a few unifying basic results. Instead, it splits immediately into many topics, each with its own fundamental theorems and methods. Indeed, even the problem of identifying those linear partial differential equations that admit any solutions at all is quite deep and solved only in special cases.

However, most problems in partial differential equations arising from physical models either have the form of evolution equations, which describe

* Sponsored in part by NSF GP-38519 and in part by the U.S. Army under Contract No. DA-31-124-ARO-D-462.

the change of a physical system in time, or result from seeking stationary solutions of some evolution problem. These evolution problems can often be regarded as ordinary differential equations in some infinite-dimensional space.

The principal goal of this chapter is to describe and hopefully to render more accessible some recent results concerning abstract Cauchy problems in infinite-dimensional spaces that are close in spirit to classical ordinary differential equations. Moreover, they cover a broad range of interesting problems in partial differential equations. In accordance with this goal, it is assumed that the reader is familiar with basic functional analysis and ordinary differential equations, but not partial differential equations, Sobolev spaces, or the Hille–Yosida theorem.

Section 1 is completely informal and is intended mainly to make the material in Section 2 seem more natural than it might otherwise appear. Section 2 describes basic abstract existence and uniqueness results for the Cauchy problem in our setting. These results are discussed for the case of general Banach spaces, which adds greatly to their complexity. However, it is this case we wish to emphasize. (The theory in, e.g., Hilbert spaces is cleaner and much more highly developed. If the reader prefers to begin with this, we recommend the book of Brezis [11].) Section 3 is intended to provide a hint of the scope and flavor of applications of the abstract existence theory to concrete problems in partial differential equations. Section 4 discusses briefly the question of continuous dependence on the equation of the solutions described in Section 2 and other perturbation questions. Section 5 indicates another setting for ordinary differential equations in infinite dimensions, which corresponds to problems in partial differential equations and the relationship of this setting to that of Section 2. Section 6 contains most of the references and various further remarks. It may be consulted while reading Sections 1–5 and is intended to help make the literature more accessible.

There are two appendixes. In Appendix 1 an estimate is established that contains a crucial part of the proof of the existence results described in Section 2. It may be used in the problems introduced in Section 5 as well. Kaplan and Yorke [30] announce results that apply to problems of the forms described in Sections 2 and 5 simultaneously, and the result in Appendix 1 is a clarification and sharpening of a result of theirs.

Appendix 2 contains an important estimate due to Benilan [3]. It is included for the reader's convenience.

1. Orientation

Let us begin by sketching one way in which a problem in partial differential equations may be transformed so that it has the appearance of an initial-value problem for an ordinary differential equation. For purposes of illustration, we choose the nonlinear parabolic equation

$$\frac{\partial u}{\partial t}(t, x) - \Delta u(t, x) + g(u(t, x)) = 0, \qquad t > 0, \quad x \in \Omega, \qquad \text{(NPE)}$$

where $\Delta = \sum_{i=1}^{N} (\partial/\partial x_i)^2$, Ω is an open subset of $\mathbb{R}^N$ with boundary $\partial\Omega$, and $g: \mathbb{R} \to \mathbb{R}$. In addition to (NPE) we impose the boundary condition

$$u(t, x) = 0 \qquad \text{for} \quad t > 0, \quad x \in \partial\Omega, \qquad \text{(BC)}$$

and the initial condition

$$u(0, x) = u_0(x) \qquad \text{for} \quad x \in \Omega, \qquad \text{(IC)}$$

on the unknown function u. The first step is to choose a Banach space X of real-valued functions on Ω [e.g., $L^p(\Omega)$] and interpret the unknown function $u(t, x)$ as the map $t \to u(t, \cdot)$, which is to have values in X. The function $t \to u(t, \cdot)$ is denoted simply by $u(t)$. Then $(\partial u/\partial t)(t, x)$ can be thought of as $(du/dt)(t)$, the derivative of the X-valued function $u(t)$. Next we define an operator $A: D(A) \subset X \to X$ by $Au = -\Delta u + g(u)$ for $u \in D(A)$, where $D(A)$ must be chosen so that the expression $Au = -\Delta u + g(u)$ has (some sort of) a meaning and lies in X for $u \in D(A)$. Moreover, the boundary condition (BC) should be built into $D(A)$ by requiring (in some sense) that $u \in D(A)$ implies $u = 0$ on $\partial\Omega$. Then (NPE), (BC), (IC) can formally be written as the abstract Cauchy problem

$$\text{(DE)} \quad (du/dt) + Au = 0, \qquad \text{(IC)} \quad u(0) = u_0. \qquad \text{(ACP)}$$

Several comments are in order at this point. To go from (NPE), (BC), (IC) to (ACP) we introduced only one space X and an operator $A: D(A) \subseteq X \to X$. There are other ways to convert problems in partial differential equations to problems that look like (ACP) but that involve more spaces. In particular, u and du/dt may take values in different spaces. (See the chapter by Tartar in this volume.) We will mostly be concerned with a "one-space" framework, i.e., $A: D(A) \subseteq X \to X$, which lies closest to classical ordinary differential equations. Let us say the reduction of a problem to the form (ACP) is "successful" if some abstract results can then be applied to obtain information about the original problem. In general, successful reductions are not unique even if they exist and before one can make any successful reduction, he must know a great deal about his original problem. For simplicity, let $\Omega = \mathbb{R}^N$ in our example. Then the choices $X = BU(\mathbb{R}^N)$ (bounded uniformly continuous functions on $\mathbb{R}^N$) or $L^p(\mathbb{R}^N)$, $1 \le p < \infty$, and $D(A) = \{u \in X : \Delta u, g(u) \in X\}$ with $Au = -\Delta u + g(u)$ are successful for (NPE), (IC) provided, e.g., that g is differentiable and g' is bounded below. (Indeed, in these cases the theory we describe in Section 2 applies.) Here Δu is understood in the sense of distributions. In this example, we see characteristic features of (ACP) corresponding to its origin in a problem in partial differential equations: X is infinite dimensional, $D(A)$ is small in the sense of category, and A is very discontinuous. These features preclude direct use of some of the standard methods from ordinary differential equations (Picard iterations, polygonal approximations) for studying the existence of solutions of (ACP), since these methods involve successive applications of A.

Thus consider instead the implicit (backwards) difference scheme

$$\frac{u_\varepsilon(t) - u_\varepsilon(t - \varepsilon)}{\varepsilon} + Au_\varepsilon(t) = 0 \qquad \text{for} \quad t \ge 0,$$

$$u_\varepsilon(t) = u_0 \qquad \text{for} \quad t < 0, \tag{BD}$$

where $\varepsilon > 0$. Not worrying about justifications at this point, we formally solve (BD) as follows: One has $u_\varepsilon(t) + \varepsilon Au_\varepsilon(t) = u_\varepsilon(t - \varepsilon)$, or $u_\varepsilon(t) = (I + \varepsilon A)^{-1}u_\varepsilon(t - \varepsilon)$. Let $J_\varepsilon = (I + \varepsilon A)^{-1}$ and iterate to find

$$u_\varepsilon(t) = J_\varepsilon^{[t/\varepsilon]+1}u_0 \qquad \text{for} \quad t > 0, \tag{1.1}$$

where $[\tau]$ is the largest integer in $(-\infty, \tau]$. This looks promising, since even if A is a differential operator, $J_\varepsilon = (I + \varepsilon A)^{-1}$ may be well behaved. (BD) is called an "implicit scheme" because it leads to formulas involving the inverses $(I + \varepsilon A)^{-1} = J_\varepsilon$. Ignoring, for the moment, our previous motivation, let us consider properties of A that might be favorable to the existence of the limit $\lim_{\varepsilon \downarrow 0} u_\varepsilon$. For one thing, it will be necessary to have u_ε defined, which requires the iterates J_ε^k of J_ε to have a large enough domain. Hence $D(J_\varepsilon) = R(I + \varepsilon A)$ should be large enough. Moreover, it seems reasonable to ask that the iterates J_ε^k be equicontinuous. The simplest way to guarantee this is to ask that $J_\varepsilon = (I + \varepsilon A)^{-1}$ be a contraction [i.e., $\|J_\varepsilon x - J_\varepsilon y\| \le \|x - y\|$ for $x, y \in D(J_\varepsilon)$] for $\varepsilon > 0$. It is a happy fact that under these rather minimal assumptions $\lim_{\varepsilon \downarrow 0} u_\varepsilon(t)$ will exist for suitable u_0. Moreover, the assumptions are satisfied in a diverse collection of problems from partial differential equations.

In the next section we will state more precisely the result hinted at above. The point here is that the implicit scheme (DE) leads naturally to the assumptions made and provides, in fact, a mnemonic device. One further thing we would like to motivate by these considerations is the following: If f is a function, f^{-1} need not be. Similarly, if $(I + \varepsilon A)^{-1}$ is a function, A need not be. Thus A will be allowed to be "multivalued," a generality very useful in applications.

Finally, let us remark that attempting to solve (ACP) via the backwards difference scheme (BD) is but one of many possible approaches. This is clear from the chapter by Tartar. However, in our context, (BD) appears a natural way to introduce a theory rich in structure and with many applications, and not simply one of a multiplicity of possibilities.

2. Accretive Operators, the Cauchy Problem, and Semigroups

For convenience, we will call a mapping $A: X \to 2^X$ (the set of subsets of X) an *operator* in X. Let $D(A) = \{x \in X : Ax \neq \varnothing\}$. An operator A is *single valued* if Ax is a singleton for $x \in D(A)$, and in this case we use Ax to denote both the singleton and its element. In this way functions $A: D(A) \subseteq X \to X$ are identified with single-valued operators. Given operators A, B, and $\alpha \in \mathbb{R}$, $A + B$, αA, and A^{-1} are defined in the obvious ways:

$$(A + B)(x) = Ax + Bx, \qquad A^{-1}y = \{x \in X : y \in Ax\},$$
$$(\alpha A)(x) = \alpha(Ax) \qquad \text{for} \quad x, y \in X.$$

One also sets $R(A) = \bigcup \{Ax : x \in D(A)\}$. The identity map of X is denoted by I, the norm by $\| \ \|$.

Definition. Let A be an operator in X. Then A is *accretive* if $(I + \varepsilon A)^{-1}$ is a contraction for $\varepsilon \geq 0$.

Let us unravel the terminology: An operator J in X is a contraction if $x_i \in Jz_i$, $i = 1, 2$, implies $\|x_1 - x_2\| \leq \|z_1 - z_2\|$. (Observe, taking $z_1 = z_2$, that contractions are single valued.) Now $x_i \in (I + \varepsilon A)^{-1}z_i$ means $z_i \in (I + \varepsilon A)x_i$ or $z_i = x_i + \varepsilon y_i$ for some $y_i \in Ax_i$. Thus $(I + \varepsilon A)^{-1}$ is a contraction if

$$\|(x_1 + \varepsilon y_1) - (x_2 + \varepsilon y_2)\| \geq \|x_1 - x_2\| \qquad \text{for} \quad y_i \in Ax_i, \quad i = 1, 2. \qquad (2.1)$$

Therefore, A is accretive iff (2.1) holds for $\varepsilon \geq 0$. There is yet a third way to say that A is accretive. Let $[\ ,\ ]_\lambda: X \times X \to \mathbb{R}$ be defined for $\lambda \neq 0$ by

$$[x, y]_\lambda = (\|x + \lambda y\| - \|x\|)/\lambda.$$

Since $\lambda \to \|x + \lambda y\|$ is convex in λ, $[x, y]_\lambda$ is nondecreasing in λ, and we may define $[\ ,\ ]_\pm: X \times X \to \mathbb{R}$ by

$$[x, y]_+ = \lim_{\lambda \downarrow 0} [x, y]_\lambda = \inf_{\lambda > 0} [x, y]_\lambda,$$

$$[x, y]_- = \lim_{\lambda \uparrow 0} [x, y]_\lambda = \sup_{\lambda < 0} [x, y]_\lambda. \tag{2.2}$$

By (2.2), $\|x + \lambda y\| \geq \|x\|$ for $\lambda \geq 0$ iff $[x, y]_+ \geq 0$. Thus A is accretive iff

$$[x_1 - x_2, y_1 - y_2]_+ \geq 0 \qquad \text{for} \quad y_i \in Ax_i. \tag{2.3}$$

The result alluded to in the previous section is stated next.

Theorem 1.   Let A be an accretive operator in X and $R(I + \lambda A) \supseteq \overline{D(A)}$ for $\lambda > 0$. If $J_\varepsilon = (I + \varepsilon A)^{-1}$, $T > 0$, and $x \in \overline{D(A)}$, then

$$S(t)x = \lim_{\varepsilon \downarrow 0} J_\varepsilon^{[t/\varepsilon]+1} x \tag{2.4}$$

exists uniformly for $0 \leq t \leq T$. Moreover, the $S(t)$ so defined is a semigroup of contractions on $\overline{D(A)}$. That is,

(i) $S(t): \overline{D(A)} \to \overline{D(A)}$ for $t \geq 0$;

(ii) $S(t)S(\tau) = S(t + \tau)$ for $t, \tau \geq 0$;

(iii) $\lim_{t \downarrow 0} S(t)x = S(0)x = x$ for $x \in \overline{D(A)}$;

(iv) $\|S(t)x - S(t)y\| \leq \|x - y\|$ for $x, y \in \overline{D(A)}$, $t \geq 0$.

Theorem 1 is proved in [22]. The proof is completely elementary. A generalization of the basic estimate of the proof is given in Appendix 1. From the discussion in the preceding section, we expect $S(t)x = u(t)$ to be a solution of $du/dt + Au = 0$, $u(0) = x$, in some sense, but the above theorem contains no assertion to that effect. Indeed, when dealing with partial differential equations, the precise notions of solution that are suitable to the problem at hand can be rather subtle, and one must always pay attention to this point. In any result for (ACP) general enough to cover examples from partial differential equations, one can expect to worry a little about the appropriate notion of solution, a problem not present in most of classical ordinary differential equations. It is worthwhile being more general while discussing this point.

Let $\mathscr{P}$ be a subinterval of $\mathbb{R}$. $L^1_{\text{loc}}(\mathscr{P}, X)$ denotes the space of functions $f: \mathscr{P} \to X$ that are (strongly) integrable on compact subsets of $\mathscr{P}$ equipped with the topology of L^1 convergence on compact subsets of $\mathscr{P}$. (The reader unfamiliar with integration of X-valued functions can simply pretend $X = \mathbb{R}^N$ when an integral appears.) Below $f, g \in L^1_{\text{loc}}(\mathscr{P}, X)$ and A is an operator in X. Consider the evolution equation:

$$(du/dt) + Au \ni f. \tag{E$_f$}$$

We want to say when a function u is a solution of (E)$_f$. Three notions (at least) are appropriate. First, u is a *strong solution* of (E)$_f$ on $\mathscr{P}$ if $u \in C(\mathscr{P}, X) \cap W^{1,1}_{\text{loc}}(\mathscr{P}, X)$ and $u'(t) + Au(t) \ni f(t)$ a.e. $t \in \mathscr{P}$. Here $C(\mathscr{P}, X)$ is the space of continuous maps from $\mathscr{P}$ into X with the topology of uniform convergence on compact subsets of $\mathscr{P}$ and $u \in W^{1,1}_{\text{loc}}(\mathscr{P}, X)$ means that there is a $v \in L^1_{\text{loc}}(\mathscr{P}, X)$ such that

$$u(t) - u(s) = \int_s^t v(\tau)\, d\tau \qquad \text{for} \quad t, s \in \mathscr{P}$$

[in which case $u'(t) = (du/dt)(t) = v(t)$ a.e. $t \in \mathscr{P}$]. The notion of a strong solution is such that one can use most of the standard arguments of calculus in studying solutions of (E)$_f$. The most obvious way to weaken this notion is to "take limits" of strong solutions in a manner so that conclusions reached by doing calculus may be preserved. Thus u is called a *weak* solution of (E)$_f$ on $\mathscr{P}$ if there is a sequence

$$\{(u_n, f_n)\}_{n=1}^\infty \subset C(\mathscr{P}, X) \times L^1_{\text{loc}}(\mathscr{P}, X)$$

such that u_n is a strong solution of (E)$_{f_n}$ on $\mathscr{P}$ and $(u_n, f_n) \to (u, f)$ in $C(\mathscr{P}, X) \times L^1_{\text{loc}}(\mathscr{P}, X)$. Finally, u is called an *integral solution* of (E)$_f$ on $\mathscr{P}$ provided $u \in C(\mathscr{P}, X)$ and

$$\|u(t) - x\| - \|u(s) - x\| \le \int_s^t [u(\alpha) - x, f(\alpha) - y]_+ \, d\alpha \tag{2.5}$$

$$\text{for} \quad t > s, \quad t, s \in \mathscr{P}, \quad x \in D(A), \quad y \in Ax.$$

The integral above is well defined by virtue of the facts that $[\ ,\ ]_+$ is upper-semicontinuous [see (2.2)] and $|[x, y]_+| \le \|y\|$ (from the definition). The notion of integral solution is appropriate only if A is accretive and it arises from the following considerations. First, using the definitions of $[\ ,\ ]_\pm$ one easily shows that if $u: \mathscr{P} \to X$ then

$$D_l\|u(t)\| = [u(t), (D_l u)(t)]_-, \qquad D_r\|u(t)\| = [u(t), (D_r u)(t)]_+ \tag{2.6}$$

where D_l, D_r are left and right differentiation, respectively. Equations (2.6) are to be understood in the sense that if $(D_l u)(t)$ exists, then so does $D_l \|u(t)\|$ and the equation holds, etc. It follows that if u *and* $\|u\|$ are differentiable at t, then

$$\frac{d}{dt} \|u(t)\| = [u(t), u'(t)]_+ = [u(t), u'(t)]_- . \tag{2.7}$$

In (2.7) the differentiability of u does not imply that of $\|u\|$ [e.g., $u(t) = t$, $X = \mathbb{R}$, so $|u(t)| = |t|$]. Now if u is a strong solution of $u' + Au \ni f$ and v is a strong solution of $v' + Av \ni g$ on $\mathscr{P}$, then $\|u(t) - v(t)\|$ is locally absolutely continuous, and we have

$$\frac{d}{dt} \|u(t) - v(t)\| = [u(t) - v(t), u'(t) - v'(t)]_-$$

$$= [u(t) - v(t), (u'(t) - f(t)) - (v'(t) - g(t)) + (f(t) - g(t))]_-$$

$$\leq [u(t) - v(t), -((g(t) - v'(t)) - (f(t) - u'(t)))]_-$$

$$+ [u(t) - v(t), f(t) - g(t)]_+$$

$$\leq [u(t) - v(t), f(t) - g(t)]_+$$

a.e. on $\mathscr{P}$. In the above calculation we used the results $[x, y + z]_- \leq [x, y]_- + [x, z]_+$ (this is easy to establish), $[x, -y]_- = -[x, y]_+$, $g(t) - v'(t) \in Av(t)$ a.e., $f(t) - u'(t) \in Au(t)$ a.e., and A satisfies (2.3) since it is accretive. The above inequality shows that

$$\|u(t) - v(t)\| - \|u(s) - v(s)\| \leq \int_s^t [u(\alpha) - v(\alpha), f(\alpha) - g(\alpha)]_+ \, d\alpha \tag{2.8}$$

if u is a strong solution of $(E)_f$ on $\mathscr{P}$, v is a strong solution of $(E)_g$ on $\mathscr{P}$, and $s < t$, $t, s \in \mathscr{P}$. If $y \in Ax$, then $v \equiv x$ is a solution of $v' + Av \ni y$. Setting $v \equiv x$, $g \equiv y$ in (2.8) yields (2.5). Thus a strong solution of $(E)_f$ is an integral solution of $(E)_f$. Moreover, since $[\ ,\ ]_+$ is upper-semicontinuous, a weak solution of $(E)_f$ is an integral solution.

Let us return to $(E)_0$ and the functions $S(t)x$ of Theorem 1. To indicate the dependence of S on A, we write $S(t) = S_A(t)$ if S is obtained from A via (2.4). A is said to be *closed* if its graph $G(A) = \{(x, y): x \in X, y \in Ax\}$ is closed. The closure of A, $\bar{A}$, is defined by $G(\bar{A}) = \overline{G(A)}$. $\bar{A}$ is accretive if A is accretive. Some basic facts follow. If A is closed, then $S_A(t)x$ is a strong solution of $(E)_0$ on $t > 0$ for $x \in \overline{D(A)}$ iff $S_A(t)x$ is differentiable a.e. The proof of this uses the fact that $S_A(t)x$ is an integral solution of $(E)_0$ for $x \in \overline{D(A)}$. Moreover, $S_A(t)x$ is Lipschitz continuous in t if $x \in D(A)$ (the proof of Theorem 1 shows this). Thus if X has the property that Lipschitz continuous functions of a real variable with values in X are differentiable a.e., then $S_A(t)x$ is a strong solution of $(E)_0$ for $x \in D(A)$, and a weak solution for $x \in \overline{D(A)}$. Reflexive spaces have this property. However, there are examples in which $S_A(t)x$ is nowhere differentiable in $t \geq 0$ for any $x \in \overline{D(A)}$. The question thus arises of whether or not integral solutions of $(E)_0$, or more generally $(E)_f$, are uniquely determined by Cauchy data. The answer is given by the following result of Benilan [3], which we state informally here and prove in Appendix 2: If $u \in C(\mathscr{P}, X)$ is obtained as the uniform limit of solutions of backward difference schemes converging to $(E)_f$, then (2.8) holds for every integral solution v of $(E)_g$ on $\mathscr{P}$. Moreover, note that the set of (u, f), $(v, g) \in C(\mathscr{P}, X) \times L^1_{loc}(\mathscr{P}, X)$ satisfying (2.8) is closed. Setting $f = g$ in (2.8), we see that if u is obtained as the limit of solutions of backward difference schemes converging to $(E)_f$, then integral solutions of $(E)_f$ with the initial value $u(0)$ coincide with u on $t \geq 0$. In particular, $S_A(t)x = u(t)$ is the *unique* integral solution of $(E)_0$ on $t \geq 0$ satisfying $u(0) = x$. Since strong solutions are integral solutions, *if* this problem has a strong solution, it must coincide with $S_A(t)x$.

The existence result Theorem 1 can be used to obtain solutions of $(E)_f$. To this end, assume for simplicity that A is *m-accretive*, i.e., A is accretive and $R(I + \lambda A) = X$ for $\lambda > 0$. Let $T > 0$, $x \in \overline{D(A)}$, and f be a step function on $[0, T]$, $f(t) = y_i$ on $a_{i-1} \leq t < a_i$, where $0 = a_0 < a_1 < a_2 < \cdots < a_n = T$. Then $u' + Au \ni f$ is the same as $u' + A_i u \ni 0$ on $a_{i-1} \leq t < a_i$, where $A_i x = Ax - y_i$. Now

$$u(t) = \begin{cases} S_{A_i}(t - a_{i-1})u(a_{i-1}), & a_{i-1} \leq t \leq a_i, \quad i = 1, 2, \ldots, n, \\ x, & t = 0, \end{cases}$$

is a solution of $(E)_f$, $u(0) = x$ obtained as the uniform limit of backwards difference schemes converging to $(E)_f$. Thus, if f and g are two step functions on $[0, T]$ and u_f, u_g are obtained as above, (2.8) holds with $u = u_f$, $v = u_g$. Setting $s = 0$, this yields

$$\|u_f(t) - u_g(t)\| \leq \int_0^t [u_f(\alpha) - u_g(\alpha), f(\alpha) - g(\alpha)]_+ \, d\alpha$$

$$\leq \int_0^T \|f(\alpha) - g(\alpha)\| \, d\alpha, \qquad \text{for} \quad 0 \leq t \leq T.$$

For general $f \in L^1([0, T], X)$, the above considerations imply that whenever $\{f_n\}$ is a sequence of step functions converging to f in $L^1([0, T], X)$ then

$u(t) = \lim_n u_{f_n}(t)$ exists and is the unique integral solution of $(E)_f$ on $[0, T]$ satisfying $u(0) = x$.

3. Examples

Demonstrating that a given evolution problem in partial differential equations falls within the scope of Theorem 1 via a reduction of the type outlined in Section 1 can be quite involved. It usually requires tools from the technical machinery of partial differential equations. Moreover, one encounters extra subtleties when the problem is set in a nonreflexive space, and we restrict our attention to this case.

An exception to these comments arises when there is only one space variable. Here the semigroup approach can be rather striking, for by it the study of a problem in partial differential equations is reduced to the study of an ordinary differential equation. Since we do not assume that the reader is familiar with Sobolev spaces and so on, the examples below are given for the case of one space variable and we shamelessly make strong unnecessary assumptions to simplify the proofs. These examples give a hint of the scope of the abstract theory and how to show a given problem may be subsumed by it.

The reader with experience in partial differential equations may prefer to go directly to the literature cited in Section 6. While this literature is steadily increasing, we want to emphasize that the study of applications is just beginning. We not only expect many new applications to be found, but also refined results for known applications.

Example 1. A quasilinear hyperbolic equation. Consider the problem

$$\begin{aligned}
&\text{(CL)} && u_t + (\phi(u))_x = 0, && t > 0, \quad 0 < x < 1, \\
&\text{(IC)} && u(0, x) = u_0(x), && \hphantom{t > 0, \,\,} 0 < x < 1, \\
&\text{(BC)} && u(t, 0) = 0, && t > 0,
\end{aligned} \qquad (3.1)$$

consisting of the conservation law (CL), the initial condition (IC), and the boundary condition (BC). In (3.1), $\phi: \mathbb{R} \to \mathbb{R}$. To have a very simple presentation, we assume ϕ is continuous, strictly increasing, $\phi(0) = 0$, and $\phi(\mathbb{R}) = \mathbb{R}$. Following the outline of Section 1, we set $X = L^1([0, 1])$, $D(A) = \{u \in C([0, 1]) : u(0) = 0 \text{ and } \phi(u) \text{ is absolutely continuous}\}$, and $Au = (\phi(u))'$ for $u \in D(A)$ (w' denotes the derivative of w). Thus A acts in accordance with (CL) and $D(A)$ consists only of functions satisfying (BC). We claim A is m-accretive in $L^1([0, 1])$.

First we show that A is accretive. Let $u, v \in D(A)$ and $\lambda > 0$. Let $p: \mathbb{R} \to \mathbb{R}$ be Lipschitz continuous and satisfy

$$p \text{ is nondecreasing}, \quad |p| \leq 1, \quad \text{and} \quad p(0) = 0. \tag{3.2}$$

If $j(s) = \int_0^s p(\tau) \, d\tau$, then

$$\int_0^1 (Au - Av)p(\phi(u) - \phi(v)) \, dx = \int_0^1 (\phi(u) - \phi(v))'p(\phi(u) - \phi(v)) \, dx$$

$$= \int_0^1 j(\phi(u) - \phi(v))' \, dx$$

$$= j(\phi(u(1)) - \phi(v(1))) \geq 0, \tag{3.3}$$

since $\phi(u(0)) = \phi(v(0))$ and $j \geq 0$. Also,

$$(u - v)p(\phi(u) - \phi(v)) = |u - v| \, |p(\phi(u) - \phi(v))|,$$

since p and ϕ are increasing. Thus, by (3.2) and (3.3)

$$\int_0^1 |u - v + \lambda(Au - Av)| \, dx \geq \int_0^1 ((u - v)p(\phi(u) - \phi(v))$$

$$+ \lambda(Au - Av)(p(\phi(u) - \phi(v)))) \, dx$$

$$\geq \int_0^1 |u - v| \, |p(\phi(u) - \phi(v))| \, dx. \tag{3.4}$$

Finally, let $p = p_n$ in (3.4), where

$$p_n(s) = \begin{cases} n^{-1}s & \text{if } |s| \leq n^{-1}, \\ \operatorname{sign} s & \text{if } |s| > n^{-1}, \end{cases}$$

and then let $n \to \infty$. Since $(u - v)p_n(\phi(u) - \phi(v)) \to |u - v|$, (3.4) then yields $\|u - v + \lambda(Au - Av)\|_1 \geq \|u - v\|_1$ [where $\| \ \|_p$ is the norm in $L^p([0, 1])$, $1 \leq p \leq \infty$]. Thus A is accretive.

It remains to show $R(I + \lambda A) = L^1([0, 1])$ and it suffices to assume $\lambda = 1$. Given $h \in L^1([0, 1])$ we seek $u \in D(A)$ such that $u + \phi(u)' = h$. If $\beta = \phi^{-1}$ and $v = \phi(u)$ the problem becomes $\beta(v) + v' = h, v(0) = 0$, and v is absolutely continuous. By standard continuation arguments of ordinary differential equations, it suffices to prove that if v is absolutely continuous on $[0, a]$, $a \geq 0$, $v(0) = 0$, and $\beta(v) - v' = h$ a.e., then

$$\int_0^a |\beta(v)| \, dx \leq \int_0^a |h| \, dx,$$

for this implies

$$|v(x)| \le \int_0^x |v'(s)|\, ds = \int_0^x |h(s) - \beta(v(s))|\, ds \le 2 \int_0^x |h(s)|\, ds$$

for $0 \le x \le a$. However, this follows as above. A is accretive, so if $u = \beta(v)$

$$\int_0^a |u|\, dx \le \int_0^a |0 + A0 - (u + Au)|\, dx = \int_0^a |h|\, dx.$$

[In the proof that A is accretive, we could have used $L^1([0, a])$]. The sketch is complete.

Thus (3.1) may be reduced to the form (ACP) with an m-accretive A, and (3.1) therefore has a unique integral solution defined by this A for all $u_0 \in \overline{D(A)}$. One can show that $\overline{D(A)} = L^1([0, 1])$ here. This type of example is especially interesting. It generalizes (with different proofs) dramatically. In particular, one can treat problems of the form

$$u_t + \sum_{i=1}^{N} \frac{\partial}{\partial x_i} (\phi_i(u)) = 0 \qquad \text{for} \quad t > 0, \quad x \in \mathbb{R}^N,$$

where $\phi: \mathbb{R} \to \mathbb{R}^N$ need only be continuous and satisfy

$$\lim_{r \to 0} |\phi(r)|/|r|^{(N-1)/N} = 0.$$

The Cauchy problem for this equation does not admit continuous solutions even if ϕ and the data are smooth [nor does (3.1)]. This *equation* is interpreted as a family of *integral inequalities* by Kruzkov [43]. Here we see that the difficulties associated with understanding $S_A(t)x$ as a solution of (ACP) appear in concrete examples. See the works cited in Section 6 for more information.

Example 2. A quasilinear parabolic problem. Consider the problem

$$u_t - (\phi(u))_{xx} = 0, \qquad t > 0, \quad 0 < x < 1,$$

$$u(0, x) = u_0(x), \qquad\qquad 0 < x < 1, \qquad (3.5)$$

$$u(t, 0) = u(t, 1) = 0, \qquad t > 0.$$

Again, let $X = L^1([0, 1])$, $\phi: \mathbb{R} \to \mathbb{R}$ be continuous, strictly increasing, $\phi(0) = 0$, and $\phi(\mathbb{R}) = \mathbb{R}$. Set $D(A) = \{u \in C([0, 1]): u(0) = u(1) = 0, \phi(u) \text{ and } \phi(u)' \text{ are absolutely continuous}\}$, and $Au = -\phi(u)''$ for $u \in D(A)$. If p satisfies

(3.2), then

$$\int_0^1 (Au - Av)p(\phi(u) - \phi(v))\, dx = \int_0^1 p'(\phi(u) - \phi(v))((\phi(u) - \phi(v))')^2 \, dx \geq 0.$$

(3.6)

The fact that A is accretive follows from (3.6) as it did from (3.3) in the first example.

Let $h \in L^1([0, 1])$. To solve $u + Au = h$ we set $v = \phi(u)$, $\beta = \phi^{-1}$ as before and solve instead $\beta(v) - v'' = h$, $v(0) = v(1) = 0$.

Step 1.   Assume β is bounded, $|\beta(v)| \leq K$ for $v \in \mathbb{R}$. Let

$$Tv = \int_0^1 g(x, y)(\beta(v(y)) - h(y))\, dy \qquad \text{for} \quad v \in L^1([0, 1]),$$

where $g(x, y) = y(x - 1)$, $0 \leq y \leq x \leq 1$, and $g(x, y) = g(y, x)$, $0 \leq x, y \leq 1$. That is, $w = Tv$ is the solution of $w'' = \beta(v) - h$, $w(0) = w(1) = 0$. A trivial estimate shows

$$\|Tv\|_\infty \leq \|\beta(v) - h\|_1 \leq \|\beta(v)\|_1 + \|h\|_1 \leq K + \|h\|_1,$$

and

$$\|(Tv)'\|_\infty \leq \|\beta(v) - h\|_1 \leq K + \|h\|_1.$$

Moreover, $T: L^1([0, 1]) \to L^1([0, 1])$ is continuous. It follows that T is a continuous mapping of $L^1([0, 1])$ into the compact convex subset $\{w \in L^1([0, 1]) : w \text{ is absolutely continuous and } \|w\|_\infty, \|w'\|_\infty \leq K + \|h\|_1\}$. (The compactness follows from Arzela–Ascoli.) T has a fixed point by Schauder's fixed-point theorem, and a fixed point of T is a solution of our problem.

Step 2.   Since A is accretive, if $\beta(v) \in D(A)$ and $\beta(v) - v'' = h$, then

$$\|\beta(v)\|_1 \leq \|\beta(v) - v'' - (\beta(0) - 0'')\|_1 = \|h\|_1.$$

Moreover, since $v(0) = v(1) = 0$, $v'(\xi) = 0$ for some $\xi \in (0, 1)$, and

$$|v'(x)| \leq \left| \int_\xi^x v''(s)\, ds \right| \leq \int_0^1 |\beta(v) - h|\, dx \leq 2\|h\|_1.$$

Finally,

$$|v(x)| \leq \left| \int_0^x v'(s)\, ds \right| \leq \int_0^1 |v'(s)|\, ds \leq 2\|h\|_1$$

for $x \in [0, 1]$. For unbounded β, set

$$\tilde{\beta}(s) = \begin{cases} 2\|h\|_1, & \text{if} \quad \beta(s) \geq 2\|h\|_1, \\ \beta(s), & \text{if} \quad |\beta(s)| \leq 2\|h\|_1, \\ -2\|h\|_1, & \text{if} \quad \beta(s) \leq -2\|h\|_1. \end{cases}$$

Now $\tilde{\beta}$ is bounded and the solution v of $\tilde{\beta}(v) - v'' = h$, $v(0) = v(1) = 0$ (which exists by Step 1) satisfies $\|v\|_\infty \leq 2\|h\|_1$ (as above). But then $\tilde{\beta}(v) = \beta(v)$ and v is the desired solution.

The study of quasilinear and semilinear parabolic equations that correspond to m-accretive operators is the most highly developed area of applications. See Section 6 for references.

Example 3. A nonlinear operator accretive in $C([0, 1])$. This final example is related to the preceding one. However, the underlying space will be $C([0, 1])$ and this time A can be multivalued. Consider the problem

$$\begin{aligned} \phi(u_t) - u_{xx} &= 0, & t &> 0, \quad 0 < x < 1, \\ u(0, x) &= u_0(x), & & 0 < x < 1, \\ u(t, 0) = u(t, 1) &= 0, & t &> 0, \end{aligned} \tag{3.7}$$

where $\phi \colon \mathbb{R} \to \mathbb{R}$ is continuous, nondecreasing, and $0 = \phi(0)$. Note that we do not assume ϕ is strictly increasing or $\phi(\mathbb{R}) = \mathbb{R}$. To define the A associated with (3.7), set

$$X = C([0, 1]),$$

$$D(A) = \{u \in C[0, 1] : u, u', u'' \text{ are continuous and } u(0) = u(1) = 0\},$$

and

$$Au = \{w \in C([0, 1]) : \phi(-w) = u''\} \quad \text{for} \quad u \in D(A).$$

The reader should ponder the relationship of (3.7) to this A for a moment.

To show A is accretive, let $u_1, u_2 \in D(A)$ and $w_1 \in Au_1$, $w_2 \in Au_2$. Let $u_1 \neq u_2$ and

$$\|u_1 - u_2\|_\infty = \max_{0 \leq x \leq 1} |u_1(x) - u_2(x)| = u_1(x_0) - u_2(x_0). \tag{3.8}$$

Since $u_1 = u_2 = 0$ at $x = 0, 1$ we have $0 < x_0 < 1$. If $w_2(x_0) \leq w_1(x_0)$ and $\lambda > 0$, then

$$\max_{0 \leq x \leq 1} |u_1(x) - u_2(x) + \lambda(w_1(x) - w_2(x))|$$

$$\geq (u_1(x_0) - u_2(x_0)) + \lambda(w_1(x_0) - w_2(x_0))$$

$$\geq u_1(x_0) - u_2(x_0) = \|u_1 - u_2\|_\infty,$$

and we are done. If $w_2(x_0) > w_1(x_0)$, the monotonicity of ϕ implies $\phi(-w_2(x)) \leq \phi(-w_1(x))$ on some interval $x_0 - \delta \leq x \leq x_0 + \delta$, where $\delta > 0$. Then

$$(u_1 - u_2)''(x) = \phi(-w_1(x)) - \phi(-w_2(x)) \geq 0$$

on this interval. Hence $u_1 - u_2$ is convex on $x_0 - \delta \leq x \leq x_0 + \delta$ and assumes its maximum at the interior point x_0. Then $u_1 - u_2$ is *constant* on $x_0 - \delta \leq x \leq x_0 + \delta$. If x_0 is the least number in $[0, 1]$ for which (3.8) holds, this cannot be. With this choice of x_0, $w_1(x_0) \geq w_2(x_0)$, and we have now shown that A is accretive.

To show, e.g., $R(I + A) = C([0, 1])$, choose $h \in C([0, 1])$ and solve $h \in (I + A)u$ or $h - u \in Au$. This means $u(0) = u(1) = 0$ and $u'' = \phi(u - h)$. This problem can be solved in a fashion similar to that of Example 2. Since A is accretive and $0 \in A0$, if u is a solution then $\|u\|_\infty \leq \|h\|_\infty$. Truncate ϕ by setting

$$\tilde{\phi}(s) = \begin{cases} 2\|h\|_\infty & \text{if} \quad \phi(s) \geq 2\|h\|_\infty, \\ \phi(s) & \text{if} \quad |\phi(s)| \leq 2\|h\|_\infty, \\ -2\|h\|_\infty & \text{if} \quad |\phi(s)| \leq -2\|h\|_\infty. \end{cases}$$

Let Tv be the solution w of $w'' = \tilde{\phi}(v - h)$, $w(0) = w(1) = 0$. T is continuous and maps $C([0, 1])$ into a compact convex subset. It has a fixed point, which is the solution of $h \in (I + A)u$.

Now we must be cautious. While it is true that A is m-accretive, consider the case $\phi(s) = 0$ for all $s \in \mathbb{R}$. In this case

$$Au = \begin{cases} X & \text{if} \quad u = 0, \\ \phi & \text{if} \quad u \neq 0, \end{cases}$$

and $D(A) = \{0\}$. Of course, if $\phi = 0$ one can only solve (3.7) for $u_0 = 0$, in which case $u = 0$. To know for which u_0 $S_A(t)u_0$ is defined we must know $\overline{D(A)}$. It is not too hard to show $\overline{D(A)} = \{u \in C([0, 1]) : u(0) = u(1) = 0\}$ iff

$\phi(\mathbb{R}) = \mathbb{R}$. The reader may consider other cases or refer to the literature cited in Section 6. If the fact that we omitted mention of $\overline{D(A)}$ in Example 2 was not noticed before, we reassure the reader that $\overline{D(A)} = L^1([0, 1])$ under the hypotheses of that example.

4. Auxiliary Results: Perturbation and Continuous Dependence

The abstract theory of Section 2 is useful in applications in that it provides existence and uniqueness theorems for, e.g., the examples of Section 3 as well as a notion of solution for certain problems that do not admit differentiable solutions. Frequently questions of existence and uniqueness may be handled by other means and one may ask why he should bother with the abstract theory. There are several answers to this. The abstract theory provides insight into the concrete problems to which it applies and exhibits common features of problems that otherwise appear to be totally unrelated. Another benefit of casting a concrete problem into a form to which the general theory applies is that one can then make use of the auxiliary results of the theory. Chief among these are the convergence theorems (which deal with questions of continuous dependence) and the perturbation theorems. Examples of these types of results are stated here.

Convergence theorems deal with the dependence of solutions of

$$(du/dt) + Au \ni f \qquad\qquad (E)_{A, f}$$

on A and f. Given a sequence $\{A_n\}_{n=1}^{\infty}$ of accretive operators, we want to define a notion of convergence $A_n \to A$. Recalling the considerations of Section 1, a natural idea is to use the resolvents $(I + \lambda A_n)^{-1}$ (there are other ways). For simplicity, assume each A_n is m-accretive. Then we say $A_n \to A$ if A is m-accretive and $(I + \lambda A_n)^{-1}x \to (I + \lambda A)^{-1}x$ for each $\lambda > 0$, $x \in X$. A typical example of a convergence theorem says the following: Let $\{A_n\}_{n=0}^{\infty}$ be a sequence of m-accretive operators $A_n \to A_0$, $\{f_n\}_{n=0}^{\infty}$ a sequence in $L^1([0, T], X)$, $f_n \to f_0$ in $L^1([0, T], X)$, $x_n \in D(A_n)$, and $x_n \to x_0 \in \overline{D(A_0)}$. If u_n is the unique integral solution of $(E)_{A_n, f_n}$, $u_n(0) = x_n$, then $\lim_n u_n = u_0$ uniformly on $[0, T]$.

In applications, convergence theorems supply information about the continuous dependence of solutions of partial differential equations on the equations. An interesting instance of this type of result is given by Kurtz [44], who showed that solutions of the scaled Carleman equations

$$u_t + \alpha u_x + \alpha^2(u^2 - v^2) = 0, \qquad v_t - \alpha v_x + \alpha^2(v^2 - u^2) = 0,$$

with

$$u|_{t=0} = v|_{t=0} \geq 0,$$

yield, in the limit as $\alpha \to \infty$, solutions of $w_t - \frac{1}{4}(\log w)_{xx} = 0$. (This does not exactly fit into the framework given above. See [44].)

Perturbation theorems supply information about the perturbed problem

$$(du/dt) + Au + Bu \ni f, \tag{PP}$$

when enough is known about the unperturbed problem $(E)_{A,f}$ and B. For example, if A is m-accretive and $B: X \to X$ may be written as $B = B_0 + B_1$, where B_0 is continuous and accretive and B_1 Lipschitz continuous, then for $x \in \overline{D(A)}, f \in L^1([0, T], X)$, there is a unique $u \in C([0, T], X)$ satisfying $u(0) = x$ such that u is an integral solution of $u' + Au \ni f - Bu$ on $[0, T]$. This result can be used to show that $A + B$ is m-accretive whenever A is m-accretive and B is continuous and accretive.

For example, the operator

$$Au = -\Delta u \qquad \text{on} \quad D(A) = \{u \in BU(\mathbb{R}^N) : \Delta u \in BU(\mathbb{R}^N)\}$$

is m-accretive in $X = BU(\mathbb{R}^N)$ (bounded uniformly continuous functions on $\mathbb{R}^N$). Now it is simple to see that $u \to \beta(u)$ is continuous and accretive in X if $\beta: \mathbb{R} \to \mathbb{R}$ is continuous and nondecreasing, while $u \to \gamma(u)$ is Lipschitz continuous in X if $\gamma: \mathbb{R} \to \mathbb{R}$ is Lipschitz continuous. Thus, by the perturbation theorem, we can solve, e.g.,

$$(\partial u/\partial t) - \Delta u + e^u + \sin u = f(t, x), \qquad u(0, x) = u_0(x),$$

for $u_0 \in X$ and suitable f, simply from the knowledge that A is m-accretive in X.

5. Tangency Conditions

Let C be a closed subset of a Banach space X and $B: C \to X$ be continuous. In order that the problem

$$(du/dt) + Bu = 0, \qquad u(0) = x \in C, \tag{5.1}$$

have a continuously differentiable solution $u: [0, T] \to C$ on some interval $[0, T]$, $T > 0$, B clearly must satisfy

$$\limsup_{\lambda \downarrow 0} d(x - \lambda Bx, C)/\lambda = 0 \qquad \text{for} \quad x \in C, \tag{5.2}$$

where $d(z, C) = \inf \{\|z - y\| : y \in C\}$. This is because then $u(t) = x - tBx + o(t) \in C$ as $t \downarrow 0$. Brezis and Bourguignon [9] show that the Euler equations of hydrodynamics may be cast in the form (5.1), where B is locally Lipschitz continuous and satisfies (5.2). It follows from the more general results of Martin [46] that under these conditions (5.1) has a solution $u: [0, T] \to C$ for some $T > 0$ provided $x \in C$. The main point here is that B is only defined on a closed set C and C may have empty interior. (Of course, B may be extended to a continuous function defined on a neighborhood of C, but the extension process cannot preserve Lipschitz continuity or accretive properties of B in general.)

We wish to make several points. First, there is the unexpected correspondence between ordinary differential equations of the form (5.1) with *continuous* B satisfying (5.2) and a difficult problem in partial differential equations. This is in contrast to the discussion in Section 1. Note, however, the problem of "small domain" persists, albeit in another guise.

Second, although this setting appears quite different from that of the previous section it is in fact closely related. To see this, note that if $x \in C$ and (5.2) holds, then for $\varepsilon > 0$ there exists x_λ in C and e_λ in X defined for small $\lambda > 0$ such that $x - \lambda Bx = x_\lambda + \lambda e_\lambda$, where $\|e_\lambda\| \leq \varepsilon$. Then

$$\|x - x_\lambda\| \leq \lambda(\|Bx\| + \varepsilon),$$

and

$$x_\lambda + \lambda Bx_\lambda = x + \lambda(Bx_\lambda - Bx) + \lambda e_\lambda = x + o(\lambda),$$

as $\lambda \downarrow 0$. Thus $x_\lambda \to x$ and

$$\left\| \frac{x_\lambda - x}{\lambda} + Bx_\lambda \right\| \leq \|Bx_\lambda - Bx\| + \varepsilon \leq 2\varepsilon, \qquad (5.3)$$

provided λ is small enough. In [47] Martin builds approximate solutions to (5.1) by starting at x, choosing λ small enough to satisfy various restrictions, interpolating linearly between x and x_λ on $[0, \lambda]$, and then repeating the process. This is *equivalent* to solving approximately a backwards difference scheme as indicated by (5.3). To make this clearer, consider the case $C = X$ so (5.2) holds for all $B: X \to X$. To construct a polygonal approximate solution to (5.1) with nodes at $0 = t_0 < t_1 < \cdots < t_N$ we set $\gamma_i = t_i - t_{i-1}$, solve the forward (explicit) difference scheme

$$\frac{x_i - x_{i-1}}{\gamma_i} + Bx_{i-1} = 0, \quad i = 1, 2, \ldots, N, \qquad x_0 = x, \qquad \text{(FS)}$$

and then interpolate the values x_i at t_i piecewise linearly. This yields the familiar expression

$$u(t) = x_i - (t - t_i)Bx_i, \qquad t_i \le t \le t_{i+1},$$

of polygonal approximation. The solution of (FS) is $x_i = (I - \gamma_i B)x_{i-1}$, so Bx_i above is $\gamma_{i+1}^{-1}(x_i - x_{i+1})$. Observe now that, using (FS), we have

$$\frac{x_i - x_{i-1}}{\gamma_i} + Bx_i = Bx_i - Bx_{i-1} = (B((I - \gamma_i B)x_{i-1}) - Bx_{i-1}),$$

so that $\{x_i\}$ *approximately* solves the implicit scheme $\gamma_i^{-1}(x_i - x_{i-1})$ $+ Bx_i = 0$. (The ith error tends to zero with γ_i if B is continuous.)

The fact that (5.2) is equivalent to being able to solve

$$x_\lambda + \lambda Bx_\lambda = x + o(\lambda) \qquad \text{and} \qquad \|x - x_\lambda\| \to 0 \qquad \text{as} \quad \lambda \downarrow 0 \quad (5.4)$$

for $x_\lambda \in C$ was observed independently by the author and J. Yorke. Construction of approximate solutions to (5.1) under (5.2) via the explicit scheme is carried out in Martin [46] and announced for the implicit scheme by Yorke and Kaplan [30] (in a setting general enough to include the case of Theorem 1 of Section 2 as well). Now it is a happy fact that the convergence of solutions of *implicit* schemes converging to (5.1) can be proved by generalizing the proof of Theorem 1 a bit. This was announced by Yorke and Kaplan [30]. A rather sharp form of the principal estimate is given in Appendix 1. Martin's convergence proof in [46] was difficult, so it is nice to have Theorem 1 and (5.1), (5.2) fit in the same framework. See also the comments on Appendix 1 in Section 6.

Before we leave this topic, we obtain another consequence of (5.4). Let $x_i, x_{i\lambda} \in C$, $i = 1, 2$, satisfy

$$\begin{aligned} x_{i\lambda} + \lambda Bx_{i\lambda} &= x_i + o(\lambda) && \text{as} \quad \lambda \downarrow 0, \\ \|x_i - x_{i\lambda}\| &\to 0 && \text{as} \quad \lambda \downarrow 0, \end{aligned} \qquad (5.5)$$

for $i = 1, 2$. Recall the functions $[\; , \;]_\pm$ of Section 2. B is accretive if $[x_1 - x_2, Bx_1 - Bx_2]_+ \ge 0$ for $x_i \in C$. Since $[\; , \;]_+ \ge [\; , \;]_-$, a stronger requirement would be $[x_1 - x_2, Bx_1 - Bx_2]_- \ge 0$. In fact, these are *equivalent* if (5.2) holds. Interest in this point arises from the results of Martin [46], in which he makes assumptions on $[\; , \;]_-$. We give the proof.

Observe that for $\lambda > 0$

$$[x_1 - x_2, Bx_1 - Bx_2]_{-\lambda}$$

$$= \frac{\|x_1 - x_2\| - \|(x_1 - x_2) - \lambda(Bx_1 - Bx_2)\|}{\lambda}$$

$$= \frac{\|(x_{1\lambda} - x_{2\lambda}) + \lambda(Bx_{1\lambda} - Bx_{2\lambda})\| - \|x_{1\lambda} - x_{2\lambda}\| + o(\lambda)}{\lambda}$$

$$= [x_{1\lambda} - x_{2\lambda}, Bx_{1\lambda} - Bx_{2\lambda}]_\lambda + o(1)$$

$$\geq [x_{1\lambda} - x_{2\lambda}, Bx_{1\lambda} - Bx_{2\lambda}]_+ + o(1)$$

as $\lambda \downarrow 0$. The relations (5.5), continuity of B, and (2.2) have been used. If $[x_{1\lambda} - x_{2\lambda}, Bx_{1\lambda} - Bx_{2\lambda}]_+ \geq \rho(x_{1\lambda} - x_{2\lambda})$, where ρ is lower semicontinuous, letting $\lambda \downarrow 0$ yields $[x_1 - x_2, Bx_1 - Bx_2]_- \geq \rho(x_1 - x_2)$. In particular, if B is continuous, accretive, and (5.2) holds, then $[x_1 - x_2, Bx_1 - Bx_2]_- \geq 0$ for $x_i \in C$. This is a new result.

6. References and Comments

General References.　This chapter can be regarded as an extended introduction to the paper of Benilan [4]. With the main exception of Theorem 1, most of the abstract results mentioned above and below are proved in greater generality in [4]. However, the emphasis on generality in [4] may not appeal to the reader at this stage and we list some other sources one could sample first. Brezis's excellent monograph [11] is useful for background material as well as the highly structured Hilbert space theory. It also contains a useful commentary and a good bibliography up to 1971.

The manuscript "Geometric theory of semilinear parabolic equations" (which is in preprint form) by Daniel Henry, University of Kentucky, also deserves a mention here. It is not directly concerned with accretive operators, but it is for the class of partial differential equations treated in this work that one finds the most striking applications of the ideas of ordinary differential equations and dynamical systems.

Our references are selective. Questions of priority are ignored and papers subsumed by later works (like [4]) are often not mentioned. However, papers not cited here will usually appear in the bibliography of either [11] or some paper we do quote.

Section 1.   This section is a bit misleading in that the theory presented here was developed internally rather than with applications in mind. The original goal was to see to what extent the linear theory of semigroups and the Hille–Yosida theorem could be carried over to nonlinear cases. It has not been necessary to refer to the linear theory before now, and this theory is not a prerequisite for understanding the nonlinear case.

Section 2

Added in proof: Recent generalizations of Theorem 1 are referenced in the comments on Appendix 1 in this section.

On multivaluedness.   We made no real attempt here to convince the reader that $A: X \to 2^X$ is a useful generality. One example is given in Section 3. The case is clearest in Hilbert spaces. See, e.g., the examples in [12]. A trivial but significant example in $\mathbb{R}$ is the operator

$$\text{sign}: x \to \begin{cases} \{1\}, & x > 0, \\ [-1, 1], & x = 0, \\ \{-1\}, & x < 0. \end{cases}$$

The theory is not simpler and the proofs are not simpler if one assumes that A is a function.

On $[x, y]_{\pm}$.   The functions $[x, y]_{\pm}$ used here are one-sided directional derivatives of the norm $\|x\|$. In many papers $\frac{1}{2}\|x\|^2$ and its derivatives appear instead, with the corresponding statements suitably modified. (This is due to the fact that $\|\ \|^2$ is nice in Hilbert spaces.) In fact, one can use $p^{-1}\|x\|^p$ and its derivatives $\|x\|^{p-1}[x, y]_{\pm}$ for $1 \le p < \infty$ and develop an equivalent set of statements. However, it is now clear that $\|\ \|$ is by far the most convenient in general Banach spaces.

There is another description of $[x, y]_{+}$. One has

$$[x, y]_{+} = \max \{x^*(y) : x^* \in X^*, \|x^*\| \le 1, \text{ and } x^*(x) = \|x\|\},$$

where we regard X as a real Banach space. Notation (especially the author's) varies wildly also and each paper must be checked for its conventions.

The reader should compute $[\ ,\ ]_{\pm}$ for the cases where X is a Hilbert space, $X = L^p([0, 1])$, $1 \le p < \infty$, and $X = C([0, 1])$ to orient himself. See also Sato [58].

Another fact, developed in [4], is that one can use more general sublinear functions in place of $\| \ \|$.

Assume now that A satisfies the hypotheses of Theorem 1.

Exponential formulas. It is the influence of the linear theory that led to the emphasis on the "exponential formula" $S_A(t) = \lim_{n \to \infty} (I + (t/n)A)^{-n}$ rather than the equivalent but more fundamental idea of scheme (BD) in, e.g., [19] and [22]. (BD) is mentioned only in passing in each case.

$S_A(t)x$ is a strong solution if it is differentiable. It follows from Appendix 2 that $S_A(t)x$ is an integral solution of $(E)_0$ for $x \in \overline{D(A)}$. This special case of Theorem A.2 is obtained more simply in Miyadera [51, proof of Lemma 1]. This is done in the form appropriate to $\frac{1}{2}\| \ \|^2$. The proof can be made very short using $\| \ \|$ instead. Given this result, one easily shows $u'(t) + Au(t) \ni 0$ whenever $u(t) = S_A(t)x$ is differentiable at $t > 0$, A is closed, and $x \in \overline{D(A)}$. See [22, pp. 277–278].

If u is a strong solution of $(E)_0$, $u(0) = x$, then $u(t) = S_A(t)x$. This result does not require Theorem A.2 either. It was proved in [14] that $\lim_{\varepsilon \downarrow 0} J_\varepsilon^{[t/\varepsilon]+1}x$ exists and is equal to the strong solution *provided* the latter exists. With the convergence known in advance, the proof is very simple. See [19, p. 162]. This comment and the one above justify *defining* $S_A(t)x$ to be a solution of $(E)_0$. However Benilan's Theorem A.2 unifies all results like this and the above and permits a much more coherent development.

Strong solutions. It is sometimes convenient to replace $W_{\mathrm{loc}}^{1,1}(\mathcal{P}, X)$ by $W_{\mathrm{loc}}^{1,1}(\text{interior } \mathcal{P}, X)$ in this definition.

Weak solutions. The term weak solution has a different meaning in [54] (see also the references). With our definition, it is not known if $S_A(t)x$ is a weak solution of $(E)_0$ for $x \in D(A)$.

Integration and differentiation. It is necessary to be familiar with the theory of integration and differentiation of vector-valued functions to pursue this subject farther. See, e.g., [65] for integration and the appendix of [11] for differentiation.

Infinitesimal generators. The notion of a semigroup of contractions on $\overline{D(A)}$ was defined in Theorem 1. If $\overline{D(A)}$ is replaced by an arbitrary $C \subseteq X$

in this definition one obtains the notion of a semigroup of contractions on C. The linear theory suggests the problem of attempting to characterize the infinitesimal generators of semigroups of contractions insofar as possible. Indeed, the nonlinear theory began with such questions. See [19] for definitions and a review of this problem, Miyadera [52] for recent results in this direction, and [4, Chapter IV] for another point of view.

The equation $du/dt + A(t)u \ni 0$.   Results for this equation, which is formally more general than the case $A(t)u = Au - f(t)$, are given in [23]. These results are not developed in terms of a notion of integral solution. L. C. Evans has (in a work in preparation) recently obtained results in the setup of [23] sufficiently general to subsume the case $A(t)u = Au - f(t)$ with $f(t) \in L^1([0, T]: X)$ directly.

Note added in proof: See L. C. Evans, Nonlinear evolution equations, Mathematical Research Center TSR #1568 (1975), Univ. of Wisconsin, Madison, Wisconsin.

Section 3
Nonlinear diffusion.   Results for mappings of both of the forms $A_1 u = -\Delta u + \beta(u)$ and $A_2 u = -\Delta\phi(u)$, where β and ϕ are accretive operators in $\mathbb{R}$, are obtained in Brezis and Strauss [16]. It is a consequence of their results that, under various assumptions, A_1 gives rise to m-accretive operators in L^p spaces, $1 \le p < \infty$, while the form A_2 defines an m-accretive operator in L^1. Actually, A_2 is not explicitly mentioned in [16]. Note, however, that the "change of variable" $v = \phi(u)$ formally transforms $u \to u - \lambda \Delta\phi(u)$ into $v \to \beta(v) - \lambda \Delta v$, where $\beta = \phi^{-1}$. This may be used to make the connection. Observe that β may be multivalued even if ϕ is not. For example, the Stefan problem (see, e.g., [10]) may be written in the form $u_t - \Delta\phi(u) = 0$, where ϕ is a function and ϕ^{-1} is not. The map $u \to -\Delta\phi(u)$ is also accretive in H^{-1}. See [12] and [13].

The equation $u_t - \Delta(u^\alpha)$ for various α arises in the study of flow through a porous medium. See the references of Konishi [40]. Difficulties not treated in [16] arise when attempting to associate an m-accretive operator with $u \to -\Delta\phi(u)$ in $L^1(\mathbb{R}^N)$ due to the infinite measure of $\mathbb{R}^N$. These problems are treated in [8]. The operator $u \to -\Delta\phi(u)$ in one space dimension is studied in [21] when $0 \notin D(\phi)$ and results of Kurtz [41] for $\phi(u) = \log u$ are generalized. One finds $u_t - \Delta\phi(u)$ studied under nonlinear boundary conditions in [3]. The problem $\phi(u_t) - \Delta u = 0$ is treated in [7] and Konishi [41]. Many of Konishi's notes [33–42] are related to parabolic problems.

Hyperbolic systems. Examples of nonlinear hyperbolic systems generating semigroups of contractions in L^1 spaces are given in [19] and [44]. These examples can be greatly generalized by abstraction, but this has not been written down yet.

Conservation laws. An excellent introduction is given by Lax [45]. The simplest paper written from our point of view is [20]. Oharu [54] and Benilan [5] treat the technically more complex case where $\phi = \phi(x, u)$ depends on x as well as u. Benilan also settles a question about $D(A)$ left open in [20]. Kruzkov's basic paper [43] provides an essential idea used in each of these presentations. See also Douglis [25].

Equations of Hamilton–Jacobi type. See Aizawa [1], Tamburro [61], Fleming [27], and Friedman [28].

Optimal control. We know of only one paper here, namely, Slemrod [59].

Functional equations. Simple functional equations correspond to m-accretive operators A perturbed by Lipschitz functions. See Webb [64] and Flaschka and Leitman [26].

Section 4
Perturbation theorems. There are only a few general known results valid in arbitrary spaces X. There is the theorem of Webb [62] as generalized by Barbu [2] to state that $A + B$ is m-accretive if A is m-accretive and B is continuous and accretive. A simpler proof and other results are given in Pierre [56]. See Martin [58] for another type of generalization of Webb's result. Calvert and Gustafson [18] deal with another class of perturbations. Results influenced by a class of applications are given in Benilan [6]. See also [63].

There are many more true theorems in the case where X is a Hilbert space. See [11]. Some of the proofs of these work if only X and X^* are uniformly convex. See, e.g., Konishi [35].

Convergence and approximation theorems.   There are many types of these proved in many places. The one we have stated is found in [4, Chapter V]. Kurtz's presentation in [44] is especially interesting and Goldstein [29] uses a similar idea in his proof. See [31, 32] for recent results of a somewhat different sort. Oharu [55] develops approximation theorems while treating a pde. We also mention [15, 52, and 53].

Asymptotic behavior.   The study of asymptotic behavior of solutions of $(E)_f$ is not highly developed. See [4, III.5] for results in Hilbert spaces. Dafermos and Slemrod [24] prove that $\{S_A(t)x : t \geq 0\}$ is precompact for $x \in \overline{D(A)}$ under certain assumptions on A and then apply the theory of dynamical systems to establish a number of interesting conclusions. See also Bruck [17].

Section 5
The tangency condition (5.2), with lim sup replaced by lim inf [let us call this (5.2)*] was used already by Nagumo [54]. Martin [47] proves that $(5.2)* \Rightarrow (5.2)$ if B is continuous. Most of the results concerning the con-sequences of (5.2) use the fact that one can build approximate solutions to (5.1) (or generalizations) in this case. In this regard, it may be helpful to read the proof of [47, Lemma 2] before pursuing other papers on this topic. This simple proof is quite close to Nagumo's, and Martin's works [46–50] delve quite deeply into the implications of (5.2) and related conditions and contain further references. See also Redheffer and Walter [57].

Comments on Appendix 1
After the preparation of this manuscript the author obtained a copy of Takahashi's paper [60]. This work contains more general results concerning existence and convergence of approximate solutions of backwards difference schemes than those previously known to the author. Takahashi uses argu-ments related to those of Appendix 2 to show convergence, and his proofs are more complex than in our Appendix 1 (which is still sufficient for known applications). In fact, however, it is possible to extend and clarify Takahashi's results by direct estimation of solutions of schemes similar to (A.9), which we will do elsewhere.

Added in proof: Significant new results were subsequently obtained by

Y. Kobayashi, Difference approximation of evolution equations and generation of nonlinear semi-groups, *Proc. Japan Acad.* (to appear), and M. G.
Crandall and L. C. Evans, On the relation of the operator $\partial/\partial s + \partial/\partial \tau$ to
evolution governed by accretive operators, *Israel J. Math.* (to appear). In
particular, Kobayashi showed that $S_A(t)$ can be defined if only the limit
inferior of $\lambda^{-1} d(x, R(I + \lambda A))$ as λ tends to $0+$ vanishes for x in the
closure of $D(A)$. A more general condition is given by M. Pierre, Un
theoreme general de generation de semi-groups non lineaires (to appear).

Appendix 1

The result proved here represents a refinement of the proof of Theorem 3
of [30]. The authors of [30] had kindly provided a rough sketch of their
proof to the author. (With respect to [30], it is presently our opinion that
there is no advantage in working with the piecewise linear functions they
use. One may as well deal with the step functions directly as is done here.)
See also the comments in Section 6.

Let A be an operator in X and $\mathscr{S} = \{0 = s_0 < s_1 < \cdots < s_N\}$ be a partition
of $[0, s_N]$ with steps $\delta_i = s_i - s_{i-1}$. Given $x_0 \in X$, consider the backwards
difference scheme

$$(x_i - x_{i-1})/\delta_i + Ax_i \ni 0, \qquad i = 1, \ldots, N. \qquad \text{(D)}_{\mathscr{S}, x_0}$$

This generalizes the scheme $\varepsilon^{-1}(u_\varepsilon(t) - u_\varepsilon(t - \varepsilon)) + Au_\varepsilon(t) \ni 0$ of Section 1,
which corresponds to $s_i = i\varepsilon$, $x_i = u_\varepsilon(i\varepsilon)$. However, here we are willing to
tolerate errors. A sequence $\{x_1, \ldots, x_N\} \subset X$ is called a (K, η)-*approximate
solution* of $\text{(D)}_{\mathscr{S}, x_0}$ if

$$\begin{aligned}
&\text{(i)} \quad \exists z_i \in (x_i - x_{i-1})/\delta_i + Ax_i, && \text{such that} \quad \|z_i\| \le \eta, \\
&&& \text{for} \quad i = 1, \ldots, N, \qquad \text{(A.1)} \\
&\text{(ii)} \quad \|x_i - x_{i-1}\|/\delta_i \le K, && \text{for} \quad i = 1, \ldots, N.
\end{aligned}$$

Here, η measures the error, while K is a Lipschitz constant for the piecewise
linear function interpolating between the values x_i at s_i. Let $\mathscr{P} = \{0 = t_0 <
t_1 < \cdots < t_M\}$ be another partition with steps $\gamma_j = t_j - t_{j-1}$ and $y_0 \in X$. We
wish to compare approximate solutions of $\text{(D)}_{\mathscr{S}, x_0}$ and $\text{(D)}_{\mathscr{P}, y_0}$. Let
$\{y_1, \ldots, y_M\}$ be a (K, η)-approximate solution of $\text{(D)}_{\mathscr{P}, y_0}$, that is,

$$\begin{aligned}
&\text{(i)} \quad \exists w_j \in (y_j - y_{j-1})/\gamma_j + Ay_j, && \text{such that} \quad \|w_j\| \le \eta, \\
&&& \text{for} \quad j = 1, \ldots, M, \qquad \text{(A.2)} \\
&\text{(ii)} \quad \gamma_j^{-1}\|y_j - y_{j-1}\| \le K && \text{for} \quad j = 1, \ldots, M,
\end{aligned}$$

Theorem A.1. Let A be accretive, (A.1) and (A.2) hold, and $\delta_i < \gamma_i$ for $1 \le i \le N$, $1 \le j \le M$. Then

$$|x_i - y_j| \le |x_0 - y_0| + 2s_i\eta + K(s_i\Delta + |t_j - s_i|^2)^{1/2}$$

for $1 \le i \le N$, $1 \le j \le M$, where

$$\Delta = \max\{\gamma_l - \delta_k : 1 \le l \le M, 1 \le k \le N\}.$$

Proof. Let $J_\lambda = (I + \lambda A)^{-1}$. Then, by (A.1i) and (A.2i)

$$x_i = J_{\delta_i}(x_{i-1} + \delta_i z_i), \qquad y_j = J_{\gamma_j}(y_{j-1} + \gamma_j w_j). \tag{A.3}$$

The relation

$$J_\lambda x = J_\mu\left(\frac{\mu}{\lambda}x + \frac{\lambda - \mu}{\lambda}J_\lambda x\right) \tag{A.4}$$

holds for $x \in D(J_\lambda)$, $0 < \lambda$, μ, as may easily be checked (or see [22, Lemma 1.2]). Set $d(j, i) = \|x_i - y_j\|$. Using (A.3) and (A.4) we have

$$d(j, i) = \|J_{\delta_i}(x_{i-1} + \delta_i z_i) - J_{\gamma_j}(y_{j-1} + \gamma_j w_j)\|$$

$$= \left\|J_{\delta_i}(x_{i-1} + \delta_i z_i) - J_{\delta_i}\left(\frac{\delta_i}{\gamma_j}(y_{j-1} + \gamma_j w_j) + \frac{\gamma_j - \delta_i}{\gamma_j}y_j\right)\right\|$$

$$\le \left\|(x_{i-1} + \delta_i z_i) - \left(\frac{\delta_i}{\gamma_j}y_{j-1} + \delta_i w_j + \frac{\gamma_j - \delta_i}{\gamma_j}y_j\right)\right\|, \tag{A.5}$$

since J_{δ_i} is a contraction. Set

$$\theta_{j,i} = \delta_i/\gamma_j, \qquad \theta_{j,i}^* = 1 - \theta_{j,i} = (\gamma_j - \delta_i)/\gamma_j. \tag{A.6}$$

Note $0 \le \theta_{j,i}, \theta_{j,i}^* \le 1$. Now (A.5) implies

$$d(j, i) \le \theta_{j,i}\|x_{i-1} - y_{j-1}\| + \theta_{j,i}^*\|x_{i-1} - y_j\| + \delta_i\|z_i - w_j\|$$

$$\le \theta_{j,i}d(j-1, i-1) + \theta_{j,i}^*d(j, i-1) + 2\delta_i\eta, \tag{A.7}$$

because $\|z_i\|, \|w_j\| \le \eta$. For notational convenience, let

$$\mathscr{R} = \{(k, l) : k = 1, \ldots, M \text{ and } l = 1, \ldots, N\},$$
$$\partial\mathscr{R} = \{(k, l) \in \mathscr{R} : l = 0 \text{ or } k = 0\},$$
$$\mathscr{R}^0 = \mathscr{R}\backslash\partial\mathscr{R}.$$

Also we know

$$d(0, i) = |x_i - y_0| \le |x_i - x_0| + |x_0 - y_0| \le Ks_i + d_{0,0},$$
$$d(j, 0) = |y_j - x_0| \le |y_j - y_0| + |x_0 - y_0| \le Kt_j + d_{0,0},$$

which may be summarized by

$$d(k, l) \leq K|t_k - s_l| + d_{0, 0} \qquad \text{for} \quad (k, l) \in \partial\mathcal{R}. \tag{A.8}$$

The estimate will be obtained from (A.7) and (A.8) by a few simple comparisons. These are, for the most part, obvious and we give no formal proofs.

First step.   Let $a: \mathcal{R} \to R$ satisfy $a \geq d$ on $\partial\mathcal{R}$ and

$$a(k, l) = \theta_{k, l} a(k - 1, l - 1) + \theta_{k, l}^* a(k, l - 1) \qquad \text{on} \quad \mathcal{R}^0. \tag{A.9}$$

Then, since $2s_l \eta = \theta_{k, l}(2s_{l-1}\eta) + \theta_{k, l}^*(2s_{l-1}\eta) + 2\delta_l \eta$,

$$d(k, l) \leq a(k, l) + 2(\delta_1 + \cdots + \delta_l)\eta = a(k, l) + 2s_l \eta.$$

Second step.   If $b, c: \mathcal{R} \to R$ both solve the same scheme (A.9) and $c \geq b^2$ on $\partial\mathcal{R}$, then $b \leq \sqrt{c}$ on $\mathcal{R}$. [If b satisfies (A.9), its values are fixed convex combinations of its values on $\partial\mathcal{R}$. Then use the Schwartz inequality.]

Third step.   The solution $c: \mathcal{R} \to R$ of (A.9) with $c(k, l) = |t_k - s_l|^2$ on $\partial\mathcal{R}$ satisfies

$$\left| c(k, l) - |t_k - s_l|^2 \right| \leq s_l \Delta \qquad \text{on} \quad \mathcal{R},$$

where $\Delta = \max_{(j, i) \in \mathcal{R}} (\gamma_j - \delta_i)$. To obtain this result, set $r(k, l) = c(k, l) - |t_k - s_l|^2$. By direct calculation [using (A.6)]

$$r(k, l) = \theta_{k, l} r(k - 1, l - 1) + \theta_{k, l}^* r(k, l - 1) + e(k, l) \qquad \text{on} \quad \mathcal{R}^0, \tag{A.10}$$

where

$$e(k, l) = \delta_l(\gamma_k - \delta_l). \tag{A.11}$$

However, (A.10) and $r = 0$ on $\partial\mathcal{R}$ imply

$$\max_{1 \leq k \leq M} |r(k, l)| \leq \max_{1 \leq k \leq M} |r(k, l - 1)| + \max_{1 \leq k \leq M} |e(k, l)|,$$

which in turn implies

$$\max_{1 \leq k \leq M} |r(k, l)| \leq \sum_{j=1}^{l} \max_{1 \leq k \leq M} |e(k, j)|. \tag{A.12}$$

Finally, (A.12) and (A.11) yield the estimate

$$|r(k, l)| \leq s_l \Delta, \qquad \Delta = \max\{(\gamma_j - \delta_i)\}. \tag{A.13}$$

To complete the proof of Theorem A.1, observe that if b is the solution of

(A.9) with $b(k, l) = |t_k - s_l|$ on $\partial \mathcal{R}$, then $a = Kb + d_{0,0}$ is a solution of (A.9) with $a \geq d$ on $\partial \mathcal{R}$ [by (A.8)]. However, $b \leq \sqrt{c}$, where c is as in the third step. Thus

$$d(k, l) \leq a(k, l) + 2s_l \eta \leq Kb(k, l) + d_{0,0} + 2s_l \eta$$
$$\leq K[c(k, l)]^{1/2} + d_{0,0} + 2s_l \eta$$
$$\leq K(|t_k - s_l|^2 + s_l \Delta)^{1/2} + d_{0,0} + 2s_l \eta,$$

and the sketch of proof is complete.

In particular, if A is accretive and $x \in D(A)$, then $x_i = J_\mu^i x$, $y_j = J_\lambda^j x$ are $(K, 0)$-approximate solutions of the schemes defined by the partitions $0 < \mu < 2\mu < \cdots < N\mu$, $0 < \lambda < 2\lambda < \cdots < M\lambda$, respectively, where $K = \{\|y\| : y \in Ax\}$ (for this last fact, see [22, Lemma 1.2]). Thus

$$\|J_\mu^n x - J_\lambda^m x\| \leq K((n\mu - m\lambda)^2 + n\lambda(\lambda - \mu))^{1/2}.$$

Compare this with (1.9) of [22] (with $\omega = 0$).

Remark 1. If A is not accretive in Theorem A.1 but $(1 - \lambda\omega)J_\lambda$ is a contraction for $\lambda > 0$ and small with a fixed ω (equivalently, $A + \omega I$ is accretive), (A.7) becomes

$$d(j, i) \leq \zeta_i(\theta_{j,i} d(j - 1, i - 1) + \theta_{j,i}^*(j, i - 1)d(j, i - 1) + 2\delta_i \eta), \quad \text{(A.14)}$$

where $\zeta_i = (1 - \delta_i \omega)^{-1}$. To handle this, simply observe that d satisfies (A.14) iff $h(j, i) = \zeta_i^{-1} \zeta_{i-1}^{-1} \cdots \zeta_1^{-1} d(j, i)$ satisfies

$$h(j, i) \leq \theta_{j,i} h(j - 1, i - 1) + \theta_{j,i}^*(j, i - 1)h(j, i - 1) + 2\zeta_{i-1}^{-1} \cdots \zeta_1^{-1} \delta_i \eta.$$

$$\text{(A.15)}$$

Assuming, without loss of generality, $\zeta_i \geq 1$ for $i = 1, \ldots, N$, then h satisfies (A.7) and (A.8) in place of d and we therefore have

$$d(k, l) \leq \zeta_1 \cdots \zeta_l(d_{0,0} + 2s_l \eta + [(t_k - s_l)^2 + s_l \Delta]^{1/2})$$
$$\zeta_i = (1 - \delta_i \omega)^{-1}, \qquad i = 1, \ldots, k. \tag{A.16}$$

Remark 2. If the partitions $\mathcal{S}$ and $\mathcal{P}$ are not comparable in the sense $\delta_i < \gamma_j$, then they may be broken into comparable "pieces" and Theorem A.1 can be used on each pair of comparable pieces. We do not make this precise here.

Of course, one can perturb these ideas in a number of other ways, which we leave to the interested reader. The author is indebted to S. Parter and

C. DeBoor for a number of insightful observations about the difference scheme (A.7) that contributed to our proof.

Remark. Be sure to read the comments on Appendix 1 in Section 6.

Appendix 2. Benilan's Uniqueness Theorem

The proof sketched here is that of Theorem 1.1 of [3] (redone with $\|x\|$ and $[x, y]_+$ in place of $\frac{1}{2}\|x\|^2$ and $\|x\|[x, y]_+$). The generalization in [4] involves somewhat more complex statements.

Let $\mathscr{P} = \{0 = a_0 < a_1 < \cdots < a_N = T\}$ be a partition of $[0, T]$ with steps $\delta_k = a_k - a_{k-1}$ and mesh $\mu(\mathscr{P}) = \max \delta_k$. Let $x_0, \ldots, x_N$ and $y_0, \ldots, y_N$ be elements of X satisfying

$$(x_k - x_{k-1})/\delta_k + Ax_k \ni y_k, \qquad k = 1, \ldots, N, \tag{A.17}$$

and $\tilde{u}, \tilde{f}$ be corresponding step functions

$$
\begin{aligned}
\tilde{u}(t) &= x_k \quad \text{for} \quad a_{k-1} \le t < a_k, \quad k = 1, \ldots, N, \\
\tilde{f}(t) &= y_k \quad \text{for} \quad a_{k-1} \le t < a_k, \quad k = 1, \ldots, N.
\end{aligned}
\tag{A.18}
$$

When the above holds we write $\tilde{u} \underset{\mathscr{P}}{\gg} \tilde{f}$.

Theorem A.2. Let $f, g \in L^1([0, T], X)$, A be an accretive operator in X, and v be an integral solution of $(E)_g$ on $[0, T]$. If $u \in C([0, T], X)$ and there is a sequence $\{\mathscr{P}_n\}$ of partitions of $[0, T]$ and functions $\tilde{u}_n \underset{\mathscr{P}_n}{\gg} \tilde{f}_n$ such that $\mu(\mathscr{P}_n) \to 0$, $\tilde{u}_n \to u$ uniformly on $[0, T]$, and $\tilde{f}_n \to f$ in $L^1([0, T], X)$, then:

$$\|v(t) - u(t)\| - \|v(s) - u(s)\| \le \int_s^t [v(\sigma) - u(\sigma), g(\sigma) - f(\sigma)]_+ \, d\sigma$$

for $0 \le s \le t \le T$.

Sketch of proof. Since v is an integral solution of $(E)_g$,

$$\|v(t) - x\| - \|v(s) - x\| \le \int_s^t [v(\tau) - x, g(\tau) - y]_+ \, d\tau \tag{A.19}$$

for $y \in Ax$ and $0 \le s \le t \le T$. Observe next that for $a, b, c \in X$ and $\lambda > 0$

$$[a, b + c]_+ \le [a, b]_+ + [a, c]_+ \le [a, b]_+ + \lambda^{-1}(\|a + \lambda c\| - \|a\|)$$

as follows easily from the definition. Hence

$$[v(\tau) - x, g(\tau) - y]_+ \le [v(\tau) - x, g(\tau) - y + w]_+$$
$$+ \lambda^{-1}(\|v(\tau) - x - \lambda w\| - \|v(\tau) - x\|) \quad \text{(A.20)}$$

for $w \in X$. Let $\tilde{u} \geqslant \tilde{f}$ as above, set

$$x = x_k, \qquad y = y_k - \delta_k^{-1}(x_k - x_{k-1}), \qquad \lambda = \delta_k, \qquad w = \delta_k^{-1}(x_k - x_{k-1})$$

in (A.20). Then plug the resulting estimate in (A.19) to find

$$\delta_k(\|v(t) - x_k\| - \|v(s) - x_k\|) \le \int_s^t (\delta_k[v(\tau) - x_k, g(\tau) - y_k]_+$$
$$+ \|v(\tau) - x_{k-1}\| - \|v(\tau) - x_k\|) \, d\tau.$$

By (A.18), the above may be written

$$\int_{a_{k-1}}^{a_k} (\|v(t) - \tilde{u}(\sigma)\| - \|v(s) - \tilde{u}(\sigma)\|) \, d\sigma$$

$$\le \int_{a_{k-1}}^{a_k} \int_s^t [v(\tau) - \tilde{u}(\sigma), g(\tau) - \tilde{f}(\sigma)]_+ \, d\tau \, d\sigma$$

$$+ \int_s^t \|v(\tau) - \tilde{u}(a_{k-2})\| - \|v(\tau) - \tilde{u}(a_{k-1})\| \, d\tau.$$

Summing over $k = m, m + 1, \ldots, n$, we obtain

$$\int_{a_{m-1}}^{a_n} \|v(t) - \tilde{u}(\sigma)\| - \|v(s) - \tilde{u}(\sigma)\| \, d\sigma$$

$$\le \int_{a_{m-1}}^{a_n} \int_s^t [v(\tau) - \tilde{u}(\sigma), g(\tau) - \tilde{f}(\sigma)]_+ \, d\tau \, d\sigma$$

$$+ \int_s^t (\|v(\tau) - \tilde{u}(a_{m-2})\| - \|v(\tau) - \tilde{u}(a_{n-1})\|) \, d\tau.$$

Letting $u \to u$ uniformly, $a_n \to \beta$, $a_{m-1} \to \alpha$, $\tilde{f} \to f$ in L^1, one finds the inequality

$$\int_\alpha^\beta (\|v(t) - u(\sigma)\| - \|v(s) - u(\sigma)\|) \, d\sigma$$

$$+ \int_s^t (\|v(\tau) - u(\beta)\| - \|v(\tau) - u(\alpha)\|) \, d\tau$$

$$\le \int_\alpha^\beta \int_s^t [v(\tau) - u(\sigma), g(\tau) - f(\sigma)]_+ \, d\tau \, d\sigma,$$

for $0 \le \alpha \le \beta \le T$, $0 \le s \le t \le T$. The second step of the proof is to show the above inequality implies the inequality of the theorem. Letting

$$\phi(t, \alpha) = \|v(t) - u(\alpha)\|, \qquad \psi(t, \alpha) = [v(t) - u(\alpha), g(t) - f(\alpha)]_+ ,$$

for $0 \le t$, $\alpha \le T$, we have

$$\int_\alpha^\beta (\phi(t, \sigma) - \phi(s, \sigma))\, d\sigma + \int_s^t (\phi(\tau, \beta) - \phi(\tau, \alpha))\, d\tau \le \int_\alpha^\beta \int_s^t \psi(\tau, \sigma)\, d\tau\, d\sigma.$$

$$(A.21)$$

Assuming ϕ, ψ are smooth, divide (A.21) by $(\beta - \alpha)(t - s) > 0$ and let $\beta \downarrow \alpha$, $s \uparrow t$, to find

$$\frac{\partial \phi}{\partial t}(t, \alpha) + \frac{\partial \phi}{\partial \alpha}(t, \alpha) \le \psi(t, \alpha).$$

Thus $(d/dt)(\phi(t, t)) \le \psi(t, t)$ and

$$\phi(t, t) \le \phi(s, s) + \int_s^t \psi(\sigma, \sigma)\, d\sigma \qquad (A.22)$$

as desired. If ϕ is not smooth, let $\rho \in C_0^\infty(\mathbb{R})$, $\rho \ge 0$, $\rho(z) = 0$ if $|z| > 1$, $\int \rho = 1$, and set

$$\phi_\varepsilon(t, \alpha) = \int_{|z| \le 1} \int_{|w| \le 1} \rho(z)\rho(w)\phi(t - \varepsilon z, \alpha - \varepsilon w)\, dz\, dw,$$

$$\psi_\varepsilon(t, \alpha) = \int_{|z| \le 1} \int_{|w| \le 1} \rho(z)\rho(w)\psi(t - \varepsilon z, \alpha - \varepsilon w)\, dz\, dw.$$

Then ϕ_ε, ψ_ε are defined and smooth on $\varepsilon \le t$, $\alpha \le T - \varepsilon$. Moreover a direct calculation verifies that (A.21) holds with ϕ, ψ replaced by ϕ_ε, ψ_ε. Hence (A.22) holds with ϕ_ε, ψ_ε on $\varepsilon \le s$, $t \le T - \varepsilon$. The theorem results from sending ε to zero. Since ϕ is continuous, the convergence $\lim_{\varepsilon \downarrow 0} \phi_\varepsilon(t, t) = \phi(t, t)$ is uniform on compact subsets of $(0, T)$. The passage to the limit in ψ_ε is not quite so obvious, but in fact

$$\overline{\lim_{\varepsilon \downarrow 0}} \int_s^t \psi_\varepsilon(\sigma, \sigma)\, d\sigma \le \int_s^t \psi(\sigma, \sigma)\, d\sigma.$$

To see this, observe that

$$\psi(\sigma - \varepsilon z, \sigma - \varepsilon w) = [v(\sigma - \varepsilon z) - u(\sigma - \varepsilon w), g(\sigma - \varepsilon z) - f(\sigma - \varepsilon w)]_+$$
$$\le [v(\sigma - \varepsilon z) - u(\sigma - \varepsilon w), g(\sigma) - f(\sigma)]_+$$
$$+ \|f(\sigma - \varepsilon w) - f(\sigma)\| + \|g(\sigma - \varepsilon z) - g(\sigma)\|,$$

and use the semicontinuity of $[\ ,\ ]_+$ together with the continuity of u, v and continuity of translations in $L^1([0, T], X)$ $[f(\sigma - \varepsilon w) \to f(\sigma)$ in L^1 as $\varepsilon w \to 0$, etc.]. Details are left to the reader. (In the important cases $f = g = 0$ and $f = 0$, $g = y$, $v \equiv x$ for $y \in Ax$, f and g are continuous. This is a more transparent case.)

REFERENCES

[1] S. Aizawa, A semigroup treatment of the Hamilton–Jacobi equation in one space variable, *Hiroshima Math. J.* **3** (1973), 367–386.

[2] V. Barbu, Continuous perturbations of nonlinear *m*-accretive operators in Banach spaces, *Boll. Un. Mat. Ital.* **6** (1972), 270–278.

[3] Ph. Benilan, Equations d'evolution dans un espace de Banach quelconque et applications, Thesis, Univ. Paris XI, Orsay, 1972.

[4] Ph. Benilan, Equations d'evolution dans un espace de Banach quelconque, *Ann. Inst. Fourier* (to appear).

[5] Ph. Benilan, Equations quasi-lineaires du 1er order, *Israel J. Math.* (to appear).

[6] Ph. Benilan, Principe du maximum et perturbation d'operateurs accretifs dans $L^1(\Omega)$ (to appear).

[7] Ph. Benilan, Sur le probléme $\Delta u \in \gamma(\cdot, -\partial u/\partial t)$ dans $L^\infty(\Omega)$ (in preparation).

[8] Ph. Benilan, H. Brezis, and M. G. Crandall, A semilinear elliptic equation in $L^1(\mathbb{R}^N)$, *Ann. Scuola Norm. Sup. Pisa* (to appear).

[9] J. P. Bourguignon and H. Brezis, Remarks on the Euler equation, *J. Functional Analysis* **15** (1974), 341–363.

[10] H. Brezis, On some degenerate nonlinear parabolic equations, *in* "Nonlinear Functional Analysis" (*Proc. Symp. Pure Math.* **18**) (F. Browder, ed.), pp. 28–38. Amer. Math. Soc., Providence, Rhode Island, 1970.

[11] H. Brezis, "Opérateurs maximaux monotones et semigroupes de contractions dans les espace de Hilbert," *Math. Stud.* **5**. North-Holland Publ., Amsterdam, 1973.

[12] H. Brezis, Monotonicity methods in Hilbert spaces and some applications to nonlinear partial differential equations, *in* "Contributions to Nonlinear Functional Analysis" (E. Zarantonello, ed.). Academic Press, New York, 1971.

[13] H. Brezis, Intégrals convexes dans les espaces de Sobolev, *Israel J. Math.* **13** (1972), 9–23.

[14] H. Brezis and A. Pazy, Accretive sets and differential equations in Banach spaces, *Israel J. Math.* **8** (1970), 367–383.

[15] H. Brezis and A. Pazy, Convergence and approximation of nonlinear semigroups in Banach spaces, *J. Functional Analysis* **9** (1971), 63–74.

[16] H. Brezis and W. Strauss, Semilinear elliptic equations in L^1, *J. Math. Soc. Japan* **25** (1973), 565–590.

[17] R. Bruck, Asymptotic convergence of nonlinear contraction semigroups, *J. Functional Analysis* **18** (1975), 15–26.

[18] B. Calvert and K. Gustafson, Multiplicative perturbation of nonlinear *m*-accretive operators, *J. Functional Analysis* **10** (1972), 149–157.

[19] M. G. Crandall, Semigroups of nonlinear transformations in Banach spaces, "Contributions to Nonlinear Functional Analysis" (E. Zarantonello, ed.), pp. 157–179. Academic Press, New York, 1971.

[20] M. G. Crandall, The semigroup approach to first order quasilinear equations in several space variables, *Israel J. Math.* **12** (1972), 108–132.

[21] M. G. Crandall and L. C. Evans, A singular semilinear equation in $L^1(\mathbb{R})$, Mathematics Research Center TSR #1566 (1975), Univ. of Wisconsin, Madison, Wisconsin.

[22] M. G. Crandall and T. M. Liggett, Generation of semigroups of nonlinear transformations on general Banach spaces, *Amer. J. Math.* **93** (1971), 265–293.

[23] M. G. Crandall and A. Pazy, Nonlinear evolution equations in Banach spaces, *Israel J. Math.* **11** (1972), 57–94.

[24] C. M. Dafermos and M. Slemrod, Asymptotic behaviour of nonlinear contraction semigroups, *J. Functional Analysis* **13** (1973), 97–106.

[25] A. Douglis, Layering methods for nonlinear partial differential equations of first order, *Ann. Inst. Fourier* **22** (1972), 141–227.

[26] H. Flaschka and M. Leitman, On semigroups of nonlinear operators and the solution of the functional differential equation $\dot{x}(t) = F(x_t)$, *J. Math. Anal. Appl.* **49** (1975), 649–658.

[27] W. Fleming, The Cauchy problem for a nonlinear first order partial differential equation, *J. Differential Equations* **5** (1969), 515–530.

[28] A. Friedman, The Cauchy problem for first order partial differential equations, *Indiana Univ. Math. J.* **23**(1) (1973), 27–40.

[29] J. Goldstein, Approximation of nonlinear semigroups and evolution equations, *J. Math. Soc. Japan* **24** (1972), 558–573.

[30] J. Kaplan and J. Yorke, Toward a unification of ordinary differential equations with nonlinear semigroup theory, *Proc. Internat. Conf. Ordinary Differential Equations, USC, 1974*, 424–433. Academic Press, New York, 1975.

[31] K. Kobayashi, Note on approximation of nonlinear semigroups (to appear).

[32] Y. Kobayashi, On approximation of nonlinear semigroups (to appear).

[33] Y. Konishi, Some examples of nonlinear semigroups in Banach lattices, *J. Fac. Sci. Univ. Tokyo, Sect. IA* **18** (1972), 537–543.

[34] Y. Konishi, On the uniform convergence of a finite difference scheme for a nonlinear heat equation, *Proc. Japan Acad.* **48** (1972), 62–66.

[35] Y. Konishi, Une remarque sur la perturbation d'opérateurs m-accrétifs dans un espace de Banach, *Proc. Japan Acad.* **48** (1972), 157–160.

[36] Y. Konishi, Some examples of nonlinear semigroups in Banach lattices, *J. Fac. Sci. Univ. Tokyo Sect. IA* **18** (1972), 537–543.

[37] Y. Konishi, On $u_t = u_{xx} - F(u_x)$ and the differentiability of the nonlinear semigroup associated with it, *Proc. Japan Acad.* **48** (1972), 281–286.

[38] Y. Konishi, Une méthod de résolution d'une équation d'évolution non linéaire dégenérée, *J. Fac. Sci. Univ. Tokyo Sect. IA* **19** (1972), 241–255.

[39] Y. Konishi, On the uniform convergence of a finite difference scheme for a nonlinear heat equation, *Proc. Japan Acad.* **48** (1972), 62–66.

[40] Y. Konishi, A remark on fluid flows through porous media, *Proc. Japan Acad.* **1** (1973), 20–24.

[41] Y. Konishi, On the nonlinear semi-groups associated with $u_t = \Delta\beta(u)$ and $\phi(u_t) = \Delta u$, *J. Math. Soc. Japan* **25** (1973), 622–628.

[42] Y. Konishi, Sur un systèm dégenéré des équations paraboliques semilineaires avec les conditions aux limites nonlineaires, *J. Fac. Sci. Univ. Tokyo* **19** (1972), 353–361.

[43] S. N. Kružkov, First order quasilinear equations in several independent variables, *Math. USSR-Sb.* **10** (1970), 217–243.

[44] T. Kurtz, Convergence of sequences of semigroups of nonlinear operators with an application to gas kinetics, *Trans. Amer. Math. Soc.* **186** (1973), 259–272.

[45] P. Lax, The formation and decay of shock waves, *Amer. Math. Monthly* **79** (1972), 227–241.

[46] R. H. Martin, Jr., Differential equations on closed subsets of a Banach space, *Trans. Amer. Math. Soc.* **179** (1973), 399–414.

[47] R. H. Martin, Jr., Approximation and existence of solutions to ordinary differential equations in Banach space, *Funkcial. Ekvac.* **16** (1973), 195–213.

[48] R. H. Martin, Jr., Invariant sets for perturbed semigroups of linear operators, *Ann. Mat. Pura Appl.* (to appear).

[49] R. H. Martin, Jr., Invariant sets for evolution equations, *Proc. Internat. Conf. Ordinary Differential Equations, USC, 1974*, 510–536. Academic Press, New York, 1975.

[50] R. H. Martin, Jr., Remarks on differential inequalities in Banach spaces (to appear).

[51] I. Miyadera, Some remarks on semigroups of nonlinear operators, *Tohoku Math. J.* **23** (1971), 245–258.

[52] J. Miyadera, Generation of semigroups of nonlinear contractions, *J. Math. Soc. Japan* **26** (1974), 389–404.

[53] I. Miyadera and S. Oharu, Approximation of semigroups of nonlinear operators, *Tohoku Math. J.* **22** (1970), 24–27.

[54] M. Nagumo, Über die lage der Intergralkurven gewöhnlicher Differentialgleichungen, *Proc. Phys.-Math. Soc. Japan* **24** (1942), 551–559.

[55] S. Oharu, On the semigroup approach to nonlinear partial differential equations of first order, I (to appear).

[56] M. Pierre, Perturbations localement Lipschitziennes et continues d'operateaurs *m*-accretifs (to appear).

[57] R. Redheffer and W. Walter, A differential inequality for the distance function in normed spaces, *Math. Ann.* **211** (1974), 299–314.

[58] K. Sato, On the generators of nonnegative contraction semigroups in Banach lattices, *J. Math. Soc. Japan* **20** (1968), 423–436.

[59] M. Slemrod, An application of maximal dissipative sets in control theory, *J. Math. Anal. Appl.* **46** (1974), 364–387.

[60] T. Takahaski, Difference approximation of Cauchy problems for quasi-dissipative operators and generation of semigroups of nonlinear contractions (to appear).

[61] M. Tamburro, On the Semigroup Solution of a Hamilton–Jacobi Type Equation, Thesis, Univ. of California at Los Angeles (1974).

[62] G. F. Webb, Continuous nonlinear perturbations of linear accretive operators in Banach spaces, *J. Functional Analysis* **10** (1972), 191–203.

[63] G. F. Webb, Nonlinear perturbations of linear accretive operators in Banach spaces, *Israel J. Math.* **12** (1972), 237–248.

[64] G. F. Webb, Autonomous nonlinear functional differential equations and nonlinear semigroups, *J. Math. Anal. Appl.* **46** (1974), 1–12.

[65] K. Yosida, "Functional Analysis," 3rd ed. Dié Grundlehren der mathematischen Wissenschaften in Einzeldarstellungen, Band 123. Springer-Verlag, Berlin and New York, 1971.

Evolution Equations in Infinite Dimensions*

LUC TARTAR†
University of Paris
Paris, France

I. Introduction: The Linear Case

Results on evolution equations in infinite dimensions are quite different from the finite-dimensional ones. Even in the linear case many differences hold.

If E is a Banach space and A a linear operator from some domain $D(A)$ of E into E, can we solve the problem

$$(du/dt) + Au = 0 \quad \text{for} \quad t \geq 0, \quad u(0) = u_0 \in E, \tag{1}$$

in such a way that the solution will be given by a semigroup

$$\begin{aligned} u(t) &= S(t)u_0 & \text{with} \quad S(t) &\in \mathscr{L}(E, E), \\ S(t + s) &= S(t)S(s) & \text{with} \quad S(0) &= I, \end{aligned} \tag{2}$$

satisfying the continuity condition

$$\text{for all} \quad e \in E, \quad S(t)e \to e \quad \text{as} \quad t \to 0? \tag{3}$$

The answer is given by a well-known theorem of Hille and Yosida (and also Philipps, Miyadera, Feller, *et al.*), which says that Eq. (1) defines a semigroup satisfying (2), (3), and

$$\|S(t)\|_{\mathscr{L}(E, E)} \leq Me^{\omega t} \quad \text{for} \quad t \geq 0, \tag{4}$$

if and only if the following conditions are satisfied:

(i) $D(A)$ is dense in E,
(ii) $\lambda I + A$ has a bounded inverse for λ real and large
    enough satisfying the estimate $\qquad\qquad\qquad\qquad\qquad$ (5)

$$\|(\lambda I + A)^{-n}\|_{\mathscr{L}(E, E)} \leq M/(\lambda - \omega)^n \quad \text{for} \quad n = 1, 2, \dots.$$

---

* Sponsored by the U.S. Army under Contract No. DA-31-124-ARO-D-462 and by the University of Paris IX.
† Present address: Université Paris-Sud Mathematiques, Orsay, Cedex, France.

Let us emphasize the differences from the finite-dimensional case by making some remarks.

Remark 1. In finite dimensions one always has $D(A) = E$. This is no longer true in infinite dimensions; the case $D(A) = E$ corresponds to ordinary differential equations in Banach spaces, whereas the case $D(A) \neq E$ leads to more difficult problems and, in most concrete examples, to partial differential equations.

Remark 2. In finite dimensions one can define the solution of (1) for $t > 0$ as well as for $t < 0$. In general, this is not true in infinite dimensions where the existence for $t < 0$ is equivalent to condition (5) for $-A$.

Example 1. $E = L^2(\mathbb{R})$, $A = \partial/\partial x$, with $D(A) = H^1(\mathbb{R})$; then problem (1) is well posed for $t > 0$ as well as for $t < 0$. In this example condition (4) is satisfied for $M = 1$ and $\omega = 0$.

Example 2. $E = L^2(\mathbb{R})$, $A = -\partial^2/\partial x^2$, with $D(A) = H^2(\mathbb{R})$; then problem (1) is well posed for $t > 0$ but not for $t < 0$. Condition (4) is also satisfied with $M = 1$ and $\omega = 0$.

Example 3. $E = L^2(\mathbb{R}) \times L^2(\mathbb{R})$, $A(u, v) = (\partial v/\partial x, \partial u/\partial x)$, with $D(A) = H^1(\mathbb{R}) \times H^1(\mathbb{R})$; then problem (1) is well posed for all t and (4) is still satisfied with $M = 1$ and $\omega = 0$.

Remark 3. Except for the case where $u_0 \in D(A)$, the solution u satisfies (1) in a generalized sense and $u(t)$ need not be in $D(A)$. However, for $u_0 \in D(A)$, (1) can be interpreted in a classical sense: u is a C^1 function of $t \geq 0$ with values in E and a C^0 function of $t \geq 0$ with values in $D(A)$ (which is a Banach space with the norm $\|u\| + \|Au\|$).

Some linear operators A have the property that, even if $u_0 \notin D(A)$, then $u(t) \in D(A)$ for $t > 0$. Examples 1 and 3 do not possess this property but Example 2 does. Most of the examples where this property holds can be stated in the following form:

H will be a Hilbert space (which plays the role of E) and V another Hilbert space continuously embedded into H and dense in it. H being identified with its dual, we have $V \subset H \subset V'$. Let A be linear and continuous from V into V' and satisfy the coercivity condition $(Au, u) \geq \alpha \|u\|_V^2$ and

$D(A)$ will be the set $\{u \in V : Au \in H\}$. Then the semigroup generated satisfies (4) with $M = 1$ and $\omega = 0$ and also satisfies

$$|Au(t)|_H \le (c/t)|u_0|_H \qquad \text{for} \quad t > 0.$$

In Example 2, V will be $H^1(\mathbb{R})$.

Remark 4. As Problem (1) for $t < 0$ is not generally well posed, it may happen that the solutions starting from 2 different points u_0 and v_0 coincide after some time t_0.

Example 4. $E = L^2(0, 1)$, $A = \partial/\partial x$, with domain $D(A) = \{u \in E$, $\partial u/\partial x \in E$, $u(0) = 0\}$; then $S(t) = 0$ for $t \ge 1$, and the solution is explicitly given by

$$u(x, t) = \begin{cases} 0, & \text{if} \quad x < t, \\ u_0(x - t), & \text{if} \quad t < x < 1. \end{cases}$$

Remark 5. Except for the case where $D(A) = E$, the proof of existence of the solution of (1) as a fixed point of the mapping $u \to u_0 - \int_0^t Au(s)\,ds$ does not work, because this mapping transforms continuous functions with values in $D(A)$ into continuous functions with values in E and thus cannot be iterated.

For the same reason the classical explicit difference scheme

$$[u((n + 1)k) - u(nk)]/k + Au(nk) = 0$$

does not generate an approximation of the solution of (1) as it is impossible in general to define $u(nk)$ for all n.

Actually the implicit difference scheme

$$[u((n + 1)k) - u(nk)]/k + Au((n + 1)k) = 0 \tag{6}$$

is well defined for small $k > 0$ and generates an approximation of the solution.

II. The Nonlinear Case: Approximation by Finite-Dimensional Problems

In this section we will consider infinite-dimensional problems as the limit of ordinary differential equations in finite dimensions; this is the method of Galerkin (associated generally with Faedo or Ritz), which we will use in some simple cases.

Let V be a separable reflexive Banach space continuously imbedded and dense in some Hilbert space H identified with its dual. Thus we have

$$V \subset H \subset V'. \tag{7}$$

Let A be a nonlinear operator, which may depend on t, from V into V'. We are interested in the equation

$$(du/dt) + A(u) = 0, \quad 0 < t \le T < +\infty, \qquad u(0) = u_0 \in H. \tag{8}$$

Step 1.   We choose a "basis" $w_1, \ldots, w_n, \ldots$ of V, that is,

for each n, $w_1, \ldots, w_n$ are linearly independent;
the subspace spanned by all the w_i is dense in V. $\tag{9}$

As V is separable, such a basis exists. Usually the following steps can be carried out for each choice of the basis, but in some cases it is useful to choose a special basis.

Step 2.   For fixed n, consider the ordinary differential equation

$$\text{find} \quad u_n = \sum_{i=1}^{n} g_i(t)w_i \quad \text{such that}$$

$$\left(\frac{du_n}{dt} + A(u_n), w_j \right) = 0 \quad \text{for} \quad j = 1, \ldots, n, \quad t \ge 0, \tag{10}$$

$$u_n(0) = u_{0n} = \sum_{i=1}^{n} \alpha_i w_i.$$

As the w_i are independent, the matrix with entries (w_i, w_j) is invertible and (10) is an ordinary differential equation for $g_1, \ldots, g_n$. To obtain local existence on an interval $[0, t_n[$ with $0 < t_n \le T$, we will assume that A satisfies

for each finite-dimensional subspace W of V, the
restriction A_W from W into W' defined by $\tag{11}$
$(A_W(u), v) = (A(u), v)$ for all $u, v \in W$ is continuous.

Step 3.   Prove that $t_n = T$ for all n and find a priori estimates on u_n independent of n. Usually one can show that if u_{0n} is bounded in H (we will suppose that $u_{0n} \to u_0$ in H), $u_n(t)$ remains in a bounded set of H; this proves that $t_n = T$ and gives the estimate $u_n \in$ bounded set of $L^\infty(0, T; H)$.

Step 4. If $u_{0n} \to u_0$ in H and if we have a solution u_n that remains in some bounded set of a reflexive Banach space [or a dual of a Banach space such as $L^\infty(0, T; H)$] then we can extract a weakly (or weakly*) convergent subsequence to some u, which we hope will be a solution of (8). There are two general arguments to carry out step 4: compactness and monotonicity. Let us begin with the results about compactness.

Compactness Lemma. If $B_0 \subset B_1 \subset B_2$ are three Banach spaces with compact injection from B_0 into B_1 and continuous injection from B_1 into B_2, then if $u_n \in$ bounded set of $L^{p_0}(0, T; B_0)$ and $du_n/dt \in$ bounded set of $L^{p_1}(0, T; B_2)$ with $T < +\infty$ and $1 < p_0, p_1 < +\infty$, then $u_n \in$ compact set of $L^{p_0}(0, T; B_1)$.

Compactness Argument. In this method we try to obtain a sufficient number of estimates on u_n and du_n/dt such that, using the compactness lemma, we can extract from u_n a strongly convergent subsequence in some Banach space (and weakly convergent in other spaces) implying the convergence of $A(u_n)$ to $A(u)$.

Let us remark that in some cases the choice of a special basis in step 1 appears to be useful to obtain estimates on du_n/dt.

Example 5

$$(\partial^2 u/\partial t^2) - \Delta u + u^3 = 0,$$

$$u(0) = u_0 \in H^1(\mathbb{R}^N) \cap L^4(\mathbb{R}^N) = V, \qquad (12)$$

$$(\partial u/\partial t)(0) = u_1 \in L^2(\mathbb{R}^N) = H.$$

This can be considered as a system (u, v) with $v = \partial u/\partial t$, but it is more convenient to work with only the unknown u and to approach (12) by

$$u_n = \sum_{i=1}^{n} g_i(t)w_i, \qquad \text{where} \quad w_j \text{ is a basis of } V,$$

$$((d^2u_n/dt^2) - \Delta u_n + u_n^3, w_i) = 0, \qquad i = 1, \ldots, n.$$

$$(12')$$

Taking in (12) the combination of w_i corresponding to du_n/dt, we obtain

$$\frac{d}{dt}\left(\frac{1}{2}\left|\frac{du_n}{dt}\right|_{L^2}^2 + \frac{1}{2}|\text{grad } u_n|_{L^2}^2 + \frac{1}{4}|u_n|_{L^4}^4\right) = 0. \qquad (13)$$

This gives the estimates

$$u_n \in \text{bounded set of } L^\infty(0, T; V),$$
$$du_n/dt \in \text{bounded set of } L^\infty(0, T; H). \tag{14}$$

At this point let us remark that if we have a weakly* convergent sequence in $L^\infty(0, T; V)$: $V_n \to V$, it does not imply that V_n^3 converges to V^3.

Now with the estimate on du_n/dt and using the compactness lemma, we have $u_n \in \text{compact set of } L^2(0, T; L^2_{\text{loc}}(\mathbb{R}^N))$. [The injection from $H^1(\mathbb{R}^N)$ into $L^2(\mathbb{R}^N)$ is not compact, but this is true if $\mathbb{R}^N$ is replaced by a bounded open set of $\mathbb{R}^N$.]

We can extract a subsequence converging in $L^\infty(0, T; V)$ weakly* and strongly in $L^2(0, T; L^2_{\text{loc}}(\mathbb{R}^N))$, which gives $u_n^3 \to u^3$ in $L^p(0, T; L^{4/3}_{\text{loc}}(\mathbb{R}^N))$ for $p < +\infty$, and so u will be a solution of (12).

Monotonicity Argument. Let us assume that A maps $L^p(0, T; V)$ into $L^{p'}(0, T; V')$ with $1 < p < +\infty$ and $(1/p) + (1/p') = 1$, and that A is monotone and hemicontinuous. A monotone means that

$$\int_0^T (A(u) - A(v), u - v)\, dt \geq 0 \qquad \text{for} \quad u, v \in L^p(0, T; V), \tag{15}$$

and A hemicontinuous means that the map

$$\lambda \to \int_0^T (A(u + \lambda v), v)\, dt \qquad \text{is continuous for} \quad \lambda \in \mathbb{R}, \quad u, v \in L^p(0, T; V). \tag{16}$$

Under these conditions on A one has:

Monotonicity Lemma. If u_n converges weakly to u in $L^p(0, T; V)$, $A(u_n)$ converges weakly to f in $L^{p'}(0, T; V')$, and

$$\limsup \int_0^T (A(u_n), u_n)\, dt \leq \int_0^T (f, u)\, dt,$$

then $A(u) = f$.

Let us remark that A may have this property without satisfying (15) and (16). If we extract a subsequence such that $u_n \to u$, $Au_n \to f$ [assuming that A maps bounded sets of $L^p(0, T; V)$ into bounded sets of $L^{p'}(0, T; V)$], and $u_n(T) \to h$, we see from (10) that

$$(du/dt) + f = 0, \qquad u(T) = h,$$

and

$$\tfrac{1}{2}\,|u_n(T)|^2 + \int_0^T (A(u_n),\, u_n)\, dt = \tfrac{1}{2}|u_n(0)|^2 \to \tfrac{1}{2}|u_0|^2.$$

So

$$\limsup \int_0^T (A(u_n),\, u_n)\, dt \le \tfrac{1}{2}|u_0|^2 - \tfrac{1}{2}|h|^2 = \int_0^T (f,\, u)\, dt,$$

and by the monotonicity lemma $f = A(u)$ and u satisfies (8).

Example 6. If

$$(\partial u/\partial t) - \Delta u + u^3 = 0, \qquad u(0) = u_0 \in L^2(\mathbb{R}^N), \tag{17}$$

then $V = H^1(\mathbb{R}^N) \cap L^4(\mathbb{R}^N)$, $H = L^2(\mathbb{R}^N)$, and $A(u) = -\Delta u + u^3$ is monotone hemicontinuous from $W = L^2(0,\, T;\, H^1(\mathbb{R}^N)) \cap L^4(0,\, T;\, L^4(\mathbb{R}^N))$ into its dual W'. With a slight modification the monotonicity argument works. The estimates are obtained in this example by taking in (10) the combination of w_i corresponding to u_n. This gives

$$\frac{1}{2}\frac{d}{dt}(|u_n|_{L^2}^2) + |\mathrm{grad}\; u_n|_{L^2}^2 + |u_n|_{L^4}^4 = 0$$

$u_n \in$ bounded set of $L^2(0,\, T;\, H^1(\mathbb{R}^N))$, $L^4(0,\, T;\, L^4(\mathbb{R}^N))$, and $L^\infty(0,\, T;\, L^\infty(\mathbb{R}^N))$.

$$\tag{18}$$

Example 7

$$(\partial^2 u/\partial t^2) - \Delta u + (\partial u/\partial t)^3 = 0,$$
$$u(0) = u_0 \in H^1(\mathbb{R}^N), \qquad (\partial u/\partial t)(0) = u_1 \in L^2(\mathbb{R}^N). \tag{19}$$

If we consider the approximation

$$u_n = \sum_{i=1}^n g_i(t) w_i \qquad \text{and} \qquad w_i \text{ is a basis of } H^1(\mathbb{R}^N) \cap L^4(\mathbb{R}^N), \tag{20}$$

$$((d^2 u_n/dt^2) - \Delta u_n + (du_n/dt)^3,\, w_j) = 0, \qquad j = 1, \ldots, n, \quad t \ge 0,$$

choosing the combination of the w_i corresponding to du_n/dt, we have

$$u_n \in \text{bounded set of } L^\infty(0,\, T;\, H^1(\mathbb{R}^N)),$$
$$du_n/dt \in \text{bounded set of } L^\infty(0,\, T;\, L^2(\mathbb{R}^N)) \cap L^4(0,\, T;\, L^4(\mathbb{R}^N)). \tag{21}$$

Then if $v_n = du_n/dt$, we can apply the monotonicity lemma to v_n and deduce that for a subsequence we will have $u_n \to u$ weakly, $du_n/dt \to du/dt$ weakly, and $(du_n/dt)^3 \to (du/dt)^3$ weakly.

Some examples may require a combination of a compactness argument and a monotonicity argument; some may be handled (with different hypotheses) by the two methods.

Example 8

$$(\partial u/\partial t) - \Delta u + u^3 = 0, \qquad u(0) = u_0 \in H^1(\mathbb{R}^N) \cap L^4(\mathbb{R}^N) \qquad (22)$$

[compare to (17)]. Taking in (10) the combination of w_i corresponding to du_n/dt, we have

$$\left| \frac{du_n}{dt} \right|_{L^2}^2 + \frac{1}{2} \frac{d}{dt} \left| \operatorname{grad} u_n \right|_{L^2}^2 + \frac{1}{4} \frac{d}{dt} \left| u_n \right|_{L^4}^4 = 0,$$

$$u_n \in \text{bounded set of } L^\infty(0, T; H^1(\mathbb{R}^N)) \cap L^\infty(0, T; L^4(\mathbb{R}^N)),$$
$$du_n/dt \in \text{bounded set of } L^2(0, T; L^2(\mathbb{R}^N)). \qquad (23)$$

Now the estimate on du_n/dt allows the use of the compactness argument as in Example 5.

III. The Nonlinear Case: Other Methods

1. ACCRETIVE OPERATORS

A simple case of (5) is where $M = 1$ and (5ii) reduces to

$$\|(\lambda I + A)^{-1}\| \le 1/(\lambda - \omega).$$

A nonlinear analog of this inequality leads to the theory of accretive and monotone operators. For these results we refer to the chapter by Crandall in this volume.

2. FIXED-POINT THEOREM FOR CONTRACTIONS

We want to find the solution u of (8) by means of the iterative method

$$(dU_{n+1}/dt) + B(U_{n+1}) = B(U_n) - A(U_n), \qquad U_{n+1}(0) = V_0, \qquad (24)$$

with a suitable B such that $u_n \overset{S}{\to} u_{n+1}$ defines a map S that is a strict contraction in some Banach space so its unique fixed point will be the desired solution.

Example 9

$$\frac{\partial u}{\partial t} - \Delta u + \varphi\left(x, u, \frac{\partial u}{\partial x}\right) = 0, \qquad u(0) = u_0 \in H^1(\mathbb{R}^N), \tag{25}$$

where φ satisfies

$$\varphi(x, 0, 0) \in L^2(\mathbb{R}^N), \qquad \left|\frac{\partial \varphi}{\partial u}(x, u, v)\right| \leq K, \qquad \left|\frac{\partial \varphi}{\partial v}(x, u, v)\right| \leq K. \tag{26}$$

Using $B(u) = -\Delta u$ and the estimate

$$(\partial w/\partial t) - \Delta w = f \qquad \text{implies} \qquad \|w(t)\|_{H^1}^2 \leq c \int_0^t |f(s)|_{L^2}^2 \, dx,$$
$$w(0) = 0, \tag{27}$$

we have the existence (and uniqueness) of u satisfying

$$u \in L^\infty(0, T; H^1(\mathbb{R}^N)), \qquad du/dt \in L^\infty(0, T; L^2(\mathbb{R}^N)). \tag{28}$$

3. FIXED-POINT THEOREM FOR COMPACT MAPPINGS

We use the same map S defined by (24) but we now require that S is continuous and maps some compact convex set of a Banach space into itself in order to apply the fixed-point theorem of Schauder.

Example 10

$$\frac{\partial u}{\partial t} - \Delta u + \varphi\left(x, u, \frac{\partial u}{\partial x}\right) = 0 \qquad \text{in} \quad \Omega \text{ a bounded open set in } \mathbb{R}^N,$$
$$u(0) = u_0 \in H_0^{\,1}(\Omega), \tag{29}$$

and φ satisfies

$$\varphi(x, u, v) \quad \text{continuous in } u, v, \quad \text{measurable in } x,$$
$$|\varphi(x, u, v)| \leq a(x) + b|u| + c|v| \quad \text{with} \quad a(x) \in L^2(\Omega), \quad b, c > 0. \tag{30}$$

We use the map defined by $B = -\Delta$ and we use the following estimate:

if $\quad (\partial w/\partial t) - \Delta w = f \in L^2(0, T; L^2(\Omega)), \qquad$ we have $\qquad w(0) = w_0 \in H_0^{\,1}(\Omega),$

(i) $\quad \|w(t)\|_{H_0^1}^2 \leq \|w(0)\|_{H_0^1}^2 + \frac{1}{4} \int_0^t |f(s)|_{L^2}^2 \, ds,$ \tag{31}

(ii) $\quad \int_0^T (|\partial w/\partial t|_{L^2}^2 + |\Delta w|_{L^2}^2) \, dt \leq c[\|w(0)\|_{H_0^1}^2 + \int_0^T |f(s)|^2 \, ds].$

Estimate (i) shows that we can find a convex set of $L^2(0, T; H_0^1(\Omega))$ that is mapped into itself. Estimate (ii) shows that, using the compactness lemma, this map is compact and we can obtain a fixed point.

4. FIXED-POINT THEOREM FOR INCREASING MAPPINGS

Using some map S, such as in (24), we now require that S is increasing in some ordered Banach space such that we can apply some special fixed-point theorem for increasing mappings.

In most cases we apply some fixed-point theorem used in Section III.2 or III.3. The order argument is used to find a suitable convex set as well as a suitable operator B.

Example 11

$$(\partial u/\partial t) - \Delta u + \varphi(x, u) = 0, \qquad u(0) = u_0 \in L^2(\mathbb{R}^N), \tag{32}$$
$$\varphi \text{ continuous in } u, \quad \text{measurable in } x.$$

We assume that we can find two functions a, b satisfying

$$a, b \in L^2(0, T; L^2(\mathbb{R}^N))$$

$$\frac{\partial a}{\partial t} - \Delta a + \varphi(x, a) \le 0 \le \frac{\partial b}{\partial t} - \Delta b + \varphi(a, b) \qquad \text{if} \quad a(x, 0) \le u_0 \le b(x, 0),$$
$$\tag{33}$$

$$\varphi(x, u) \in L^2(0, T; L^2(\mathbb{R}^N)) \qquad \text{if} \quad a \le u \le b,$$

$$\frac{\partial \varphi}{\partial x}(x, u) \le K \qquad \text{if} \quad a(x) \le u \le b(x).$$

Then

$$\text{there is a solution of (32) satisfying } a \le u \le b, \text{ and}$$
$$(\partial u/\partial t) - \Delta u \in L^2(0, T; L^2(\mathbb{R}^N)). \tag{34}$$

For this we consider $v = Su$ defined by

$$(\partial v/\partial t) - \Delta v + Kv = Ku - \varphi(x, u), \qquad v(0) = u_0, \tag{35}$$

and remark that K has been chosen so that $Ku - \varphi(x, u)$ is increasing on the interval $[a, b]$.

Now the crucial point is the maximum principle for the operator $-\Delta + K$:

$$\text{if} \quad (\partial w/\partial t) - \Delta w + Kw \ge 0, \quad w(0) \ge 0, \qquad \text{then} \qquad w \ge 0. \tag{36}$$

The hypotheses on a, b imply that S is increasing and maps the interval $[a, b]$ into itself, and so S has a fixed point.

Example 12

$$\frac{\partial u}{\partial t} + \frac{\partial u}{\partial x} + u^2 - v^2 = 0, \qquad \frac{\partial v}{\partial t} - \frac{\partial v}{\partial x} + v^2 - u^2 = 0, \tag{37}$$

$$u(0) = u_0 \in L^\infty(\mathbb{R}), \qquad v(0) = v_0 \in L^\infty(\mathbb{R}), \qquad 0 \le u_0, \ v_0 \le M.$$

For $K \ge 2M$ we consider the iterative process:

$$\frac{\partial u^{n+1}}{\partial t} + \frac{\partial u^{n+1}}{\partial x} + K u^{n+1} = K u^n - (u^n)^2 + (v^n)^2,$$

$$\frac{\partial v^{n+1}}{\partial t} + \frac{\partial v^{n+1}}{\partial x} + K v^{n+1} = K v^n - (v^n)^2 + (u^n)^2, \tag{38}$$

$$U^{n+1}(0) = u_0, \qquad V^{n+1}(0) = v_0.$$

Note that K has been chosen so that $Ku - u^2 + v^2$ and $Kv - v^2 + u^2$ are increasing for $0 \le u, v \le M$.

The crucial point is that

$$\text{if} \quad \frac{\partial w}{\partial t} + a\frac{\partial w}{\partial x} + Kw \ge 0 \qquad \text{then} \quad w \ge 0;$$

$$w(0) \ge 0, \tag{39}$$

then (38) defines an increasing mapping $(u^n, v^n) \to (u^{n+1}, v^{n+1})$, which is Lipschitz continuous and maps $0 \le u, v \le M$ into itself. So (37) has a unique solution such that $0 \le u, v \le M$. This approach proves that the solution u, v increases if u_0, v_0 increases. Using this, one can also show that $\|u(t)\|_{L^\infty} + \|v(t)\|_{L^\infty}$ decreases.

5. OTHER METHODS

Of course there are many other methods. One always has to improve some classical methods in order to fit a particular problem.

The examples given here are not given in the most general form and we have restricted ourselves to questions of existence. Problems of regularity, periodic solutions, and asymptotic behavior of the solutions can be solved for these examples.

For more details and examples let us give a single reference:

J. L. Lions, "Quelques Méthodes de Résolution des Problèmes aux Limites non Linéaires." Dunod–Gauthier–Villars, Paris, 1969.

Chapter 4: FUNCTIONAL DIFFERENTIAL EQUATIONS

Functional Differential Equations of Neutral Type*

JACK K. HALE
Lefschetz Center for Dynamical Systems
Division of Applied Mathematics
Brown University, Providence, Rhode Island

I had the good fortune of knowing Professor Lefschetz for a number of years. As with so many other young people, he had a profound influence on my outlook toward life and the direction of my research. Although I had an interest in functional differential equations before going to RIAS in 1958, it was during a conversation with Lefschetz that he told me, "You should read the Russian book on stability by Krasovskii." Naturally, he was right and my serious research on this subject began at this point. The book of Krasovskii has had a tremendous impact on later developments. And, once more, the deep perception of Lefschetz showed through even in areas not of direct interest to his research. It is indeed an honor to dedicate this chapter to Professor Lefschetz.

1. Introduction

The number of stimulating chapters on functional differential and integral equations presented in this volume is a clear indication of the rapid development of the theory over the past few years. The literature has become so extensive that it is now impossible in a few pages to give a clear picture of the spectrum of current research. Therefore, I will confine my attention to a special class of neutral equations (which will include retarded ones with finite delay) and to those results for this class that are of particular interest to me.

* This research was supported by the National Science Foundation under GP-28931X2 and in part by the U.S. Army Research Office, Durham, under DA-ARO-D-31-124-73-G-130.

In general terms, a neutral functional differential equation (NFDE) is a differential relation in which the derivative of the unknown function may depend on past values of the function as well as its derivative. In contrast to retarded equations, even the basic problem of existence, uniqueness, and continuous dependence does not have a straightforward solution. The manner in which the norm on the initial data depends on the derivative plays a significant role. This was pointed out in the fundamental paper [8], which contains many interesting examples as well as a solution of the above problems for equations linear in the derivatives and initial functions absolutely continuous. Extensions were given in [30, 31]. A glance through Mathematical Reviews will show that there is a tremendous amount of work being done, especially in the USSR, on neutral equations with initial data that have derivatives whose pth power is integrable. However, a qualitative theory has not yet been developed.

In [21] in 1967, a particular class of neutral equations was introduced for which initial data could be specified in the space of continuous functions. In subsequent years, this class was further restricted in such a way that a qualitative theory now seems possible. It is this class that will concern us in this chapter. With the knowledge we now have, it seems possible that a similar theory could be developed for initial data in other spaces.

2. Notation

Suppose $r \geq 0$ is a given real number, E^n an n-dimensional linear vector space with norm $|\cdot|$, and $C = C([-r, 0], E^n)$ the space of continuous functions mapping the interval $[-r, 0]$ into E^n, with $|\phi| = \sup_{-r \leq \theta \leq 0} |\phi(\theta)|$ for $\phi \in C$. Let $\mathscr{B}$ be the space of bounded linear operators from C to E^n with $\|B\| = \inf\{k : |B\phi| \leq k|\phi|, \phi \in C\}$ for any $B \in \mathscr{B}$. For any $B \in \mathscr{B}$, there is an $n \times n$ matrix function μ on $(-\infty, \infty)$ of bounded variation such that $\mu(\theta) = \mu(-r)$, $\theta \leq 0$, $\mu(\theta) = \mu(0)$, $\theta \geq 0$, and

$$B\phi = \int_{-r}^{0} [d\mu(\theta)]\phi(\theta)$$

for all $\phi \in C$. Furthermore, there is a continuous, nonnegative scalar function γ on $[0, \infty)$ such that $\gamma(0) = 0$,

$$\left| \int_{-s}^{0} [d\mu(\theta)]\phi(\theta) - [\mu(0) - \mu(0_-)]\phi(0) \right| \leq \gamma(s) \sup_{-s \leq \theta \leq 0} |\phi(s)|$$

for $s \in [0, r]$, $\phi \in C$. We say B is *atomic at* 0 if $\det[\mu(0) - \mu(0_-)] \neq 0$ and B is *nonatomic at* 0 if $\mu(0) - \mu(0_-) = 0$. Notice that to say B is atomic is not to negate the statement that B is nonatomic except for $n = 1$. In the same way, one can define atomic and nonatomic at any point $\theta \in [-r, 0]$.

Let $\overline{\Omega}$ be the closure of an open set Ω of C, $k \geq 0$ an integer,

$$\mathscr{F}^k = \{f : \overline{\Omega} \to E^n, f^{(j)}, j = 0, 1, \ldots, k, \text{ continuous and bounded in } \overline{\Omega}\},$$

where $f^{(j)}$ is the jth Fréchet derivative of f. For any $f \in \mathscr{F}^k$, we define the norm of f by $\|f\| = \sup_{\phi \in \Omega} \{\sum_{j=0}^{k} \|f^j(\phi)\|\}$, where $\|f^0(\phi)\| = |f(\phi)|$ and, for $j \geq 1$, the double bars represent the norms for multilinear operators.

If $A \geq 0$ and $x : [\sigma - r, \sigma + A] \to E^n$, then for any $t \in [\sigma, \sigma + A]$ we let $x_t : [-r, 0] \to E^n$ be defined by $x_t(\theta) = x(t + \theta)$, $-r \leq \theta \leq 0$.

For any $(D, f) \in \mathscr{B} \times \mathscr{F}^k$ with D atomic at zero, we define a *neutral functional differential equation* by a relation

$$\frac{d}{dt} D(x_t) = f(x_t) \tag{2.1}$$

and designate it by NFDE (D, f). For any $\phi \in \Omega$, a solution $x(\sigma, \phi)$ through $(\sigma, \phi) \in R \times C$ is a continuous function $x : [\sigma - r, \sigma + A) \to E^n$ for some $A > 0$ such that $x_t(\phi) \in \Omega$, $t \in [\sigma, \sigma + A)$, $x_\sigma = \phi$, and Dx_t is continuously differentiable on $[\sigma, \sigma + A)$ and satisfies the equation on $[\sigma, \sigma + A)$. If $\sigma = 0$, we designate the solution by $x(\phi)$.

By far the most interesting equations have the form

$$\frac{d}{dt} \left[x(t) - \sum_{k=1}^{N} B_k x(t - r_k) \right] = f(x_t), \tag{2.2}$$

where each B_k is an $n \times n$ matrix and each $r_k > 0$.

3. Existence, Uniqueness, Continuous Dependence

If $(D, f) \in \mathscr{B} \times \mathscr{F}^0$, then for any $\phi \in \Omega$ there exists a solution of the NFDE (D, f) through $(0, \phi)$ on a maximal interval $[0, A_\phi)$ such that either $A_\phi = +\infty$ or $\lim_{t \to \infty} |x_t| = +\infty$ or $\lim_{t \to A_\phi} x_t = \psi$ exists and $\psi \in \partial\Omega$. If, in addition, $(D, f) \in \mathscr{B} \times \mathscr{F}^1$, then the solution $x(\phi, D, f)(t)$ through $(0, \phi)$ depends continuously on $(\phi, D, f, t) \in \Omega \times \mathscr{B} \times \mathscr{F}^1 \times [-r, A_\phi)$ (see [4, 10, 30, 31]).

A function $x : (-r - A, 0] \to E^n$, $A > 0$, is said to be a *backward continuation* through $(0, \phi)$ for the NFDE (D, f) if $x_0 = \phi$ and Dx_t is continuously differentiable on $(-A, 0]$ and satisfies the equation on $(-A, 0]$.

If D is atomic at $-r$, $(D, f) \in \mathcal{B} \times \mathcal{F}^0$, $\phi \in \Omega$, then there is a backward continuation through $(0, \phi)$ and the solution $x(\phi, D, f)(t)$ depends continuously on $(\phi, D, f, t) \in \Omega \times \mathcal{B} \times \mathcal{F}^1 \times (-r - A_\phi, 0]$, where A_ϕ is defined by the maximal interval of existence (see [10]).

Define the solution operator of an NFDE (D, f) by

$$T_{D, f}(t): \Omega \to \Omega, \qquad T_{D, f}(t)\phi = x_t(\phi, D, f),$$

for $t \in [0, A_\phi)$. From the above remarks, if D is atomic at $-r$, then $T_{D, f}(t)$ is a homeomorphism. Since the set of $D \in \mathcal{B}$ for which D is atomic at $-r$ is an open dense set in $\mathcal{B}$, it follows that the set of NFDE (D, f) for which $T_{D, f}(t)$ is a homeomorphism is open and dense in $\mathcal{B} \times \mathcal{F}^1$ (see [10]). Consequently, any retarded functional differential equation

$$\dot{x}(t) = f(x_t) \tag{3.1}$$

can be approximated in $\mathcal{B} \times \mathcal{F}^1$ by an NFDE for which $T_{D, f}(t)$ is a homeomorphism; in particular, by

$$\frac{d}{dt}[x(t) - \varepsilon x(t - r)] = f(x_t). \tag{3.2}$$

It would be interesting to know how the solution operator $T_{D, f}(t)$ of (3.2) approaches the solution operator $T_{D, f}(t)$ of (3.1) as $\varepsilon \to 0$.

The above results on continuous dependence are not sufficiently general to handle many practical problems. More specifically, if

$$D\phi = \phi(0) - \sum_{k=1}^N B_k \phi(-r_k) \overset{\text{def}}{=} \phi(0) - g(\phi), \tag{3.3}$$

where each B_k is an $n \times n$ matrix and $r_k > 0$, the operator $D \in \mathcal{B}$ is not continuous in the parameters r_k, $k = 1, 2, \ldots, N$. On the other hand, it is shown in [8, 30] that the solution is continuous in the r_k provided each $r_k > 0$. Furthermore, this continuity property even holds as each $r_k \to 0$ provided that $\|g\| < 1$. For $n = 1$, this latter condition is equivalent to $\sum_{k=1}^N |B_k| < 1$. In [32], it is shown that for $n = 1$ the inequality $\sum_{k=1}^N |B_k| > 1$ implies that in any neighborhood of $0 \in R^n$ there are positive $(r_1, \ldots, r_N)$ and a root λ of

$$1 - \sum_{k=1}^N B_k \exp(-\lambda r_k) = 0 \tag{3.4}$$

with positive real part. Using this fact, one can show that for $n = 1$ the condition $\sum_{k=1}^N |B_k| \leq 1$ is necessary to have continuous dependence of the solutions of the NFDE (D, f) as the $r_k \to 0$. Analogous results for the matrix case seem to be difficult to obtain.

Throughout the remainder of this paper, we assume the solution operator $T_{D, f}(t): \Omega \to \Omega$ is defined for all $t \geq 0$.

4. Stable D-Operators

A few years ago (see [5]), a class of neutral functional differential equations was defined for which a rather general theory has evolved. It is the purpose of this section to define and give some properties of this class.

For $D \in \mathscr{B}$, D atomic at 0, let $C_D = \{\phi \in C : D\phi = 0\}$. On C_D, the functional equation

$$Dy_t = 0, \qquad t \geq 0, \quad y_0 = \psi \in C_D,$$

generates a strongly continuous semigroup of linear transformations $T_D(t)$: $C_D \to C_D$, $t \geq 0$, defined by $T_D(t)\psi = y_t(\psi)$. Let a_D be the order of the semigroup $T_D(t)$, $t \geq 0$; that is,

$$a_D = \inf \{a \in R : \text{there is a } K = K(a) \text{ such that } \|T_D(t)\| \leq Ke^{at}, t \geq 0\}.$$

Definition   [5]. D is stable if $a_D < 0$.

The simplest operator D that is stable is $D\phi = \phi(0)$. This operator corresponds to retarded functional differential equations.

Theorem 4.1   [5].   The operator D is stable if and only if there is an $a > 0$, $b > 0$, such that for any $\sigma \in R$, $h \in C([\sigma, \infty), E^n)$, any solution y of

$$Dy_t = h(t), \qquad t \geq \sigma, \tag{4.1}$$

must satisfy

$$|y_t| \leq be^{-a(t-\sigma)}|y_\sigma| + b \sup_{\sigma \leq u \leq t} |h(u)|. \tag{4.2}$$

In particular, if y is a bounded solution of (4.1) on $(-\infty, 0]$, then

$$|y_t| \leq b \sup_{-\infty < u \leq t} |h(u)|, \qquad -\infty < t \leq 0. \tag{4.3}$$

Theorem 4.2 **[23].** Let $D \in \mathscr{B}$, $D\phi = \int_{-r}^{0} d\mu(\theta)\phi(\theta)$, and suppose μ has no singular part. Then D is stable if and only if there is a $\delta > 0$ such that Re $\lambda < -\delta$ for all λ satisfying the characteristic equation

$$\det D(e^{\lambda}I) = 0. \tag{4.4}$$

In particular, if $D\phi = \phi(0) - \sum_{k=1}^{N} B_k \phi(-r_k)$, $0 < r_k \le r$, then D is stable if and only if there is a $\delta > 0$ such that Re $\lambda < -\delta$ for all λ satisfying

$$\det \left(I - \sum_{k=1}^{N} B_k e^{-\lambda r_k} \right) = 0. \tag{4.5}$$

If the r_j/r_k in (4.5) are rational, the proof of Theorem 4.2 is trivial. The complications arise when some of these ratios are irrational and the proof is very recent [6, 23]. A less complete discussion of the asymptotic behavior of solutions has been given by many authors. In fact, in [1] it was proved that individual solutions of difference equations approach zero exponentially as $t \to \infty$ under the above hypotheses. In [21], it was shown that the hypotheses imply there are $\alpha > 0$, $K > 0$, such that

$$|T_D(t)\psi| \le Ke^{-\alpha t}|\psi|_1, \qquad t \ge 0,$$

for all $\psi \in C^1([-r, 0], E^n) \cap C_D$ and $|\psi|_1$ is the C^1 norm on ψ.

The following pages will be concerned with NFDE (D, f) with D stable. A few years after stable D-operators were discussed in some detail, Melvin considered the following problem: Suppose D is the difference operator given by (3.3) and D is stable. Do there exist intervals (with interior points) I_k containing r_k, $k = 1, \ldots, N$, such that

$$\bar{D}\phi = \phi(0) - \sum_{k=1}^{N} B_k \phi(-\bar{r}_k) \tag{4.6}$$

is stable for each $\bar{r}_k \in I_k$? If such a situation holds, we say D is stable locally in the delays (sld).

To be sld is a necessary requirement in a physical problem and a computable characterization of this property is very important. Some results have been obtained. For $n = 1$, it was shown in [32] that D is sld if and only if $\sum_{k=1}^{N} |B_k| < 1$. This result is also given implicitly in [23].

The matrix case seems to be extremely difficult. The reason for this is that the characteristic equation (4.5) is of the form

$$1 = \sum_{j=1}^{p} a_j \exp \left(-\lambda \sum_{k=1}^{N} n_{jk} r_k \right), \tag{4.7}$$

where all the n_{jk} are nonnegative integers. We say that (4.7) is stable globally in the delays (sgd) if for any nonnegative numbers r_k, $k = 1, 2, \ldots, N$, there is a $\delta > 0$ such that the solutions λ of (4.7) satisfy Re $\lambda < -\delta$. It is then shown in [11] that sld is equivalent to sgd. This result is far from the characterization that one would like, but it is useful to obtain the regions of stability in the parameters a_j in specific examples.

5. ω-Limit Sets

In this section we summarize some of the implications for NFDE (D, f) of the hypothesis that D is stable. For a solution $x(\phi)$ of an NFDE (D, f), let $\gamma^+(\phi) = \bigcup_{t \geq 0} x_t(\phi)$ be the positive orbit through ϕ and let $\omega(\phi)$, $\alpha(\phi)$ be the ω- and α-limit sets, respectively, of the orbit through ϕ. A set U in C is invariant for the NFDE (D, f) if for any $\phi \in U$, there is a function $x: (-\infty, \infty) \rightarrow E^n$ such that $x_0 = \phi$, $x_t \in U$, $t \in (-\infty, \infty)$, and x satisfies the equation (D, f) for all $t \in (-\infty, \infty)$.

Theorem 5.1 [5]. If D is stable, then $\gamma^+(\phi)$ is relatively compact if and only if $\gamma^+(\phi)$ is bounded. If $\gamma^+(\phi)$ is bounded, then $\omega(\phi)$ is a nonempty compact connected invariant set.

To understand the manner in which stable D-operators simplify problems, we give a proof of the first part of this theorem. Obviously $\gamma^+(\phi)$ relatively compact implies $\gamma^+(\phi)$ bounded. Conversely, suppose $\gamma^+(\phi)$ is bounded. There exists a constant M such that $|f(\gamma^+(\phi))| \leq M$. Since for any $\tau \geq 0$, $t \geq 0$, x_t satisfies

$$D(x_{t+\tau} - x_t) = \int_t^{t+\tau} f(x_s) \, ds,$$

it follows from Theorem 4.1 that

$$|x_{t+\tau} - x_t| \leq b|x_\tau - \phi| + b\tau M.$$

Consequently, $x(\phi)(t)$ is bounded and uniformly continuous on $[-r, \infty)$. Thus, $\gamma^+(\phi)$ is relatively compact and the result is proved.

The author conjectured some time ago that D stable implies the backward extension on ω-limit sets of bounded orbits are unique. For analytic systems this conjecture is true. In fact, the following result is known.

Theorem 5.2 [33]. Suppose x is a solution on $(-\infty, 0]$ of the retarded functional differential equation (D, f), $D\phi = \phi(0)$, f analytic. Then x is analytic in a region containing $(-\infty, 0]$.

With D any stable operator, the proof in [33] can be modified to show that the same result is true for any NFDE (D, f). Consequently, there cannot be two backward extensions on ω-limit sets of bounded orbits.

Mallet-Paret has communicated an example to the author that shows that the above conjecture is false for general NFDE (D, f) with f in C^∞. Consequently, it is desirable to prove that this result is true generically.

For a retarded functional differential equation

$$\dot{x}(t) = f(x_t),$$

any bounded solution on $(-\infty, \infty)$ obviously must be C^{k+1} if f is C^k. Surprisingly, this same result is true for NFDE (D, f) with D stable. A less general fact for $k = 0$ is in [12] and the general case is treated in the thesis [27]. For later reference, this is stated as

Theorem 5.3 [27]. For an NFDE $(D, f) \in \mathcal{B} \times \mathcal{F}^k$, D stable, any solution bounded on $(-\infty, \infty)$ must be C^{k+1}. Even more, the ω-limit set of a bounded orbit consists of C^{k+1} functions such that for each $j \leq k + 1$, the family consisting of the jth derivative of each element of the ω-limit set is uniformly bounded and equicontinuous.

In particular, any periodic solution of an NFDE $(D, f) \in \mathcal{B} \times \mathcal{F}^k$ must be in C^{k+1}. Theorem 5.3 exhibits a very important property of an NFDE (D, f) with a stable D-operator. Bounded orbits tend to "smooth" at ∞.

The basic ideas of the proof of Theorem 5.3 are as follows. We need the following lemma.

Lemma 5.1. Suppose D is stable, $h: (-\infty, a] \to E^n$ and $\dot{h}$ are uniformly continuous and bounded. If $x(t)$ is a bounded solution of

$$Dx_t = h(t)$$

on $(-\infty, a]$, then $\dot{x}(t)$ exists and is uniformly continuous and bounded on $(-\infty, a]$.

Proof.   Since $D(x_{t+\tau} - x_t) = h(t + \tau) - h(t)$, Theorem 4.1 implies

$$|x_{t+\tau} - x_t| \leq b \sup_{u \in (-\infty, a]} |h(u + \tau) - h(u)|$$

for $t \in (-\infty, a], t + \tau \in (-\infty, a]$. Thus x is uniformly continuous on $(-\infty, a]$. Also, using the same argument,

$$\left| \frac{x_{t+\tau} - x_t}{\tau} - \frac{x_{t+s} - x_t}{s} \right| \leq b \sup_{u \in (-\infty, a]} \left| \frac{h(u + \tau) - h(u)}{\tau} - \frac{h(u + s) - h(u)}{s} \right|$$

$$\leq b \sup_{u \in (-\infty, a]} \left| \int_0^1 [\dot{h}(u + v\tau) - \dot{h}(u + vs)] \, dv \right|.$$

Since $\dot{h}$ is uniformly continuous, $[x(t + \tau) - x(t)]/\tau$ approaches a continuous limit $\dot{x}(t)$ as $\tau \to 0$. Obviously, $D(\dot{x}_t) = \dot{h}(t)$ and, as before, $\dot{x}(t)$ is bounded and uniformly continuous on $(-\infty, a]$.

With this lemma, we prove Theorem 5.3 for $k = 0$. If $h(t) = Dx_t$, then $h(t)$ is bounded and continuous on $(-\infty, a]$. Also, $dDx_t/dt = \dot{h}(t) = f(x_t)$ is bounded and continuous on $(-\infty, a]$. Thus, h is uniformly continuous and D stable implies x uniformly continuous on $(-\infty, a]$, and so $\{x_t, t \in (-\infty, a]\}$ is precompact. Thus, $\dot{h}(t)$ is uniformly continuous on $(-\infty, a]$. Lemma 5.1 implies x is C^1 and $\dot{x}$ uniformly continuous and bounded on $(-\infty, a]$. The remainder of the proof of the theorem follows from similar reasoning.

6. Representation of Solution Operator

In a retarded functional differential equation, $\dot{x}(t) = f(x_t), f \in \mathscr{F}^k, k \geq 0$, the solution operator $T(t)$ satisfies the following property: For any $\omega \geq r$ and any bounded set $A \subset C$ for which $\{T(t)A, 0 \leq t \leq \omega\}$ is bounded, the set $T(\omega)A$ is precompact. It is a fundamental result that the solution operator of an NFDE is the sum of a bounded linear operator and an operator of the above type.

To state this result precisely, some notation is needed. If D is atomic at zero, there is a matrix $\Phi = (\phi_1, \ldots, \phi_n)$ with each $\phi_j \in C$ such that $D(\Phi) = I$. Let $C_D = \{\phi \in C: D\phi = 0\}$, $\Psi = I - \Phi D(\cdot)$, and let $T_D(t)$ be the semigroup on C generated by the equation $D(y_t) = D\phi$, $y_0 = \phi$. Then $T_D(t)|C_D$ is also a semigroup on C_D and $\Psi: C \to C_D$.

Theorem 6.1 **[13].** The solution operator $T_{D, f}(t)$ of the NFDE (D, f) can be written as

$$T_{D, f}(t) = T_D(t)\Psi + U(t), \qquad t \geq 0, \tag{6.1}$$

where $U(t) : C \to C$ satisfies the following property: For any $\omega > 0$ and any bounded set $A \subset C$ for which $\{T(t)A, 0 \leq t \leq \omega\}$ is bounded, the set $U(\omega)A$ is precompact. If, in addition, D is stable, then there is an $\alpha > 0$ such that the spectral radius $\rho(T_D(t)\Psi) \leq \exp(-\alpha t)$, $t \geq 0$.

We indicate the proof of the representation formula (6.1) in Theorem 6.1. There exist positive constants $K > 0$, $\alpha > 0$, such that $\|T_D(t)\Psi\| \leq Ke^{-\alpha t}$, $t \geq 0$. Define $U(t) = T_{D, f}(t) - T_D(t)\Psi$. Then

$$\begin{aligned}
D(U(t)\phi) &= D(T_{D, f}(t)\phi - T_D(t)\Psi\phi) \\
&= D(T_{D, f}(t)\phi) - D(T_D(t)\Psi\phi) \\
&= D\phi + \int_0^t f(T_{D, f}(s)\phi)\, ds - D(\Psi\phi) \\
&= D\phi + \int_0^t f(T_{D, f}(s)\phi)\, ds.
\end{aligned}$$

Therefore,

$$D(U(t + \tau)\phi - U(t)\phi) = \int_t^{t+\tau} f(T_{D, f}(s)\phi)\, ds.$$

Theorem 4.1 implies

$$|U(t + \tau)\phi - U(t)\phi| \leq b|U(\tau)\phi - \Phi D\phi| + b \sup \left| \int_v^{v+\tau} f(T_{D, f}(s)\phi)\, ds \right|.$$

Since $U(0)\phi = \Phi D\phi$ lies in a finite-dimensional space, it is now clear that U satisfies the properties stated in the theorem. This proves (6.1).

Suppose $T_{D, f}(t)$ has the following property:

(B) For each t, the set $\{T_{D, f}(\tau), 0 \leq \tau \leq t\}$ is bounded. Then Theorem 6.1 asserts that $T_{D, f}(t)$ is the sum of a bounded linear operator $T_D(t)\Psi$ and a completely continuous operator. Furthermore, if D is stable, then there is an $\alpha > 0$ and an equivalent norm in C so that $|T_D(t)\Psi| \leq e^{-\alpha t}$, $t \geq 0$.

Consequently, for each $t > 0$, the solution operator is the sum of a contraction operator and a completely continuous operator. This is stated as

Corollary 6.1. If $T_{D,f}(t)$ has property (B) and D is stable, then $U(t)$ in (6.1) is completely continuous and there is an $\alpha > 0$ and an equivalent norm in C such that $|T_D(t)\Psi| \leq \exp(-\alpha t)$.

For D stable and $T_{D,f}(t)$ satisfying property (B), the map $T_{D,f}(t)$, $t > 0$, is an α-contraction in the sense of Kuratowski [26], with $\alpha(T_{D,f}(t)) \leq \exp(-\alpha t)$.* We can therefore bring to bear on these problems all of the known results on α-contractions [36]. The most important applications so far have been in the understanding of the spectral theory for linear systems and the existence of periodic solutions of periodic NFDEs. Some aspects of this subject will be mentioned in later sections. At the present time, we remark only that the functions (D, f) can depend on the independent variable t and the previous theory remains valid.

7. Linear Equations

In this section we consider linear autonomous NFDE $(D, L) \in \mathscr{B} \times \mathscr{B}$ and perturbations of such systems. The solution operator $T_{D,L}(t)$, $t \geq 0$, is a strongly continuous semigroup of linear operators with infinitesimal generator A given by $A\phi(\theta) = \dot\phi(\theta)$, $-r \leq \theta \leq 0$, and $D(A) = \{\phi \in C : \dot\phi \in C, D\dot\phi = L\phi\}$. The spectrum $\sigma(A)$ of A contains only the point spectrum $P\sigma(A)$ and $\lambda \in \sigma(A)$ if and only if λ satisfies the characteristic equation

$$\det[D(e^{\lambda}I) - L(e^{\lambda}I)] = 0. \tag{7.1}$$

For a proof of these facts see [21].

If Q is a bounded linear operator on C, then a complex number μ is called a normal eigenvalue of Q if $\mu \in P\sigma(Q)$ and there is an integer k such that $C = \mathfrak{N}(Q - \mu I)^k \oplus \mathcal{R}(Q - \mu I)^k$ and $\mathfrak{N}(Q - \mu I)^k$ has finite dimension. A point μ is a normal point of Q if it is either a normal eigenvalue or a point in the resolvent set of Q. We need the following result from [9]:

Lemma 7.1. Suppose P, Q are bounded linear operators on C with Q compact. If U is an unbounded connected component of normal points of P, then U consists only of normal points of $P + Q$.

---

* For any bounded set B in a Banach space X, $\alpha(B) = \inf\{d : B$ has a finite cover of diam $< d\}$. If $T : S \subset X \to X$, then $\alpha(T) = \inf\{k : \alpha(TB) \leq k\alpha(B)$ for all bounded sets $B \subset X\}$.

From Lemma 7.1, Theorem 6.1, and the fact that $P\sigma(T_{D,L}(t))\backslash\{0\} =$ exp $\sigma(A)t$, we have the following important result, which is a special case of [23].

Theorem 7.1. For any $a > a_D$ and any λ satisfying (7.1) with Re $\lambda \geq a$, it follows that $\mu(t) = e^{\lambda t}$ is a normal eigenvalue of $T_{D,L}(t)$.

In particular, if D is stable, then $a_D < 0$, the set $\Lambda = \{\lambda : \text{Re } \lambda \geq 0, \lambda \in \sigma(A)\}$ is finite, and there is a natural decomposition of C as $C = P \oplus Q$, where P, Q are subspaces invariant under $T_{D,L}(t)$ with the spectrum of $T_{D,L}(t)$ restricted to P being exp Λt and P is finite dimensional.

This basic result allows one to begin the discussion of the behavior of solutions in a neighborhood of an equilibrium point for nonlinear equations; for example, the saddle-point property [4a], stability in critical cases [14, 22], bifurcation [2, 15]. A recent treatment of some of these latter questions for retarded equations in the space W_∞^1 rather than C has been given in [35]. With this latter space, one can discuss bifurcation caused by the delays.

8. Linear Periodic Systems

In this section we consider periodic linear NFDE (D, L); that is, $D(t, \phi)$, $L(t, \phi)$ are both linear in ϕ and ω-periodic in t. The solution operator $T_{D,L}(t, \tau)$ is defined for all $t \geq \tau$ and is a bounded linear operator on C. The following definition is a modification of the one in [19].

Definition 8.1. A complex number ρ is a characteristic multiplier of NFDE (D, L) if ρ is a normal point of $T_{D,L}(\tau + \omega, \tau)$.

As in [16, 37], one shows that the characteristic multipliers are independent of τ. There is a decomposition of C as $C = P \oplus Q$, $P = \mathfrak{N}(T(\tau + \omega, \tau) - \mu I)^k$, $Q = \mathfrak{R}(T(\tau + \omega, \tau) - \mu I)^k$. If Φ is a basis for P with P of dimension d, then there is a $d \times d$ constant matrix B such that the spectrum of $e^{B\omega}$ is ρ and an $n \times d$ ω-periodic matrix $C(t)$ such that if $\phi \in P$, $\phi = \Phi b$, then

$$T(t, \tau)\phi = C(t)e^{Bt}b.$$

In this sense, there is a Floquet representation on the generalized eigenspace P.

From Theorem 6.1, if exp $a_D \tau$ is the spectral radius of $T_D(\tau + \omega, \tau)$, it follows that any point $\mu \in \sigma(T(\tau + \omega, \tau))$ with $|\mu| > \exp a_D \tau$ must be a characteristic multiplier. In particular, if D is stable, then $\exp a_D \tau < 1$ and the stability properties of the zero solution are determined by the characteristic multipliers.

This latter remark allows one to prove the Fredholm alternative for nonhomogeneous linear systems [17], and therefore develop a theory of nonlinear oscillations for weakly nonlinear equations. The Fredholm alternative is also a consequence of the very general theory of two-point boundary-value problems in [24].

Suppose p is a nonconstant ω-periodic solution of an autonomous NFDE $(D, f) \in \mathscr{B} \times \mathscr{F}^k$. For $k \geq 1$, it is meaningful to consider the linear variational equation for p,

$$\frac{d}{dt} Dy_t = f_\phi(p(t))y_t \overset{\text{def}}{=} L(t, y_t), \tag{8.1}$$

and $L(t, \phi)$ is ω-periodic in t.

From Theorem 5.1, p is a C^{k+1} function. Therefore, $D(\dot{p}_t) = f(p_t)$. Differentiating this expression with respect to t, we have that $\dot{p}$ is a nontrivial periodic solution of (8.1). Consequently, one is a characteristic multiplier. One should now be able to extend to NFDEs the results on RFDEs concerning orbital stability of periodic orbits and more generally the behavior of solutions near a periodic orbit [18, 38, 39]. Another very interesting result that should be extended to NFDEs is the existence of "useful" coordinate systems near a periodic orbit [40].

9. Periodic Solutions

Consider an NFDE (D, f) with $D(t, \phi)$, $f(t, \phi)$ ω-periodic in t, $\omega > 0$, and D stable. We may assume from Section 6 that the solution operator

$$T_{D, f}(t) \overset{\text{def}}{=} T_{D, f}(t, 0)$$

is a weak α-contraction for each $t > \sigma$. Therefore, an immediate consequence of the fixed-point theorem in [25] or [7] is the following

Theorem 9.1. If there exists a closed bounded convex set $\Gamma \subset C$ such that $\{T_{D, f}(t), 0 \leq t \leq \omega\}$ is bounded and $T_{D, f}(\omega)$: $\Gamma \to \Gamma$, then there is an ω-periodic solution of the NFDE (D, f).

Although this result is interesting, it is not very useful because such a *convex* Γ almost never exists. Usually, one is only able to prove some type of ultimate boundedness of the solution. Fixed-point theorems sufficiently general to cover these situations have only recently been proved in [20, 34]. To describe the result, we say that the NFDE (D, f) is *compact* dissipative if there is a bounded set $B \subset C$ such that for any compact set $K \subset C$ there is $\tau = \tau(K)$ such that $T_{D, f}(t)K \subset B$ for $t \geq \tau$.

Theorem 9.2 [20]. If the NFDE (D, f) is compact dissipative, then there is an ω-periodic solution.

To apply Theorem 9.2, one must have a method for determining when a system is compact dissipative. It turns out that the use of Liapunov functionals generally puts some restrictions on the delays (see [41]). In [28], sufficient conditions for ultimate boundedness were given using the Razumikhin-type functions, and applications were made to transmission line problems.

The development of fixed-point theorems in [20] was motivated directly from the application to NFDEs. Other fixed-point theorems are in this paper and an application to the existence of periodic solutions of equations that are perturbations of one with an asymptotically stable equilibrium point is given in [29]. The proofs in [20] also were the motivation for the introduction of strongly limit compact maps in [3].

REFERENCES

[1] Bellman, R., and Cooke, K., "Differential-Difference Equations." Academic Press, New York, 1963.

[2] Chafee, N., A bifurcation problem for a functional differential equation of finitely retarded type, *J. Math. Anal. Appl.* **35** (1971), 312–348.

[3] Chow, S., and Hale, J. K., Strongly limit compact maps, *Funkcial. Ekvac.* **17** (1974), 31–38.

[4] Cruz, M. A., and Hale, J. K., Existence, uniqueness and continuous dependence for hereditary systems, *Ann. Mat. Pura Appl.* **85** (1970), 63–82.

[4a] Cruz, M. A., and Hale, J. K., Exponential estimates and the saddle point property for neutral functional differential equations, *J. Math. Anal. Appl.* **34** (1971), 267–288.

[5] Cruz, M. A., and Hale, J. K., Stability of functional differential equations of neutral type, *J. Differential Equations* **7** (1970), 334–355.

[6] Datko, R., Stability in neutral functional differential equations, Personal communication (1974).

[7] Darbo, G., Punti uniti in trasformazione a condominio non compatto, *Rend. Sem. Mat. Univ. Padova* **24** (1955), 84–92.

[8] Driver, R., Existence and continuous dependence of solutions of a neutral functional differential equation, *Arch. Rational Mech. Anal.* **19** (1965), 147–166.

[9] Gokberg, I. C., and M. G. Krein, "Introduction to the Theory of Linear Nonselfadjoint Operators," *Amer. Math. Soc. Transl. Math. Monogr.* **18** (1969).

[10] Hale, J. K., Forward and backward continuation for neutral functional differential equations, *J. Differential Equations* **9** (1971), 168–181.

[11] Hale, J. K., Parametric stability in difference equations. *Boll. Un. Mat. Italia* **16** (1974).

[12] Hale, J. K., Smoothing properties of neutral equations, *An. Acad. Brasil Ci.* **45** (1973), 49–50.

[13] Hale, J. K., A class of neutral equations with the fixed point property, *Proc. Nat. Acad. Sci. U.S.A.* **67** (1970), 136–137.

[14] Hale, J. K., Critical cases for neutral functional differential equations, *J. Differential Equations* **10** (1971), 59–82.

[15] Hale, J. K., Behavior near constant solutions of functional differential equations, *J. Differential Equations* **15** (1974), 278–294.

[16] Hale, J. K., "Functional Differential Equations," *Appl. Math. Sci.* **3**. Springer-Verlag, Berlin, 1971.

[17] Hale, J. K., Oscillations in neutral functional differential equations, *in* "Nonlinear Mechanics," C.I.M.E., June, 1972.

[18] Hale, J. K., Solutions near simple periodic orbits of functional differential equations, *J. Differential Equations* **9** (1970), 126–138.

[19] Hale, J. K., α-contractions and differential equations, "Equations Différentielles et Fontionelles Nonlinéaires," pp. 15–42. Herman, Paris, 1973.

[20] Hale, J. K., and Lopes, O., Fixed point theorems and dissipative process, *J. Differential Equations* **13** (1973), 391–402.

[21] Hale, J. K., and Meyer, K., A class of functional differential equations of neutral type, *Memoirs Amer. Math. Soc.* No. 76 (1967).

[22] Hausrath, A., Stability in critical cases of purely imaginary roots for neutral functional differential equations, *J. Differential Equations* **13** (1973), 329–357.

[23] Henry, D., Linear autonomous neutral functional differential equations, *J. Differential Equations* **15** (1974), 106–128.

[24] Henry, D., Adjoint theory and boundary value problems for neutral functional differential equations, preprint.

[25] Krasnoselskii, M. A., "Topological Methods in the Theory of Nonlinear Integral Equations." Macmillan, New York, 1964.

[26] Kuratowskii, C., Sur les espaces completes, *Fund. Math.* **15** (1930), 301–309.

[27] Lopes, O., Asymptotic fixed point theorems and forced oscillations in neutral equations, Ph.D. thesis, Brown Univ. (1973).

[28] Lopes, O., Forced oscillations in nonlinear neutral differential equations, submitted to *SIAM*.

[29] Lopes, O., Periodic solutions of perturbed neutral differential equations, *J. Differential Equations* **15** (1974), 70–76.

[30] Melvin, W. R., A class of neutral functional differential equations, *J. Differential Equations* **12** (1972), 524–534.

[31] Melvin, W. R., Topologies for neutral functional differential equations, *J. Differential Equations* **13** (1973), 24–32.

[32] Melvin, W. R., Stability properties of functional differential equations, *J. Math. Anal. Appl.* **48** (1974), 749–763.

[33] Nussbaum, R., Periodic solutions of analytic functional differential equations are analytic, *Michigan Math. J.* **20** (1973), 249–255.

[34] Nussbaum, R., Some asymptotic fixed point theorems, *Trans. Amer. Math. Soc.* **171** (1972), 349–375.

[35] Ruiz-Claeyssen, J., Effect of delays on functional differential equations, Ph.D. thesis, Brown Univ. (1974).

[36] Sadovskii, B. N., Limit compact and condensing operators, *Usp. Mat. Nauk* **271** (1972), 81–146 [*English transl.: Russian Math. Surveys* 85–146].

[37] Stokes, A., A Floquet theory for functional differential equations, *Proc. Nat. Acad. Sci. U.S.A.* **48** (1962), 1330–1334.

[38] Stokes, A., On the stability of a limit cycle of an autonomous functional differential equation, *Contrib. Differential Equations* 3 (1964), 121–140.

[39] Stokes, A., On the stability of integral manifolds of functional differential equations, *J. Differential Equations* 9 (1971), 405–419.

[40] Stokes, A., Local coordinates around a limit cycle of functional differential equations. *J. Appl. Anal.* (to appear).

[41] Yoshizawa, T., "Stability Theory by Liapunov's Direct Method." Math. Soc. Japan, Tokyo (1966).

Functional Differential Equations—Generic Theory

WALDYR M. OLIVA *
Instituto de Matemática e Estatística
Universidade de São Paulo, Sao Paulo, Brazil

Introduction

The purpose of this chapter is to present recent results on functional differential equations from the generic point of view. Although most results obtained so far have been given for the retarded case, it seems to us that neutral equations need to be treated soon. Keeping this in mind, in Section I a general definition is given for autonomous functional differential equations on a manifold M, which includes retarded and neutral cases. Examples and generic results extending the Kupka–Smale theorem for vector fields to retarded and some differential–delay equations are presented in Section II; the one-to-oneness of the flow for retarded equations, the global unstable manifolds of hyperbolic critical points and periodic orbits, and the invariance of nonwandering sets make sense generically and are also considered in this section. Generic linear nonautonomous equations have one-to-oneness in the solution operator and appear in Section III.

I. Autonomous Functional Differential Equations

Let M be a C^∞ (or analytic) finite-dimensional manifold, I the closed interval $[-r, 0]$, $r \geq 0$, and $C^0(I, M)$ the Banach manifold of all continuous maps ϕ of I into M. The differentiable structure of $C^0(I, M)$, the smoothness of the evaluation map $\rho : \phi \to \phi(0)$, and other related properties can be found in [12]. We follow closely the notation used in [1] for calculus on manifolds.

We will now give a geometric definition for an autonomous functional differential equation in a way that includes some important and well-known cases.

* Research supported in part by FAPESP, Fundação de Amparo à Pesquisa do Estado de São Paulo, Brazil, Proc. 11-74/603.

An autonomous functional differential equation (FDE) on M is a pair (D, F) of continuous functions

$$F: C^0(I, M) \to TM, \qquad D: C^0(I, M) \to M,$$

such that $\tau_M \cdot F = D$. Here TM denotes the tangent bundle of M and τ_M is the C^∞ (or analytic) canonical projection of TM onto M. Locally one can think of TM as a product; then for each $\phi \in C^0(I, M)$,

$$F(\phi) = (D(\phi), f(\phi)).$$

A solution of (D, F) with initial condition ϕ at t_0 is a continuous function $x(t)$ defined on $t_0 - r \le t < t_0 + A, 0 \le A \le \infty$, with values on M, such that if $x_t \in C^0(I, M)$ is defined by $x_t(\theta) = x(t + \theta), \theta \in I, t_0 \le t < t_0 + A$, one has

    (i) $x_{t_0} = \phi$;
    (ii) $D(x_t)$ is a C^1 function for $t \in [t_0, t_0 + A)$;
    (iii) $d(D(x_t))/dt = F(x_t)$ for all $t \in [t_0, t_0 + A)$, where $d(D(x_t))/dt$ denotes the tangent vector to the curve $D(x_t)$ at the point t.

Locally one can write on TM:

$$[D(x_t), d(D(x_t))/dt] = [D(x_t), f(x_t)] \qquad \text{or} \qquad d(D(x_t))/dt = f(x_t).$$

The solution $x(t)$ is sometimes denoted by $x(t; t_0, \phi)$ and x_t by $x_t(t_0, \phi)$ or $x_t(\phi)$ if $t_0 = 0$.

A C^1 function $D: C^0(I, M) \to M$ is said to be *atomic at zero* if for any ϕ in $C^0(I, M)$ the derivative $T_\phi D$ given by

$$T_\phi D(\Psi) = \int_{-r}^{0} d\eta(\phi, \theta)\Psi(\theta)$$

has the continuous jump $A(\phi) = [\eta(\phi, 0) - \eta(\phi, 0_-)]$ nonsingular, say $\det A(\phi) \ne 0$ [6, p. 279].

An autonomous neutral functional differential equation (NFDE) is an FDE (D, F) with D atomic at zero. When D is the evaluation map ρ, the pair (ρ, F) is called a *retarded functional differential equation (RFDE)* (see [12, Def. 1]). Note that any RFDE is an NFDE. In fact, we only need to prove that ρ is atomic at zero; but

$$T_\phi \rho(\Psi) = \Psi(0) = \int_{-r}^{0} d\eta(\phi, \theta)\Psi(\theta),$$

where $\eta(\phi, \theta) = 0$ for $\theta \in [-r, 0)$ and $\eta(\phi, 0) = I$. This shows that in this case $A(\phi) = I$.

If (D, F) is an NFDE with F continuous and locally Lipschitzian, it can be proved that one obtains existence, uniqueness, and continuous dependence of solutions; when $M = R^n$ most of the results on existence, uniqueness, and continuous dependence can be seen in [7]. For the case of an RFDE on a manifold M some of these questions were considered in [12] and the case $M = R^n$ is exhaustively discussed in [4]. We will return to this subject later in this section.

An *equilibrium* (*or critical*) *point* of an FDE (D, F) is a constant function $c \in C^0(I, M)$ such that $F(c) = 0$. A function $g(t)$, $-\infty < t < +\infty$, is said to be a solution of (D, F) on $(-\infty, +\infty)$ if for every $\sigma \in (-\infty, +\infty)$, the solution $x(t; \sigma, g_\sigma)$ of (D, F) exists and satisfies $x_t(\sigma, g_\sigma) = g_t$, $t \geq \sigma$. As in the case of ordinary differential equations, the *constant* and *periodic solutions* $g(t)$ of (D, F) on $(-\infty, +\infty)$ are of fundamental importance in the qualitative theory of functional differential equations.

Many questions can be considered when we try to understand the qualitative behavior of the solutions of FDEs with respect to the variation of the pairs (D, F) in some topological space. The closed condition $\tau_M \cdot F = D$ shows that the smoothness of D is at least equal to the smoothness of F.

The generic theory for RFDEs on compact manifolds has been developed considering F in the Banach space $\mathscr{BX}^\kappa(I, M)$ of all C^κ maps that are bounded with bounded derivatives up to the order $\kappa \geq 1$. In the neutral case, to each fixed D corresponds the Banach space of all C^κ maps

$$F: C^0(I, M) \to TM$$

such that $\tau_M \cdot F = D$ and that are bounded with bounded derivatives up to the order $\kappa \geq 1$.

In the case $M = R^n$ the FDEs can be written as

$$dD(x_t)/dt = f(x_t).$$

It now makes sense to consider special NFDEs as pairs $(D, f) \in \mathscr{B} \times F^\kappa$, with D atomic at zero, $\mathscr{B}$ being the Banach space of the bounded linear operators from $C = C^0(I, R^n)$ into R^n with the usual norm and F^κ the Banach space of all C^κ maps from $\overline{\Omega}$ (the closure of an open set Ω of C) into R^n that are bounded with bounded derivatives up to the order $\kappa \geq 1$ (see [6, p. 279]).

II. Autonomous Retarded Functional Differential Equations

We will start this section with some examples of RFDEs on a C^∞ manifold M.

Example 1. *Any C^κ-vector field on M defines a C^κ-RFDE on M, $\kappa \geq 1$.* If $X: M \to TM$ is a C^κ-vector field on M, it is easy to see that $F = X \cdot \rho$ is a C^κ-RFDE on M. All solutions of X are also solutions of F. The solutions of F on $(-\infty, +\infty)$ are solutions of X; in particular, constant and periodic solutions are the same for X and F.

Example 2. *Any C^κ-vector field on $C^0(I, M)$ defines a C^κ-RFDE on M,* $\kappa \geq 1$. If $V: C^0(I, M) \to TC^0(I, M) = C^0(I, TM)$ is a C^κ-vector field on $C^0(I, M)$, then $F = T\rho \cdot V$ is a C^κ-RFDE on M, since $\tau_M \cdot T\rho = \rho \cdot \tau_{C^0(I, M)}$ and then $\tau_M \cdot F = \tau_M \cdot (T\rho \cdot V) = \rho$.

Example 3. Fix $\kappa \geq 1$. If $\gamma: C^0(I, M) \to R$ is a C^κ-real-valued function and $X: M \to TM$ a C^κ-vector field on M, then F given by

$$F(\phi) = \gamma(\phi) \cdot X(\phi(0))$$

is a C^κ-RFDE on M. The constant solutions (critical points) of X are constant solutions of F. In the special case $\gamma(\phi) = 1 + \int_{-r}^0 \phi^2(\theta) \, d\theta$, the critical points of F correspond to the critical points of X.

Example 4. *Equation of first variation.* Fix $\kappa \geq 1$ and let $F: C^0(I, M) \to TM$ be a $C^{\kappa+1}$-RFDE on M. The double tangent space of M, T^2M, admits a canonical involution $\omega: T^2M \to T^2M$, $\omega^2 = id T^2M$, and following [1, p. 18] ω is a C^∞-diffeomorphism of T^2M such that $\tau_{TM} \cdot \omega = T\tau_M$ and $T\tau_M \cdot \omega = \tau_{TM}$. Using the above relations one can see that the map

$$\omega \cdot TF: C^0(I, TM) = TC^0(I, M) \to T^2M$$

is a C^κ-RFDE on TM called the first variation equation associated to F. Locally, if $(x(t), y(t)) \in TM$ is a solution of $\omega \cdot TF$ and $F(\phi) = (\phi(0), f(\phi))$, one can write

$$\dot{x}(t) = f(x_t), \qquad \dot{y}(t) = Tf(x_t)y_t.$$

Example 5. *Second-order retarded functional differential equations on M^n.* The double tangent space T^2M of M^n contains a $3n$-dimensional sub-manifold J^2 such that an element $\bar{v}$ of T^2M belongs to J^2 if and only if

$\omega(\bar{v}) = \bar{v}$, ω being the canonical involution. It is clear that τ_{TM} and $T\tau_M$ coincide on J^2. A second-order RFDE on M is a C^κ-RFDE G: $C^0(I, TM) \to T^2M$ such that G has values on J^2. Locally, if $(x(t), v(t)) \in TM$ is a solution of G then

$$\dot{x} = v, \qquad \dot{v} = g(x_t, v_t).$$

In such a problem we look for basic solutions [curves $x(t)$ on M] that obviously satisfy $\ddot{x} = g(x_t, \dot{x}_t)$, $t \ge r$.

Example 6. Consider an imbedding of a compact manifold M as a closed submanifold of R^N, for a large enough N, and let U be a tubular neighborhood of M, U an open set of R^N, and $\alpha: U \to M$ the C^∞-canonical projection. The maps

$$q, p: C^0(I, M) \to TM$$

defined by

$$p(\phi) = T\alpha_{\phi(0)}[\phi(-r)] \qquad \text{and} \qquad q(\phi) = T\alpha_{\phi(0)}[\phi(0)]$$

are C^∞-RFDEs. Here $\phi(0)$ and $\phi(-r)$ are considered tangent vectors of U at the point $\phi(0) \in M$.

Example 7. *Differential–delay equations.* Consider a C^κ-function g: $M \times M \to TM$ such that $\tau_M g = \pi_1$, $\pi_1: M \times M \to M$ being the first canonical projection. The map g defines a C^κ-RFDE F given by $F(\phi) = g(\phi(0), \phi(-r))$. Locally if $x(t)$ is a solution of this equation for F, one obtains $\dot{x}(t) = g(x(t), x(t - r))$. An analogous geometric definition can be given for equations of the type $\dot{x} = g(x(t), x(t - r_1), \ldots, x(t - r))$.

Example 8. *An RFDE on R^n. Linear RFDE.* Any C^κ-function $F: C^0(I, R^n) \to R^n \times R^n$ of the form $F(\phi) = (\phi(0), f(\phi))$ is a C^κ-RFDE on R^n. It is clear that, in this particular case, we only need to give the C^κ-function $f: C^0(I, R^n) \to R^n$. If $x(t)$ is a solution, one obtains $\dot{x} = f(x_t)$. In this case, $C^0(I, R^n)$ is a Banach space; then it makes sense to consider linear equations, viz., choose f as a linear continuous map. Examples of RFDEs on R^n can be found in [4].

Example 9. Assume that for a given RFDE on R^n, $\dot{x} = f(x_t)$, there exists a submanifold $S \subseteq R^n$ that is invariant with respect to the equation, in the sense that for any $\phi \in C^0(I, R^n)$, $\phi(\theta) \in S$, $\theta \in I$, the solution $x(t; 0; \phi)$

remains on S for all $t \geq 0$. It is easy to see that the restriction of f to $C^0(I, S)$ defines the following RFDE on S:

$$F: C^0(I, S) \to TS$$

by $F(\phi) = (\phi(0), f(\phi))$. The next two examples will be special cases of this situation.

Example 10. *An RFDE on the sphere S^2.* Let $M = S^2$ be the set of all $(x, y, z) \in R^3$ such that $x^2 + y^2 + z^2 = 1$. Consider in R^3 the following differential–delay system of equations:

$$\dot{x} = -x(t - 1)y(t) - z(t),$$
$$\dot{y} = x(t - 1)x(t) - z(t),$$
$$\dot{z} = x(t) + y(t).$$

If $(x(t), y(t), z(t))$ is a solution of such a system it is easy to see that

$$x\dot{x} + y\dot{y} + z\dot{z} = 0 \qquad \text{for all} \quad t \geq 0.$$

Then $x^2(t) + y^2(t) + z^2(t) = a^2$, and then, if an initial condition ϕ is such that $\phi(\theta) \in S^2$ for all $\theta \in [-1, 0]$, one concludes that the solution $x(t; 0, \phi)$ has values on S^2 for all $t \geq 0$. The above considerations show that F is a well-defined C^∞-RFDE on S^2. The critical points of F are

$$v = ((\sqrt{2} - 1)^{1/2}, -(\sqrt{2} - 1)^{1/2}, \sqrt{2} - 1)$$

and

$$v = (-(\sqrt{2} - 1)^{1/2}, (\sqrt{2} - 1)^{1/2}, \sqrt{2} - 1).$$

Example 11. *An RFDE on the circle S^1.* A similar procedure shows that the system in R^2

$$\dot{x} = -x(t - 1)y(t), \qquad \dot{y} = x(t - 1)x(t),$$

induces a C^∞-RFDE on the set S^1 of all $(x, y) \in R^2$ such that $x^2 + y^2 = 1$. The critical points in this case are $v = (0, 1)$ and $v = (0, -1)$.

Example 12. This example is a combination of Examples 1 and 6. Take in S^2 the vector field Y given by an infinitesimal rotation around the vertical z-axis. Let $r = \pi/2$ and F be given by $F(\phi) = 2Y(\phi(0)) + p(\phi)$. Assume that if a point $x \in S^2$ belongs to the equator, $|Y(x)| = 1$. The critical points of F are $\pm v = (0, 0, 1)$ and $(\cos 2t, \sin 2t, 0)$ is a periodic solution.

After the above examples we will try to describe for the RFDE on a compact manifold M *the existence and uniqueness of solutions* and the *behavior* of them *with respect to initial data and to the equation.*

For simplicity of notation we will denote $C = C^0(I, M)$ and $\mathscr{X}^\kappa = \mathscr{B}\mathscr{X}^\kappa(I, M)$ for a fixed $\kappa \geq 1$.

For $F \in \mathscr{X}^\kappa$ and $\phi \in C$, consider the following problem of existence and uniqueness of solution:

$$\dot{x}(t) = F(x_t), \qquad t \geq t_0, \quad x_{t_0} = \phi.$$

The answer is that there exists a unique solution $x(t)$ with such an initial condition ϕ at $t = t_0$; the function $x(t) = x(t; t_0, \phi)$ is defined on $t_0 - r \leq t < +\infty$ and has values on M for all $t \geq (t_0 - r)$ (see [10] and [12]). If the continuous function F is only supposed to be locally Lipschitzian in $\phi \in C$, the same result is true [12] (recall that M is compact).

Following [10] define now the semiflow map of $F \in \mathscr{X}^\kappa$

$$\Phi: \mathscr{X}^\kappa \times C \times [0, \infty) \to C,$$

where $\Phi(F, \phi, t) = x_t = x_t(\phi)$. It can be proved that

(a) Φ is continuous on $\mathscr{X}^\kappa \times C \times [0, \infty)$,
(b) for each $\tau \geq 0$, Φ is C^κ on $\mathscr{X}^\kappa \times C \times \{\tau\}$,
(c) for any s, $0 \leq s \leq \kappa$, Φ is C^s on $\mathscr{X}^\kappa \times C \times (sr, \infty)$.

Now fix $F \in \mathscr{X}^\kappa$ and define for each $t \geq 0$ the C^κ-map $\Phi_t: C \to C$ by $\Phi_t(\phi) = x_t(\phi)$. Then

$$\Phi_{t_1 + t_2} = \Phi_{t_1} \cdot \Phi_{t_2} = \Phi_{t_2} \cdot \Phi_{t_1} \qquad \text{and} \qquad \Phi_0 = idC.$$

The *solution operator* Φ_t is also called the *flow* of F; Φ_t need not be one-to-one, but if $\phi \neq \Psi$ and $\Phi_{t_1}(\phi) = \Phi_{t_2}(\Psi)$ then

$$\Phi_{t_1 + \sigma}(\phi) = \Phi_{t_2 + \sigma}(\Psi) \qquad \text{for all} \quad \sigma \geq 0.$$

For the equations given by Example 1 the solution operator is never one-to-one for $t = r$.

Example 13.   Consider the restriction to S^1 of the following system of differential–delay equations:

$$\dot{x} = -x(t - 1)y(t)(1 - x(t)), \qquad \dot{y} = x(t - 1)x(t)(1 - x(t)).$$

The critical points in this case are $\pm v = (0, 1)$ and $v = (1, 0)$. The initial conditions $t_0 = 0$ and $\phi(\theta) = (\cos (\pi/2)\theta, \sin (\pi/2)\theta)$, $\theta \in [-1, 0]$ define a unique solution $x(t; 0, \phi)$ that for $t \geq 0$ is the constant map $(1, 0)$. The solution operator Φ_1 is not one-to-one on $C^0(I, S^1)$.

Example 14. Take $M = R$ and consider the scalar equation $\dot{x} = x(t - 1)$. It is easy to see that the solution operator is one-to-one in this case.

An important property of the solution operator Φ_t is that for $t \geq r$, Φ_t is a compact map [i.e., Φ_t maps bounded subsets of $C^0(I, M)$ into relatively compact subsets of $C^0(I, M)$]. This is a consequence of the boundedness of F in $\mathscr{B}\mathscr{X}^\kappa(I, M)$ and the classical Arzela theorem. Then even if ϕ_t is one-to-one, it can never be a homeomorphism.

The behavior at infinity of the solutions is another very important qualitative concept. An *invariant set* of a given RFDE F is any subset S of $C^0(I, M)$ such that for every $\phi \in S$, there is a solution x on $(-\infty, +\infty)$ with $x_0 = \phi$ and $x_t \in S$ for all $t \in (-\infty, +\infty)$. It is clear that the set $A(F)$ of all x_t of all solutions $x(t)$ on $(-\infty, +\infty)$ is the largest invariant set of F. When M is compact $A(F)$ is compact in $C^0(I, M)$ and

$$A(F) = \bigcap_{s=1}^{\infty} \Phi_{sr}(C^0(I, M)).$$

The set $A(F)$ is called the *attractor of F*. An element $\Psi \in C^0(I, M)$ is said to be in the *ω-limit set* $\omega(\phi)$ of an *orbit* $\gamma^+(\phi) = \bigcup_{t \geq 0} x_t(\phi)$ *through* ϕ if there is a sequence of $t_n \to \infty$ as $n \to \infty$ such that $x_{t_n} \to \Psi$ as $n \to \infty$. An element $\Psi \in C^0(I, M)$ is said to be in the *α-limit set* $\alpha(\phi)$ of an orbit through ϕ if there is a solution $x(t)$ on $(-\infty, +\infty)$, $x_0(\phi) = \phi$, and a sequence $t_n \to -\infty$ as $n \to \infty$ such that $x_{t_n}(\phi) \to \Psi$ as $n \to \infty$. The following result is proved in [4]: any ω-limit set $\omega(\phi)$ is nonempty, compact, connected, and invariant [then contained in $A(F)$]. Again we need the boundedness of $F \in \mathscr{B}\mathscr{X}^\kappa(I, M)$ and M. Also if $\phi \in A(F)$ and if Φ_t is one-to-one on $A(F)$ then $\alpha(\phi)$ is nonempty, compact, connected, and invariant.

A point $\Psi \in A(F)$ is a *nonwandering point* of F if for any neighborhood U of Ψ in $A(F)$ and any real number $T > 0$ there exists $t = t(U, T) > T$ and $\overline{\Psi} \in U$ such that $\phi_t(\overline{\Psi}) \in U$. The *nonwandering* set $\Omega(F)$ is the set of all non-wandering points of F. If $\phi \in A(F)$, then $\omega(\phi) \subseteq \Omega(F)$ and $\alpha(\phi) \subseteq \Omega(F)$. The nonwandering set $\Omega(F)$ is closed and if Φ_t is one-to-one on $A(F)$ then $\Omega(F)$ is an invariant set. As in the ordinary case, points in $\Omega(F)$ can exist that do not belong to the ω- or α-limit sets of $\phi \in C^0(I, M)$.

Example 15. Consider in a vertical torus T^2 the Hamiltonian vector field X of its height function (see [15]); the orbits of X belong to the intersections of T^2 with horizontal planes. This vector field has four critical points (two centers and two saddles). Use Example 1 to construct the RFDE $X \cdot \rho$ on T^2. Any nonconstant solution $x(t)$ of X in the level of a saddle is a solution of $X \cdot \rho$ such that x_t is a nonwandering point and does not belong to the ω- or α-limit set of elements $\phi \in C^0(I, M)$.

Given a critical point ϕ of an RFDE F we get $\Phi_t(\phi) = \phi$ for all $t \geq 0$. The derivative $T_\phi \Phi_t$ of Φ_t at the point ϕ (for fixed $t \geq 0$) is a linear continuous map

$$T_\phi \Phi_t : T_\phi C^0(I, M) \to T_\phi C^0(I, M),$$

$T_\phi C^0(I, M)$ being the tangent space of the manifold $C^0(I, M)$ at the point ϕ. This derivative $T_\phi \Phi_t$ is given by

$$T_\phi \Phi_t(\Psi) = z_t(\Psi),$$

where $\Psi \in T_\phi C^0(I, M)$ and $z_t(\Psi)$ is the flow of $\omega \cdot TF$, the equation of first variation of F (see Example 4). The critical point ϕ is said to be non-degenerate or *transversal* (see [10, 12]) if 1 does not belong to the spectrum $\sigma(T_\phi \Phi_t)$ of the operator $T_\phi \Phi_t$; and ϕ is said to be *elementary* or *hyperbolic* if $\sigma(T_\phi \Phi_t) \cap S^1 \neq \Phi$, where S^1 is the unit circle in the complex plane. In [4] some properties of the compact operator $T_\phi \Phi_r$ are described (see also [9] and [4, Chapter 26]). A hyperbolic critical point has the saddle point property. The solutions defined for all $t \leq 0$ and remaining near ϕ form the finite-dimensional *local unstable manifold*; the solutions remaining near ϕ for all $t \geq 0$ form the finite codimensional *local stable manifold*. The critical point ϕ is the intersection of these transversal smooth manifolds and their tangent spaces split at ϕ.

Since the flow Φ_t is, in general, not one-to-one, it is not clear if we can talk about a global unstable manifold of ϕ. But if Φ_t is one-to-one on the attractor $A(F)$ and $T_\phi \Phi_t$ is also one-to-one when restricted to the tangent vectors of the local unstable manifold of the point ϕ, then the global unstable manifold of ϕ is smooth.

As in the case of critical points, we can take a nonconstant T-periodic solution $x(t)$ of F $(T > 0$ is any period) and if $\phi = x_0$ we get $\Phi_T(\phi) = \phi$; choose T sufficiently large in order to have $T \geq r$, and again the map

$$T_\phi \Phi_T : T_\phi C^0(I, M) \to T_\phi C^0(I, M)$$

is a compact operator [4, p. 196]. The description of the spectrum $\sigma(T_\phi \Phi_T)$ is also given in [4]. It is easy to see that

$$T_\phi \Phi_T(\dot{x}_0) = \dot{x}_0 = [dx_t/dt]_{t=0},$$

so 1 belongs to $\sigma(T_\phi \Phi_T)$. The period T is called a *transversal period* or the periodic solution is called *nondegenerate* [10, 13] if 1 has multiplicity one

(1 is a simple multiplier [2, 3]). If $x(t)$ is nondegenerate and $\sigma(T_\phi \Phi_T) \cap S^1 = \{1\}$, then $x(t)$ is said to be an *elementary* or *hyperbolic* periodic solution. The concepts of transversality and hyperbolicity depend only on the orbit $\Gamma = \{x_t | t \in R\}$ and not on the point ϕ that we considered.

Hale [3] showed that in a neighborhood of the orbit of a hyperbolic periodic solution there are *local stable* and *unstable manifolds* as a kind of saddle property.

The behavior of solutions near a hyperbolic periodic orbit was considered by Stokes [17] in the case where $x(t)$ has all multipliers different from 1 with modulus less than one. In some sense he showed the existence of the local stable manifold (in this case the orbit is the unstable manifold) and described the asymptotic behavior of the solutions near Γ. He also considered the stable manifold for a periodic orbit Γ in the case where there exist multipliers different from 1 in S^1 but with (what he called) simple elementary divisors. Later Hale [3] showed that the hyperbolicity condition implies that the orbit Γ exhibits a saddle structure in the sense that the unstable manifold is locally diffeomorphic either to the unstable manifold in the equation of first variation crossed with a circle or a generalized Möbius band, and the stable manifold is a special union of sets each of which are locally homeomorphic to the asymptotically stable manifold in the equation of first variation. Under the weaker hypothesis that Γ is nondegenerate, Hale also showed that there is a periodic solution of a system that is a perturbation of the one given provided that the perturbation is small.

It is quite evident that we need to extend the techniques for obtaining the qualitative behavior of solutions near other invariant sets. A good definition of hyperbolicity for the nonwandering set $\Omega(F)$ and general theorems analogous to Smale's spectral decomposition theorem for diffeomorphisms [16] will certainly give a good description of the behavior of solutions at infinity.

GENERIC RESULTS FOR RFDE

In the case where M is a compact manifold, let $G_i{}^\kappa$ be the set of all elements in $\mathscr{B}\mathscr{X}^\kappa(I, M)$ such that all critical points are nondegenerate (case $i = 0$) or hyperbolic (case $i = 1$). The first generic result says that $G_i{}^\kappa$ is open and dense in $\mathscr{B}\mathscr{X}^\kappa(I, M)$, $\kappa \geq 1$, $i = 0, 1$ (see [12]).

Let us denote by $G_\Delta{}^\kappa(a)$ the set of all $F \in G_1{}^\kappa$ such that all nonconstant periodic solutions of F with period $T \in (0, a]$ have T as a transversal period; and let $G_2{}^\kappa(a)$ be the set of all $F \in G_1{}^\kappa$ such that all nonconstant periodic solutions of F with period $T \in (0, a]$ are hyperbolic. In [13] it is proved that $G_\Delta{}^\kappa(a)$ and $G_2{}^\kappa(a)$ are open sets of $\mathcal{B}\mathcal{X}^\kappa(I, M)$, $\kappa \geq 1$. The proof of the openness of $G_2{}^\kappa(a)$ was given by the author together with Hale, following the same lines of Peixoto's version of the Kupka–Smale theorem [14]. Peixoto uses a classical Hartman theorem for ordinary differential equations, but this can be avoided in the retarded case. The proof of the openness of $G_\Delta{}^\kappa(a)$ was based on a transversal openness theorem [1, p. 47].

For the case $M = R^n$, Mallet-Paret [10] considers the Banach space $\mathcal{X}$ of C^κ-functions $f \colon C = C^0(I, R^n) \to R^n$ such that all derivatives of f of order at most κ are bounded on C; for technical reasons related to the utilization of the Sard–Smale theorem, he required $\kappa \geq 2$. Calling $\mathcal{G}_2 \subseteq \mathcal{X}$ the set of all $f \in \mathcal{X}$ such that all critical points and all periodic solutions of $\dot{x} = f(x_t)$ are hyperbolic, Mallet-Paret proved that $\mathcal{G}_2$ is a residual subset of $\mathcal{X}$. For doing that he defines four classes of subsets of $\mathcal{X}$, and proves that they are open; for $E \subseteq R^n$ compact, and $A \in (0, \infty)$:

$\mathcal{G}_0(E)$ $\{f \in \mathcal{X} \,|\, \text{all critical points } a \in E \text{ are transversal}\}$;

$\mathcal{G}_1(E)$ $\{f \in \mathcal{X} \,|\, \text{all critical points } a \in E \text{ are hyperbolic}\}$;

$\mathcal{G}_{3/2}(E, A)$ $\{f \in \mathcal{G}_1(E) \,|\, \text{all nonconstant periodic solutions of } f \text{ lying in } E$ and with period $T \in (0, A]$ have T as a transversal period$\}$;

$\mathcal{G}_2(E, A)$ $\{f \in \mathcal{G}_1(E) \,|\, \text{all nonconstant periodic solutions of } f \text{ lying in } E$ and with period $T \in (0, A]$ are hyperbolic$\}$.

With a very interesting and nontrivial proof he showed that $\mathcal{G}_2(E, A)$ is open and dense in $\mathcal{X}$. Then, taking sequences E_m whose union is R^n, and $A_m \to \infty$ as $m \to \infty$, he found that $\mathcal{G}_2 = \bigcap_{m=1}^\infty \mathcal{G}_2(E_m, A_m)$ is residual in $\mathcal{X}$. Mallet-Paret also told the author in a private communication that he can achieve the same result by staying in the subset $\mathcal{D}$ of $\mathcal{X}$ of all differential–delay equations of the type $\dot{x} = g(x(t), x(t - 1))$. In this case

$$\mathcal{D} = \{f \in \mathcal{X} \,|\, f(\phi) = g(\phi(0), \phi(-1)); \, \phi \in C\}.$$

An open question is to prove the same result staying on the set

$$\tilde{\mathcal{D}} = \{f \in \mathcal{X} \,|\, f(\phi) = g(\phi(-1)); \, \phi \in C\}.$$

Call $\mathcal{G}_3$ the set of all $f \in \mathcal{G}_2$ such that the stable and unstable manifolds of all critical points and periodic solutions are transversal. The following conjecture could be an extension of the Kupka–Smale theorem: $\mathcal{G}_3$ is residual.

It seems also reasonable that following [10] one can get $G_2{}^\kappa(a)$ dense in $\mathcal{B}\mathcal{X}^\kappa(I, M)$, for compact M, $\kappa \geq 2$. Is the Kupka–Smale result true for an RFDE on a compact manifold?

Another natural question is to introduce a definition of local equivalence between two RFDEs in the neighborhood of a hyperbolic critical point and try to prove the local equivalence between a given RFDE and its linear part. This corresponds to a kind of Hartman theorem. This and many other questions can be considered in the qualitative and generic theory of RFDE with regard to the nice corresponding theories for ordinary differential equations.

BACKWARD CONTINUATION AND ONE-TO-ONENESS

In [4, Theorem 6.1] it is proved that if the initial condition $\phi \in C^0(I, M)$ of the solution $x(t; 0, \phi)$ of an RFDE F is C^1 in a neighborhood of (0_-), satisfies $\dot{\phi}(0) = F(\phi)$, and F verifies properties (a) and (b) below, then the solution has a small unique backward continuation. To write conditions (a) and (b) we need to describe F locally, say, in a neighborhood U of $\phi \in C^0(I, M)$, F is given by $F(\phi) = (\phi(0), f(\phi))$ and

$$T_\phi f(\Psi) = \int_{-r}^{0} d\eta(\phi, \theta)\Psi(\theta).$$

(a) F is atomic at $(-r)$ at ϕ, which means

$$\det \mathscr{B}(\phi) = \det (\eta(\phi, -r) - \eta(\phi, -r+)) \neq 0.$$

(b) F has smoothness on the measure, which means the existence of a continuous function $\gamma(\phi, s)$, $\phi \in U$, $s \geq 0$, $\gamma(\phi, 0) = 0$, such that

$$\left| \int_{-r}^{-r+s} d\eta(\phi, \theta)\Psi(\theta) - \mathscr{B}(\phi)\Psi(-r) \right| \leq \gamma(\phi, s) \sup |\Psi(\theta)|,$$

$$\theta \in [-r, -r + s].$$

It can be proved that any C^1-RFDE F has smoothness on the measure. The only sufficient condition to be verified for the uniqueness of the backward continuation is that F must be atomic at $-r$ at the given ϕ.

If $F \in \mathscr{B}\mathscr{X}^\kappa(I, M)$ is atomic at $-r$ at all ϕ in some compact invariant subset S of $C^0(I, M)$, the flow Φ_t, $t \geq 0$, restricted to S will be one-to-one and then a homeomorphism of S onto S. It was conjectured in [5] that Φ_t is always one-to-one on ω-limit sets; the conjecture is true for F analytic, because in this case the solutions defined on $(-\infty, +\infty)$ are bounded (M compact), thereby analytic [11].

Example 16.   Consider the scalar equation

$$\dot{x}(t) = f(x_t) = [x(t - r)]^3.$$

Since $x(t - r) = [\dot{x}(t)]^{1/3}$, the flow Φ_t is one-to-one on $C^0(I, R)$, $t \geq 0$, but the function f is not atomic at $-r$ at $\phi = 0$; in fact, $f(\phi) = [\phi(-r)]^3$ and

$$T_\phi f(\Psi) = 3[\phi(-r)]^2 \Psi(-r).$$

This example shows that to be atomic at $-r$ at any ϕ is not a necessary condition for one-to-oneness of the flow.

A very important question is: Under what conditions is the attractor $A(F)$ a smooth manifold? When M is compact and $F \in \mathscr{BX}^\kappa(I, M)$, $\kappa \geq 2$, $A(F)$ as a compact (thereby finite-dimensional) manifold implies that the solutions $x(t)$ of F on $(-\infty, +\infty)$ will be given by a vector field on $A(F)$; in particular, the flow of this vector field will be the restriction of Φ_t to $A(F)$, and thus a diffeomorphism.

Note that if the compact manifold M is analytic and F is also an analytic RFDE on M, all solutions on $A(F)$ must be disjoint and the solution operator Φ_t is a homeomorphism of $A(F)$ onto $A(F)$. The nonwandering set $\Omega(F)$ is invariant and the global unstable manifolds of hyperbolic critical points and hyperbolic nonconstant periodic orbits are analytic finite-dimensional manifolds. If $\Omega(F)$ is the union of a finite number of hyperbolic critical points with a finite number of hyperbolic periodic orbits, the attractor $A(F)$ is the union of all (in finite number) unstable manifolds of critical points and periodic orbits. The globally defined solutions of F are then solutions of a finite number of vector fields. Is $A(F)$ a manifold in this case?

The procedure of approximating a C^κ-RFDE by an analytic one gives some information about the $A(F)$, since in the analytic case the flow is one-to-one. It will be very important to know if, at least generically, $A(F)$ is a manifold. Also conditions on $\Omega(F)$ need to be considered in order to get a kind of continuity for the maps $F \to \Omega(F)$ and $F \to A(F)$.

A Morse–Smale RFDE can be defined as retarded equations F for which the $\Omega(F)$ is equal to the union of a finite number of critical points and periodic orbits, all hyperbolic, and the corresponding stable and unstable manifolds are in a transversal position. Is the set of all Morse–Smale RFDEs an open subset of $\mathscr{BX}^\kappa(I, M)$?

The equation of first variation relative to a solution $x(t)$ of a nonlinear equation $\dot{x} = f(x_t)$ is given by $\dot{y} = Tf(x_t)y_t$ (see Example 4). The one-to-oneness of its flow is needed to define globally invariant manifolds.

III. Nonautonomous Retarded Functional Differential Equations

All the definitions we need to consider nonautonomous retarded functional differential equations of the type $\dot{x} = f(t, x_t)$ can be found in [4]. In order to point out some recent generic results of Hale and the author [8] we will restrict ourselves to the linear case. Let $\mathcal{U}$ be the Banach space of all linear continuous functions $A: C^0(I, R^n) \to R^n$ with the usual norm. *A linear retarded functional differential equation* LRFDE is defined by any function $L: R \to \mathcal{U}$ and the solutions needed to satisfy

$$\dot{x}(t) = L(t)x_t.$$

If $C^1(R, \mathcal{U})$ is the space of C^1-LRFDEs with the C^1-uniform topology on compact sets of R, it is proved in [8] that the set of all LRFDEs such that the solution operator is one-to-one on $C^0(I, R^n)$ is a dense subset of $C^1(R, \mathcal{U})$ and is not an open set. For a fixed-time initial condition, they showed that the set of all $L \in C^1(R, \mathcal{U})$ such that the solution operator is one-to-one, except at a countable number of points with no finite accumulation point, is residual.

Let $\mathcal{D}(R) \subseteq C^1(R, \mathcal{U})$ the set of all L such that there is an integer N such that the corresponding measure has at most N discontinuities in $\theta \in [-r, 0]$ for all $t \in R$ and is a step function in θ. For $K \subseteq R$ compact, define $\mathcal{D}(K)$ in a similar way. Then the set of L in $\mathcal{D}(R)$ for which the solution operator is one-to-one is residual. Also, the corresponding set is open in $\mathcal{D}(K)$.

ACKNOWLEDGMENTS

The author wishes to thank the organizing committee for giving him the opportunity to present this paper at the International Symposium on Dynamical Systems at the Lefschetz Center for Dynamical Systems, Brown University, August, 1974.

REFERENCES

[1] R. Abraham and J. Robbin, "Transversal Mappings and Flows." Benjamin, New York, 1967.
[2] J. K. Hale, Behavior near a periodic orbit of functional differential equations, *Stud. Appl. Math.* **5** (1968), 71–75.

[3] J. K. Hale, Solutions near simple periodic orbits of functional differential equations, *J. Differential Equations* **7** (1970), 126–138.

[4] J. K. Hale, "Functional Differential Equations." Springer-Verlag, Berlin, 1971.

[5] J. K. Hale, Some infinite-dimensional dynamical systems, *in* "Dynamical Systems" (J. K. Hale and J. P. LaSalle, eds.), pp. 129–133. Academic Press, New York, 1973.

[6] J. K. Hale, Behavior near constant solutions of functional differential equations, *J. Differential Equations* **15** (1974), 278–294.

[7] J. K. Hale and M. A. Cruz, Existence, uniqueness and continuous dependence for hereditary systems, *Ann. Mat. Pura Appl. (IV)* **85** (1970), 63–82.

[8] J. K. Hale and W. M. Oliva, One-to-oneness for linear retarded functional differential equations, *J. Differential Equations* (to appear).

[9] J. K. Hale and C. Perello, The neighborhood of a singular point of functional differential equations, *Contrib. Differential Equations* **3** (1964), 351–375.

[10] J. Mallet-Paret. Fixed points and periodic solutions of generic retarded functional differential equations, Thesis, Univ. of Minnesota, 1974.

[11] R. D. Nussbaum, Periodic solutions of analytic functional differential equations are analytic, *Michigan Math. J.* **20** (1973), 249–255.

[12] W. M. Oliva, Functional differential equations on compact manifolds and an approximation theorem, *J. Differential Equations* **5** (1969), 483–496.

[13] W. M. Oliva, Functional differential equations on manifolds, *Atlas Soc. Bras. Mat.* **1** (1971), 103–116.

[14] M. M. Peixoto, On an approximation theorem of Kupka and Smale, *J. Differential Equations* **3** (1967), 214–227.

[15] R. C. Robinson, Global properties of Hamiltonian systems, *Proc. AMS Sum. Inst. Global Anal. Berkeley* (1968).

[16] S. Smale, Differentiable dynamical systems, *Bull. Amer. Math. Soc.* **73** (1967), 747–817.

[17] A. Stokes, On the stability of a limit cycle of an autonomous functional differential equation, *Contrib. Differential Equations* **3** (1964), 121–139.

Chapter 5: TOPOLOGICAL DYNAMICAL SYSTEMS

Stability Theory and Invariance Principles*

JOSEPH P. LASALLE
Lefschetz Center for Dynamical Systems
Division of Applied Mathematics
Brown University, Providence, Rhode Island

1. Introduction

I do not know the origin of the terminology "invariance principle" except that I picked it up from others around 1965. The principle states that, if the positive limit sets of a dynamical system have an invariance property, then Liapunov functions can be used to obtain information on the location of positive limits sets. The application of this principle then gives a generalization of the classical Liapunov theory of stability and instability. The term invariance principle has also been used at times to refer to the results obtained by the application of the principle. (Propositions 1 and 2 in Section 2 are general invariance principles.) For applications it has the advantage of providing more general results, making applications easier, and of enlarging the class of useful Liapunov functions.

This exploitation of the invariance property of the positive limit sets of solutions of autonomous ordinary differential equations to extend and unify Liapunov stability theory can be dated back to 1960 [17, 18]. Barbashin and Krassovski in 1952 gave a partial result, but they overlooked the importance of invariance. Primitive forms of invariance principles can be found in many applications. The basic ideas are simple and natural, and perhaps for this reason invariance principles have proved to be useful both for theory and applications. Applications of invariance principles have now gone beyond autonomous ordinary differential equations and are known for many types of general dynamical systems. It would be rash to say that

---

* This research was supported in part by the Air Force Office of Scientific Research Grant #AF-AFOSR 71-2078C, National Science Foundation Grant #GP-28931X2, Office of Naval Research Grant #NONR N00014-67-A-0191-000906, U.S. Army Grant #DA-ARO-D-31-124-73-G-130.

the theory is complete. In spite of the simplicity of the basic ideas, discovering just how to associate a dynamical system with a particular class of functional equations, integral equations, partial differential equations, or evolutionary equations can be difficult, and it can be nontrivial and nonelementary to obtain a theory that gives direct methods for studying stability and asymptotic behavior.

Among the people who have contributed to these developments are LaSalle, Hale, Yoshizawa, Miller, Infante, Sell, Slemrod, Dafermos, Hurt, Cruz, Chafee, Wakeman, and Walker. There are also specific applications too numerous to mention. (A few can be found in the references).

Originally, I had thought I might survey some of these more recent developments and perhaps discuss some specific applications. This turns out to be too difficult to do in any reasonable fashion within space limitations, and there are many papers and a number of surveys easily accessible.

What I have decided to do is to give a simple and abstract formulation that shows that one need impose, insofar as invariance principles and stability theory are concerned, very little topological structure on the state space. This brings out the generality of the invariance principle and illustrates quite well the simple and elementary nature of the basic ideas. In itself the abstraction is not uninteresting, and stripping the theory to its bare bones makes it possible to see clearly the role of certain assumptions. At the same time the structure is sufficient for discussing the difficulties involved in associating dynamical systems with general types of equations. The difficulties arise in obtaining a theory that can actually be applied to the solution of real problems. In particular, a general formulation is given that shows the relationship between the work of Sell ([38] and Miller and Sell [27, 29–32], and also their joint paper in this volume) and the work of Dafermos [5–8] for associating dynamical systems with nonautonomous ordinary differential equations and integral equations and processes (nonautonomous flows) in general. This is discussed in Section 3.

2. Abstract Dynamical Systems

Here the state space will be a Fréchet space (these are the spaces of class (L) studied by Fréchet in his dissertation around 1906). A set P is a Fréchet space if each sequence of elements $p_n \in P$ either has a limit $p \in P$ associated with it or not. If it has a limit p, we denote this by $p_n \to p$, and we say that the sequence converges. This convergence has the following properties:

(1) $p_n \to p$ and $p_n \to q$ implies $p = q$;
(2) $p_n = p$, $n = 1, 2, \ldots$, implies $p_n \to p$; and
(3) $p_n \to p$ implies $p_{n_i} \to p$ for each subsequence p_{n_i} of p_n.

In this chapter, I shall always mean sequential continuity, sequential compactness, etc., and the adjective "sequential" is omitted throughout. Note that in such a space P the closure of a set may not be closed. However, compact sets are always closed.

Let $\phi: [0, \omega) \to P$ $(0 < \omega \leq \infty)$. Following Birkhoff [2], q is said to be a positive *limit point* of ϕ if there is a sequence $t_n \to \omega^-$ such that $\phi(t_n) \to q$. Throughout we limit ourselves to what happens as $t \to \omega^-$, and from this point omit everywhere the adjective "positive." (For instance, "invariance" will always mean "positive invariance.") The set of all such limit points will be denoted by $\Omega(\phi, \omega)$ and is called the *limit set* of $\phi([0, \omega))$. We say that $[0, \omega)$ is *maximal* if $\omega = \infty$ or $\Omega(\phi, \omega)$ is empty ($[0, \omega)$ is the maximal positive domain of definition of ϕ). Throughout P is a Fréchet space.

Definition 1 [Dynamical Systems of Type (i)].   With each $p \in P$ there is associated an interval $I(p) = [0, \omega(p))$, $0 < \omega(p) \leq \infty$. Let $\chi = \{(t, p); p \in P, t \in I(p)\}$. A map $\pi: \chi \to P$ is called a D_i-system on P if

A$_1$.   Relative to $\pi(\cdot, p)$ each $I(p)$ is maximal.
A$_2$. $\pi(0, p) = p$.
A$_3$. s, $t \in R^+ = [0, \infty)$ and $s + t \in I(p)$ implies $s \in I(\pi(t, p))$ and $\pi(s + t, p) = \pi(s, \pi(t, p))$.
A$_4$. $\pi \in C_i$ [some continuity conditions on $I(p)$ and $\pi(t, p)$].

Systems such as these have been called "local semiflows" or "local dynamical systems." Usually the state space P is given more topological structure (see, for instance, [29] and [1]).

Immediate consequences of A$_3$ are

P$_{1.1}$. $\omega(p) \leq t + \omega(\pi(t, p))$ for all $t \in I(p)$.
P$_{1.2}$. $t \in I(p) \cap I(q)$ and $\pi(t, p) = \pi(t, q)$ implies $\pi(t + s, p) = \pi(t + s, q)$ for all $s \geq 0$ and $t + s \in I(p) \cap I(q)$.

Property 1.2 is the uniqueness of the flow in the positive direction of t.

It is also easy to see that A_3 could be replaced by

A_3*.   Given $p \in P$ and $s \in I(p)$ there is a $q \in P$ such that $t \geq 0$ and $s + t \in I(p)$ implies $t \in I(q)$ and $\pi(t + s, p) = \pi(t, q)$.

A_3* states that the translate of a motion is a motion [of course, $q = \pi(s, p)$ for a D_i-system].

Definition 2. $\pi \in C_1$ if $p_n \to q$ and $t \in I(q)$ imply $t \in I(p_n)$ for all n sufficiently large and $\pi(t, p_n) \to \pi(t, q)$.

Thus $\pi \in C_1$ means lower semicontinuity of $I(p)$ and continuity of $\pi(t, p)$ with respect to p for fixed t.

Definition 3.   (a)   A set M in P is said to be *weakly invariant* (with respect to π) if $\pi(I(p), p) \subset M$ for all $p \in M$.
(b)   A set M is said to be *invariant* if M is weakly invariant and $I(p) = R^+$ for each $p \in M$.
(c)   We say that $\pi(t, p)$ is compact if $\pi(I(p), p)$ is precompact (contained in a compact set of P).

For D_i-systems it is easy to see that $[\Omega(p)$ is the limit set of $\pi(I(p), p)]$:

$P_{1.3}$. $\Omega(p)$ is weakly invariant.
$P_{1.4}$.   If $\pi(t, p)$ is compact, then $\Omega(p)$ is nonempty, precompact, and invariant.

Even though $\pi(t, p)$ is compact, it may be that $\Omega(p)$ is not closed, and $\Omega(p)$ is closed if and only if it is compact. Also, it is not difficult to construct a D_1-system, where $\omega(p) < t + \omega \, (\pi(t, p))$ for some $t \in I(p)$.

We shall now introduce a general concept of a Liapunov function for a D_i-system π and show for D_1-systems that Liapunov functions give information about the location of limit sets. (See Proposition 1.) This establishes the invariance principle in this general and abstract setting. How well the Liapunov function locates limit sets depends on the Liapunov function itself, and the difficulty in applications is to find "good" Liapunov functions. For instance, a constant function is a Liapunov function but provides no useful information.

Let π be a D_i-system on P. Throughout, G denotes an arbitrary subset of P and V is a real-valued function on P. We say that V is a *Liapunov function for* π *on* G if (i) V is continuous on $\overline{G}$, and (ii) $\pi([0, \beta], p) \subset G$ implies $V(\pi(t, p))$ is nonincreasing with respect to t on $[0, \beta)$. Let

$$V_c = \{p;\ V(p) = c,\ p \in \overline{G}\} = V^{-1}(c) \cap \overline{G},$$

and define M_1 to be the union of all $q \in \overline{G}$ with the property that (i) $I(q) = R^+$, and (ii) $\pi(R^+, q) \subset V_c$ for some $c = c(q)$.

Proposition 1. Let π be a D_1-system and let V be a Liapunov function for π on G. If $\pi(t, p)$ is compact and $\pi(R^+, p) \subset G$, then $\Omega(p) \subset M_1 \cap V_c$ for some $c = c(p)$.

Proof. The assumptions on $\pi(t, p)$ and V imply that $V(\pi(t, p))$ is nonincreasing and bounded from below on R^+. Hence, $V(\pi(t, p)) \to c$ as $t \to \infty$, and $\Omega(p) \subset V_c$. It then follows from $P_{1.4}$ (and this is the simple observation behind the invariance principle) that $q \in \Omega(p)$ implies $I(q) = R^+$ and $\pi(R^+, q) \subset V_c$. Therefore, $\Omega(p) \subset M_1$. This completes the proof.

It is usually not pointed out explicitly that, in addition to $\Omega(p)$ being in M_1, it also lies on a $V = \text{const}$ surface. Although an obvious fact, it is worth noting. For example, if it is known that $M_1 \cap V_c$ is either empty or a single point for each c, then M_1 is the set of all equilibrium points in $\overline{G}$ and each compact solution that remains in G approaches an equilibrium point as $t \to \infty$. If R^+ is replaced by the nonnegative integers, then the result is an invariance principle for discrete dynamical systems (difference quations; see [16]).

We now consider a slightly stronger continuity property for π and see what changes this makes in the above results.

Definition 4. $\pi \in C_2$ if (i) $\pi \in C_1$ and (ii) $t_n \to t \in I(p)$ implies $\pi(t_n, p) \to \pi(t, p)$.

It then follows easily for a D_2-system π that

$P_{2.1}$. $\omega(p) = t + \omega(\pi(t, p))$, $t \in I(p)$.

$P_{2.2}$. $t \in I(p) \cap I(q)$ and $\pi(t, p) = \pi(t, q)$ implies $I(p) = I(q)$ and $\pi(t + s, p) = \pi(t + s, q)$ for all $s \geq 0$ and $t + s \in I(p)$.

For a D_2-system π, we can come closer to the direct methods of Liapunov theory. Let V be a real-valued function on P and define

$$\dot{V}(p) = \lim_{t \to 0+} \inf \frac{V(\pi(t, p)) - V(p)}{t}.$$

If $V(\pi(t, p))$ is nonincreasing on $[0, \beta) \subset I(p)$, then

$$\dot{V}(\pi(t, p)) = D_+ V(\pi(t, p)) \leq 0$$

on $[0, \beta)$. Conversely, if π is a D_2-system and V is lower semicontinuous on P, then $\dot{V}(\pi(t, p)) \leq 0$ on $[0, \beta)$ implies $V(\pi(t, p))$ is nonincreasing on $I(p)$. We have therefore.

Lemma 1. Let π be a D_2-system on P. Then V is a Liapunov function of π on G if and only if V is continuous on $\overline{G}$ and $\dot{V}(p) \leq 0$ for all $p \in G$. If V is a Liapunov function of π on G, then

$$M_1 \subset M,$$

where M is the largest invariant set in $E = \{q; \dot{V}(q) = 0, q \in \overline{G}\}$ [that is, M is the union of all $\pi(I(p), p)$ for which $I(p) = R^+$ and $\pi(R^+, p) \subset E$].

As a consequence of this lemma and Proposition 1, we have

Proposition 2. Let π be a D_2-system and let V be a Liapunov function of π on G. If $\pi(t, p)$ is compact and remains in G for all $t \geq 0$, then

$$\Omega(p) \subset M \cap V_c \qquad \text{for some} \quad c = c(p).$$

If $V(\pi(t, q))$ is absolutely continuous with respect to t for each $q \in E$, then, of course, $M = M_1$. In any case in the application of this type of result, it is the set M that can be located by direct methods, and we see in this abstract setting the role of continuity properties of $\pi(t, p)$ with respect to t. In applications, the space P will be a metric space and the motions $\pi(t, p)$ correspond to solutions of some sort of evolutionary equation. For ordinary differential equations, the space P is n-dimensional Euclidean space, but more generally P is a Banach space and is not locally compact. This raises a practical difficulty. To use Proposition 2 requires being able to verify the compactness of solutions, and this may not be possible or obvious by direct methods. Getting around this difficulty is the major problem to be overcome in developing a practical stability theory. This is well illustrated by the examples discussed in [14].

Let us note one more difference between D_1- and D_2-systems. Another definition used for the positive limit set is

$$\Gamma(p) = \bigcap_{\beta > 0} \overline{\pi([\beta, 0), p)}.$$

It is clear that Birkhoff's $\Omega(p)$ is always contained in $\Gamma(p)$, and for D_1-systems $\Omega(p)$ can be smaller than $\Gamma(p)$. However, for D_2-systems we have

$P_{2.3}$. If π is a D_2-system, then $\Omega(p) = \Gamma(p)$.

3. Associating Dynamical Systems with Nonautonomous Flows (Processes)

Following a technique of Miller [27], Sell [38] showed how to associate in a significant manner a dynamical system with a nonautonomous system of ordinary differential equations. (Today this associated dynamical system is being called a "skew-product" dynamical system.) Then from the invariance property of the positive limit sets of the dynamical system (see $P_{1.3}$ and $P_{1.4}$), Sell obtained all the invariance properties known at that time for the limit sets of the solutions of nonautonomous systems (see, for example, [19, 24, 25, 27, 33, 49]). This is a very fruitful idea. Dafermos obtained similar invariance results in his study [5] of compact processes. In [8] Dafermos relates his ideas to those of Miller and Sell. Here I shall place these ideas of Miller, Sell, and Dafermos in an abstract setting, also borrowing the terminology they have introduced.

Definition 5. Let X be a Fréchet space. With each $(t_0, x) \in R \times X$ there is associated an interval $I(t_0, x) = [0, \omega(t_0, x))$, $0 < \omega(t_0, x) \leq \infty$. Let

$$\chi = \{(t, t_0, x); (t_0, x) \in R \times X, t \in I(t_0, x)\}.$$

A map $u: \chi \to X$ is called a *process* if

(1) relative to u each $I(t_0, x)$ is maximal,
(2) $u(0, t_0, x) = x$,
(3) $s, t \in R^+$ and $s + t \in I(t_0, x)$ implies

$s \in I(t_0 + t, u(t, t_0, x))$ and $u(s + t, t_0, x) = u(s, t_0 + t, u(t, t_0, x))$.

(4) [some continuity conditions on $I(t_0, x)$ and u].

For a nonautonomous differential equation, $\dot{x} = f(t, x)$ with $\phi(t, t_0, x^0)$ the solution (for $t \geq t_0$) satisfying $\phi(t_0, t_0, x^0) = x^0$. Note that u corresponds to

$$u(t, t_0, x^0) = \phi(t + t_0, t_0, x^0).$$

Although it is important to consider specific continuity conditions (4), I wish here to discuss only general ideas behind associating a dynamical system with a process and the difficulties behind doing this insofar as stability theory is concerned. Also, for simplicity, I will assume that $I(t_0, x) = R^+$ for all $(t_0, x) \in R \times X$ $(\chi = R^+ \times R \times X)$. The changes necessary for a local theory are not difficult to carry out.

Let W denote the set of all mappings of $R^+ \times R \times X \to X$ and let U be the subset of all processes in W. The *translate* u_τ of a process u is defined by $u_\tau(t, t_0, x) = u(t, t_0 + \tau, x)$.

Define $\alpha: R^+ \times X \times U \to X$ by

$$\alpha(t, x, u) = u(t, 0, x).$$

It then follows from (3) that

$$\alpha(s + t, x, u) = \alpha(t, \alpha(s, x, u), u_s) \qquad \text{for all} \quad s, t \geq 0.$$

Then with $p = (x, u) \in X \times U = P$ define

$$\pi(t, x, u) = (\alpha(t, x, u), u_t).$$

Clearly π satisfies A_1–A_3 of Definition 1 for a dynamical system. The remaining and difficult question (if π is to be at least a D_1-system) is the continuity property A_4. It is too much to expect that π will be a dynamical system on $P = X \times U$ but we can hope that π will be a dynamical system on $P^* = X \times U^*$, where U^* is a subset of U. Suppose that W is a Fréchet space and that convergence in X and the continuity property (4) for processes have been fixed. We then want to have the invariance property of compact motions $\pi(t, p)$ induce an invariance property for compact motions $\alpha(t, x, u) = u(t, 0, x)$.

This means that U^* must be translation invariant ($u \in U^*$ implies $u_t \in U^*$) and, in fact, that $u \in U^*$ implies $H(u) = \overline{u_{R^+}} \subset U^*$ [$H(u)$ is called the *hull* of u]. These are Sell's *regular processes* [38]. Moreover, for induced invariance the motion u_t in W must be compact in order that compactness of $u(t, 0, x)$ in X correspond to compactness of $\pi(t, p)$ in P. These are Dafermos's *compact processes* [5, 8].

Let $\Gamma(x, u) = \Gamma(p) \subset X$ be the limit set of $u(t, 0, x)$, and let $\Omega(p) \subset P$ be the limit set of $\pi(t, p)$. $\Omega(p) = \Gamma(p) \times H_\infty(u)$, where $H_\infty(u)$ is the limit set of u_t in W [$H_\infty(u)$ is called the asymptotic *hull* of u]. We then have

Proposition 3 (**Induced Invariance**). If u is a compact regular process and $u(t, 0, x)$ is compact in X, then $\Gamma(x, u)$ is invariant in the sense that, if $y \in \Gamma(x, u)$, there is a $v \in H_\infty(u)$ such that $v(R^+, 0, y) \subset \Gamma(x, u)$.

In the processes considered by Dafermos, his convergence is such that every process is regular but this will not, in general, be true. Proposition 3 is simply a statement of a general method for obtaining invariance properties of nonautonomous flows defined by processes. The processes themselves may be defined by the solutions of nonautonomous evolutionary equations. The difficulty is to find a "good" convergence for processes. A "best" convergence, if one exists, maximizes the class of regular compact processes. Weakening the convergence may make the class of regular processes larger but may decrease the class of compact processes.

In the case of ordinary differential equations $\dot{x} = f(t, x)$, Sell [38] took the compact open topology for the functions f. He then showed that the processes defined by f are regular if f satisfies a local Lipschitz condition with Lipschitz constant independent of t. However, then f is compact if and only if f is bounded and uniformly continuous on each $R^+ \times K$, K a compact set of R^n. This covered all the invariance principles for nonautonomous ordinary differential equations known at that time. In 1972, Peng [36], motivated by the practical problem of controlling a system with unknown but bounded parameters, obtained directly an invariance property for his systems, which is not included in the above-mentioned result of Sell. In 1974, Wakeman [44] used a weaker topology for the functions f [the compact open topology for $F(x, t) = \int_0^t f(s, x)\, ds$] and obtained an invariance property for the limit sets of bounded solutions for a much wider class of f. His class of compact regular f is

(1) f satisfies a Carathéodory condition, and
(2) for each compact K in R^n there is an l such that
    (a) $|f(t, x)| \leq l$,
    (b) $|f(t, x) - f(t, y)| \leq l(x - y)$,
for all (t, x) in $R^+ \times K$.

By the invariance principle this gives a stability theory analogous to that for autonomous ordinary differential equations, and this should be a much more powerful tool for the stability analysis of nonautonomous systems than the corresponding classical Liapunov theory, which rules out many natural Liapunov functions. A possible type of application can be found in [11], where the problem is to control a space station in an unstable periodic orbit. One way of doing this is to determine feedback control to stabilize the motion (make the orbit asymptotically stable). Fearnsides and Levine [11], using the same notation as in Section 2, selected instead a sufficiently large G of initial conditions and a set E (a sufficiently small set about the periodic orbit). Then using a Liapunov function V they determined controls so that G is positively invariant and E corresponds to $V = 0$ in $\overline{G}$. This then ensures that motions starting in G approach E (in fact, by the invariance property for nonautonomous systems and Proposition 2 they could have concluded that the solutions approach the smaller set M). Digital computer simulation showed that the control (thrusters) kept the actual motion close to the desired motion with a 50% reduction in fuel and the number of thruster firings compared with making the orbit itself asymptotically stable.

One difficulty in connection with Wakeman's results is that he obtains sufficient conditions for asymptotic stability (with the stability being uniform) but not for uniform asymptotic stability. In applications one wants to assure stability under perturbations, and this is the reason for wanting uniform asymptotic stability. Of course, another question here is, can Wakeman's results be improved and is there in any sense a "best" topology (or convergence) for the functions f?

REFERENCES

[1] Bhatia, N. P., and O. Hajek, "Local Semi-Dynamical Systems." Springer-Verlag, Berlin, 1969.
[2] Birkhoff, G. D., "Dynamical Systems," Amer. Math. Soc. Colloq. Publ. Vol. 9. Amer. Math. Soc., Providence, Rhode Island, 1927.
[3] Chaffee, N., and Infante, E. F., A bifurcation problem for a nonlinear partial differential equation of parabolic type, *CDS Tech. Rep.* **74-5**, Brown Univ. (March 1974).
[4] Cruz, M. A., and Hale, J. K., Stability of functional differential equations of neutral type, *J. Differential Equations* **7** (1970), 334–355.
[5] Dafermos, C. M., An invariance principle for compact processes, *J. Differential Equations* **9** (1971), 239–252.
[6] Dafermos, C. M., Applications of the invariance principle for compact processes. I. Asymptotically dynamical systems, *J. Differential Equations* **9** (1971), 291–299; II. Asymptotic behavior of solutions of a hyperbolic conservation law, *ibid.* **11** (1972), 416–424.
[7] Dafermos, C. M., Uniform processes and semi-continuous Liapunov functionals, *J. Differential Equations* **11** (1972), 401–405.
[8] Dafermos, C. M., Semiflows generated by compact and uniform processes, *Math. Systems Theory* **8** (1975), 142–149.
[9] Dafermos, C. M., and Slemrod, M., Asymptotic behavior of nonlinear contraction semigroups, *J. Functional Analysis* **13** (1973), 97–106.
[10] DiPasquantonio, F., Stability in the first approximation and a critical case relating to nuclear reactor kinetics equations, *Nukleonik* **11** (1968), 276–282.
[11] Fearnsides, J. J., and Levine, W. S., On the determination of the asymptotic behavior of an inertially oriented space station, *IEEE Trans. Automatic Control* **AC-19** (1974), 186–191.

[12] Hale, J. K., A stability theorem for functional differential equations, *Proc. Nat. Acad. Sci. U.S.A.* **50** (1963), 942–946.

[13] Hale, J. K., Sufficient conditions for stability and instability of autonomous functional differential equations, *J. Differential Equations* **1** (1965), 452–482.

[14] Hale, J. K., Dynamical systems and stability, *J. Math. Anal. Appl.* **26** (1969), 39–59.

[15] Hale, J. K., and Infante, E. F., Extended dynamical systems and stability theory, *Proc. Nat. Acad. Sci. U.S.A.* **58** (1967), 405–409.

[16] Hurt, J., Some stability theorems for ordinary difference equations, *SIAM J. Numer. Anal.* **4** (1967), 582–596.

[17] LaSalle, J. P., The extent of asymptotic stability, *Proc. Nat. Acad. Sci. U.S.A.* **46** (1960), 363–365.

[18] LaSalle, J. P., Some extensions of Liapunov's second method, *IRE Trans. Circuit Theory* **CT-7** (1960), 520–527.

[19] LaSalle, J. P., Asymptotic stability criteria, *in Proc. Symp. Appl. Math., Hydrodynamic Instability* **13**, 299–307. Amer. Math. Soc., Providence, Rhode Island (1962).

[20] LaSalle, J. P., Liapunov's second method, stability problems of solutions of differential equations, *Proc. NATO Advan. Study Inst., Padua, Italy* pp. 95–106. Edizioni "Oderisi," Gubbio (1966).

[21] LaSalle, J. P., An invariance principle in the theory of stability, *in* Differential Equations and Dynamical Systems, *Proc. Int. Symp., Puerto Rico* pp. 277–286. Academic Press, New York (1967).

[22] LaSalle, J. P., Stability theory for ordinary differential equations, *J. Differential Equations* **4** (1968), 57–65.

[23] LaSalle, J., and Lefschetz, S., "Stability by Liapunov's Direct Method with Applications." Academic Press, New York, 1961.

[24] Markus, L., Asymptotically autonomous differential systems, *in* "Contributions to the Theory of Nonlinear Oscillations," Vol. 3, pp. 17–29. Princeton Univ. Press, Princeton, New Jersey, 1956.

[25] Miller, R. K., On almost periodic differential equations, *Bull. Amer. Math. Soc.* **70** (1964), 792–795.

[26] Miller, R. K., Asymptotic behavior of nonlinear delay–differential equations, *J. Differential Equations* **1** (1965), 293–305.

[27] Miller, R. K., Almost periodic differential equations as dynamical systems with applications to the existence of a. p. solutions, *J. Differential Equations* **1** (1965), 337–345.

[28] Miller, R. K., Asymptotic behavior of solutions of nonlinear differential equations, *Trans. Amer. Math. Soc.* **115** (1965), 400–416.

[29] Miller, R. K., The topological dynamics of Volterra integral equations, *Stud. Appl. Math.* **5** (1969), 82–87.

[30] Miller, R. K., and Sell, G. R., A note on Volterra integral equations and topological dynamics, *Bull. Amer. Math. Soc.* **74** (1968), 904–908.

[31] Miller, R. K., and Sell, G. R., Existence, uniqueness and continuity of solutions of integral equations, *Ann. Mat.* **80** (1968), 135–152.

[32] Miller, R. K., and Sell, G. R., Volterra integral equations and topological dynamics, *Mem. Amer. Math. Soc.* **102** (1970).

[33] Opial, Z., Sur la dépendance des solutions d'un système d'équations différentielles de leurs seconds membres. Application aux systèmes presque autonomes, *Ann. Polon. Math.* **8** (1960), 75–89.

[34] Parks, P. C., A stability criterion for a panel flutter problem via the second method of Liapunov, *in* Differential Equations and Dynamical Systems, *Proc. Int. Symp., Puerto Rico* pp. 287–298. Academic Press, New York, 1967.

[35] Pazy, A., On the applicability of Liapunov's theorem in Hilbert space, *SIAM J. Math. Anal.* **3** (1972), 291–294.

[36] Peng, T. K. L., Invariance and stability for bounded uncertain systems, *SIAM J. Control* **10** (1972), 679–690.

[37] Plaut, R. H., Asymptotic stability and instability criteria for some elastic systems by Liapunov's direct method, *Quart. Appl. Math.* (1972), 535–540.

[38] Sell, G. R., Nonautonomous differential equations and topological dynamics, *Trans. Amer. Math. Soc.* **127** (1967), 241–283.

[39] Slemrod, M., Asymptotic behavior of a class of abstract dynamical systems, *J. Differential Equations* **7** (1970), 584–600.

[40] Slemrod, M., Asymptotic behavior of periodic dynamical systems on Banach spaces, *Ann. Mat. Pura Appl.* **86** (1970), 325–330.

[42] Slemrod, M., and Infante, E. F., An invariance principle for dynamical systems on a Banach space; application to the general problem of thermoelastic stability, *in* "Instability of Continuous Systems," pp. 215–221. Springer-Verlag, Berlin, 1971.

[43] Slemrod, M., and Infante, E. F., Asymptotic stability criteria for linear systems difference-differential equations of neutral type and their discrete analogues, *J. Math. Anal. Appl.* **38** (1972), 339–415.

[44] Wakeman, D. R., An application of topological dynamics to obtain a new invariance property for nonautonomous ordinary differential equations, Ph.D. dissertation, Brown Univ., June 1973 *J. Differential Equations* **17** (1975), 259–295.

[45] Walker, J. A., Liapunov analysis of the generalizer Pflueger problem, *ASME J. Appl. Mech.* **39** (1972), 935–938.

[46] Walker, J. A., Energy-like Liapunov functionals for linear elastic systems in a Hilbert space, *Quart. Appl. Math.* **30** (1973), 465–480.

[47] Walker, J. A., On the application of Liapunov's direct method to linear dynamical systems, *J. Math. Anal. Appl.* (to appear).

[48] Walker, J. A., and E. F. Infante, Some results on the precompactness of orbits of dynamical systems, *CDS Tech. Rep.* **74-2**, Brown Univ. (February 1974) (to appear in *J. Math. Anal. Appl.*).

[49] Yoshizawa, T., Asymptotic behavior of solutions of a system of differential equations, *Contrib. Differential Equations* **1** (1963), 371–387.

[50] Yoshizawa, T., "Stability Theory by Liapunov's Second Method," Publ. No. 9, Math. Soc. of Japan, Tokyo, 1966.

Topological Dynamics and Its Relation to Integral Equations and Nonautonomous Systems

RICHARD K. MILLER*
Mathematics Department
Iowa State University, Ames, Iowa

GEORGE R. SELL*
School of Mathematics
University of Minnesota, Minneapolis, Minnesota

This paper is dedicated to the memory of Professor Solomon Lefschetz.

I. Introduction

Our main objective in this chapter is to investigate the role of the theory of topological dynamics in the study of nonautonomous ordinary differential equations and Volterra integral equations. The application of topological dynamics to autonomous differential equations is a classical theory with its origin in the works of Henri Poincaré and George Birkhoff. However, the applications to nonautonomous processes and integral equations are rather recent with most of the developments occurring within the last decade.

In this chapter we hope to cover the main developments in this area within the last five years. Other contributions to this symposium cover such important topics as differentiable dynamics, dynamical systems in a Banach space, and the theory of the dynamics of functional differential equations. So we shall not touch on these related topics here. For references to other results prior to 1970, see [36, 62, 64, 66].

The main problem in studying nonautonomous systems of ordinary differential equations is the embedding of the time parameter t in a suitable space. The well-known artifice of replacing the nth-order nonautonomous system $x' = f(t, x)$ with an $(n + 1)$st-order autonomous system by setting $x_{n+1} = t$, or $x'_{n+1} = 1$, is of little use in studying stability questions or other dynamical properties. For periodic equations the circle S^1 is a natural space for embedding time. However, for aperiodic equations one must seek a

* Supported in part by the National Science Foundation under grants No. GP-311 84X (RKM) and GP-38955 (GRS).

different space for the embedding. The coefficient f contains all of the essential information about time dependence. Moreover, time translation of f determines a flow that is compatible with translation of solutions. Thus time translates of f determine a natural embedding space. This observation is certainly simple but has profound consequences. We shall examine this in more detail in Sections II and III.

In Section IV we shall look at the corresponding dynamical structure of integral equations. Section V will be devoted to the study of the asymptotic properties of solutions and the limiting equations. The LaSalle invariance principle is very useful in this context.

Section VI will be devoted to an analysis of the theories of the existence of periodic and almost periodic solutions. In Section VII we will study the role of generic theories in the qualitative study of ordinary differential equations.

The last three sections are concerned primarily with linear equations. Section VIII covers the theory of exponential dichotomies and invariant splittings, and Section IX deals with the structure of the Liapunov-type numbers and possible extensions of the Floquet theory to aperiodic linear differential equations. Finally, in Section X we shall present two points of view that illustrate how the solutions of linear Volterra integrodifferential equations can be used to construct one-parameter semigroups on a suitable Banach space.

II. Local Dynamical Systems

Our starting point will be the notion of a local dynamical system on a Hausdorff space X.

Definition 1. Given any point p in X, let $I(p) = (\alpha_p, \omega_p)$ be a given interval with $\alpha_p < 0 < \omega_p$. Let

$$S = \{(t, p) \in R \times X : t \in I(p)\}.$$

A function $\pi: S \to X$ is a *local dynamical system* or *flow* on X if the following hold:

   (i) $\pi(0, p) = p$ for all p in X.
   (ii) If $t \in I(p)$ and $s \in I(\pi(t, p))$ then $t + s \in I(p)$ and $\pi(s, \pi(t, p)) = \pi(t + s, p)$.

(iii)   Each $I(p)$ is maximal in the sense that either $\omega_p = +\infty$ or the closure of the semitrajectory

$$\gamma^+(p) = \{\pi(t, p) : 0 \le t < \omega_p\}$$

is not compact; and similarly either $\alpha_p = -\infty$ or the closure of the semi-trajectory

$$\gamma^-(p) = \{\pi(t, p) : \alpha_p < t \le 0\}$$

is not compact.

(iv) $\pi : S \to X$ is continuous. Moreover, if $p_n \to p$ in X, then $\lim \inf I(p_n) \subset I(p)$ $(n \to \infty)$. (We shall occasionally use the notions of generalized sequences or nets to describe various topological properties of dynamical systems; cf. [63] for more details.)

This definition of a local dynamical system reduces to the usual definition of a *dynamical system* (or *flow*) whenever $I(p) = (-\infty, \infty)$ for all p in X. In this case (iii) and the last part of (iv) are automatically true. If we restrict the intervals $I(p)$ to have the form $I(p) = [0, \omega_p)$, then with the definition of S and with (iii) appropriately modified, π defines a *local semiflow* on X. If $I(p)$ is always $[0, \infty)$, π is a semiflow. For more details see [31, 36, 62, 63].

We shall note that the definition of a local dynamical system $\pi : S \to X$ implies that the domain S is open in $R \times X$. In fact, an equivalent definition can be formulated in terms of an open domain for S (cf. [69], for example). Also we should note that the use of the term "local" in the definition of a local dynamical system is a miscasting. The mapping π is not defined only locally. One can show that the domain S is maximal in the sense that π has no proper extension in $R \times X$ to a mapping that is also a local dynamical system.

We shall refer to the function of t, $\pi(t, p)$ as the *motion through p*. We shall call the point set

$$\gamma(p) = \{\pi(t, p) : t \in I(p)\}$$

the *trajectory through p* and $H(p) = \text{Cl } \gamma(p)$ the *hull of p* (where Cl denotes the closure operation). Also $H^+(p) = \text{Cl } \gamma^+(p)$ and $H^-(p) = \text{Cl } \gamma^-(p)$ will denote the *positive and negative hulls* of p. The α- and ω-*limit sets* are then defined by

$$A(p) = \bigcap_{t \in R} H^-(\pi(t, p)), \qquad \Omega(p) = \bigcap_{t \in R} H^+(\pi(t, p)).$$

A set $M \subseteq X$ is *invariant* if for all $p \in M$ one has $I(p) = (-\infty, \infty)$ and $\gamma(p) \subseteq M$. A set $M \subseteq X$ is said to be a *minimal set* if M is nonempty, closed, and invariant, and contains no proper subset with these three properties. M is said to be an *a.p. minimal set* if M is minimal and the motions in M are almost periodic (cf. [63] for more details). A motion $\pi(t, p)$ is said to be *positively compact* if $\gamma^+(p)$ is precompact, i.e., if $H^+(p)$ is compact. The following lemmas are easily established.

Lemma 2. For each $p \in X$ with $\omega_p = +\infty$, the ω-limit set $\Omega(p)$ is closed and invariant. If the motion $\pi(t, p)$ is positively compact, then $\omega_p = +\infty$ and $\Omega(p)$ is nonempty, compact, and connected.

Lemma 3. If $\pi(t, p)$ is positively compact and if $\pi(t, p)$ approaches a set E as $t \to +\infty$, then $\pi(t, p) \to M$ as $t \to +\infty$, where M is the largest invariant subset of Cl E.

Let $G \subseteq X$ and let $V : \mathrm{Cl}\, G \to R$ be a continuous mapping. Following LaSalle [14] we say that V is a *Liapunov function* on G (with respect to a local dynamical system π) if $V(\pi(t, p))$ is nonincreasing in t whenever $\gamma^+(p) \subset G$. The following result is an abstract formulation of the LaSalle invariance principle:

Theorem 4. Let V be a Liapunov function on G with respect to π. Let $\pi(t, p)$ be a positively compact motion and assume that $\gamma^+(p) \subseteq G$. Then $\pi(t, p) \to M$ as $t \to \infty$, where M is the union of all trajectories $\gamma(q)$ with the property that $I(q) = (-\infty, \infty)$, $\gamma(q) \subset G$, and $V(\pi(t, q)) \equiv V(q)$ for all $t \in R$.

This theorem together with the construction described in the next section can be combined to obtain results about the limiting behavior of solutions of nonautonomous differential equations, as in the chapter by LaSalle.

There do exist many variations of the concept of a flow. These occur when the real line R is replaced by a topological group T. In this case, the flow is a continuous mapping $\pi : T \times X \to X$ satisfying $\pi(0, x) = x$ and $\pi(s, \pi(t, x)) = \pi(s + t, x)$. Applications where $T = Z$, the integers, are quite common. A similar extension of a semiflow is possible where T is now a topological semigroup.

One very interesting application of this more general viewpoint appears in Dafermos [5, 6], where he studies the partial differential equation

$$v_t + f(v)_x = 0.$$

Without going into detail, let us note that Dafermos introduces the concept of a *process*, which can be viewed as a semiflow

$$\pi: R \times X \times R^+ \to R \times X,$$

where R and $R^+ = [0, \infty)$ have the discrete topology and X is a complete metric space.

III. Nonautonomous Ordinary Differential Equations

Consider the family of nonautonomous differential equations

$$x' = f(t, x), \tag{DE1}$$

where f belongs to a function space $\mathfrak{F}$ of functions $g: R \times W \to R^n$ (where W is an open set in R^n). We shall assume that $\mathfrak{F}$ is a Hausdorff space that
   (i)   *is translation invariant* (i.e., if $f \in \mathfrak{F}$, then $f_\tau \in \mathfrak{F}$ for all $\tau \in R$, where $f_\tau(t, x) = f(\tau + t, x)$), and
   (ii)   *satisfies the Carathéodory property* (i.e., for each $f \in \mathfrak{F}$ and $x_0 \in W$ there is a unique noncontinuable solution $\varphi(t, x_0, f)$ of

$$x' = f(t, x), \qquad x(0) = x_0,$$

defined on an interval $I(x_0, f) = (\alpha(x_0, f), w(x_0, f))$.

The following theorem, which is basic for our theory, has been discussed in a number of papers (cf. [36, 40, 62, 63]).

Theorem 5. Let $\mathfrak{F}$ be translation invariant and satisfy the Carathéodory property. Then the mapping

$$\pi(t, x, f) = (\varphi(t, x, f), f_t) \tag{1}$$

defined on $\{(t, x, f) : t \in I(x, f)\}$ defines a local dynamical system on $W \times \mathfrak{F}$ if and only if
   (iii)   the mapping $(\tau, f) \to f_\tau$ is continuous, and
   (iv) π is properly continuous, i.e., the mapping $(t, x, f) \to \varphi(t, x, f)$ is continuous and $I(x, f) \subseteq \liminf I(x_n, f_n)$ whenever $(x_n, f_n) \to (x, f)$ in $W \times \mathfrak{F}$.

Remarks. Several extensions of Theorem 5 are possible. First one does not need W to be open in R^n but rather only that the solution of the initial value problem $x' = f(t, x), x(0) = x_0$, remains in W. For example, W could be a manifold (cf. [26]). Also, it is possible to eliminate the uniqueness of solutions of (DE1), but in this case the formulation of the flow π is somewhat different than that given by (1) (cf. [66] for details). The above formulation of the flow π is an example of a skew-product flow (cf. [53, 57]).

Since $\mathfrak{F}$ is translation invariant, the continuity property (iii) means that the mapping $\sigma(\tau, f) = f_\tau$ defines a flow on $\mathfrak{F}$. Therefore, if $f \in \mathfrak{F}$, the trajectory $\gamma(f)$, hull $H(f)$, and ω-limit set $\Omega(f)$ are as defined in Section II. It should be emphasized here that the sets $H(f)$ and $\Omega(f)$ do depend on the space $\mathfrak{F}$. If one changes $\mathfrak{F}$ (makes it larger or smaller), then the sets $H(f)$ and $\Omega(f)$ may change, too. The trajectory $\gamma(f)$ is, of course, determined by f alone and not the space $\mathfrak{F}$.

In order to apply Theorem 5, and the corresponding theory of topological dynamics, one must determine a function space $\mathfrak{F}$ satisfying conditions (i)–(iv). Often one is given a specific equation

$$x' = f(t, x).$$

In this case the trajectory $\gamma(f) = \{f_t : t \in R\}$ must lie in $\mathfrak{F}$ since $\mathfrak{F}$ is to be translation invariant. The Carathéodory property (ii) is, of course, the conclusion of the basic existence and uniqueness theorems. Next the topology on $\mathfrak{F}$ must be chosen so that (iii) and (iv) hold. In practice the continuity of translations (iii) is not very restrictive. Property (iv) is the hardest to satisfy, but it too is a standard problem arising in the theory of ordinary differential equations. Some recent research of Z. Artstein and D. R. Wakeman has given new insight into continuity property (iv).

For $t \neq 0$, let I_t denote the closed interval in R with endpoints t and 0, and let $C_t = C(I_t, R^n)$ be the Banach space of continuous functions from I_t to R^n with sup-norm. The following result is basically due to Artstein [2]:

Theorem 6.   Let $\mathfrak{F}$ be translation invariant and satisfy the Carathéodory property. Then continuity property (iv) is satisfied if for every $t \neq 0$ the mapping

$$G(\varphi, f) = \int_0^t f(s, \varphi(s))\, ds \tag{2}$$

is a continuous mapping from $C_t \times \mathfrak{F}$ to R^n.

For example, if $\mathfrak{F}$ is a translation invariant subset of $C_L(R \times W, R^n)$, the space of continuous functions from $R \times W$ to R^n that are locally Lipschitz in x, and $C_L(R \times W, R)$ has the topology of uniform convergence on compact subsets of $R \times W$, then with the help of Theorems 5 and 6 it is easy to see that (1) defines a local dynamical system on $W \times \mathfrak{F}$. This is the first example studied by Miller [28] and Sell [62].

Wakeman [70] studied a more general class of problems. His generalization was needed in order to study certain problems in control theory. He considered $\mathfrak{F}$ the collection of all functions $f: R \times W \to R^n$ such that the following hold:

(a) $f(t, x)$ is continuous in x for each t and measurable in t for each x.

(b) Given any compact set $K \subseteq W$ there exist $M > 0$ and a locally L^1-function $m(t)$ such that

$$|f(t, x)| \le M, \qquad |f(t, x) - f(t, y)| \le m(t)|x - y|, \qquad \int_t^{t+1} m(s)\, ds \le M,$$

for all $t \in R$ and $x, y \in K$. The topology on $\mathfrak{F}$ is defined by saying that a generalized sequence $\{f_n\}$ converges to f $(f_n \to f)$ if for every $T > 0$ and every compact set $K \subseteq W$ one has

$$\sup\left\{ \left| \int_0^t \{f_n(s, x) - f(s, x)\}\, ds \right| : -T \le t \le T, \; x \in K \right\} \to 0. \tag{3}$$

Under these conditions, Wakeman showed that (1) defines a local dynamical system on $W \times \mathfrak{F}$.

The construction of the flow π described in (1) does include the standard theory for autonomous and periodic differential equations as special cases. If $x' = f(x)$ is an autonomous equation, then one could choose $\mathfrak{F}$ to be the set $\{f\}$, containing the single function f. In the periodic case $x' = f(t, x)$, where f is T-periodic in t, the trajectory $\gamma(f)$ is homeomorphic to a circle. Also the corresponding flow π on $W \times \gamma(f)$ is equivalent to the $(n + 1)$st-order autonomous system

$$x_0' = 1, \qquad x' = f(x_0, x),$$

where we identify $x_0 = x_0 + T$ (modulo T).

Another important class of differential equations that we shall return to shortly are the almost periodic equations (cf. [9, 28, 53, 63]).

Definition 7. A continuous function $f: R \times W \to R^n$ is almost periodic in t (uniformly for x in compact subsets of W) if given any real sequence

$\{t_n{}'\}$ there is a subsequence $\{t_n\}$ and a function g such that $f(t + t_n, x) \to g(t, x)$ as $n \to \infty$ uniformly in (t, x) for $t \in R$ and x in compact subsets of W.

It is known that $(t, f) \to f_t$ defines a flow on the family of all almost periodic functions and further that the hull $H(f)$ is a compact minimal set; in fact, $H(f)$ is a compact abelian group of functions (cf. [63]). In the almost periodic case one often chooses $\mathfrak{F}$ to be the hull $H(f)$ of an almost periodic function. Since $\mathfrak{F}$ is then compact, it happens that the phase space $W \times \mathfrak{F}$ for the flow π is locally compact, a simple mathematical property that pays countless dividends.

IV. Integral Equations and Semiflows

A set of Volterra integral equations can be used to construct a local semiflow. The construction is similar to construction (1) for (DE1). Let $\varphi(t, f, g)$ be a solution of

$$x(t) = f(t) + \int_0^t g(t, s, x(s)) \, ds \tag{IE1}$$

on the maximal interval $[0, \alpha(f, g))$. Then formally

$$\pi(t, f, g) = \left(f_t(\cdot) + \int_0^t g(t + \cdot, s, \varphi(s, f, g)) \, ds, g_t \right) \tag{4}$$

on $0 \le t < \alpha(f, g)$, where $g_t(u, s, x) = g(t + u, s + t, x)$. For appropriate spaces of pairs (f, g), this will define a local semiflow (see [31, 36] for details). Moreover, if f is continuous, then results of Artstein [2] give necessary and sufficient conditions for proper continuity of π.

Similarly, if $\varphi(t) = \varphi(t, x_0, f, g)$ solves

$$x'(t) = f(t, x(t)) + \int_0^t g(t, s, x(s)) \, ds, \qquad x(0) = x_0, \tag{IE2}$$

then under appropriate hypotheses π defined by

$$\pi(t, x_0, f, g) = \left(\varphi(t), f_t(\cdot) + \int_0^t g(t + \cdot, s, \varphi(s)) \, ds, g_t \right) \tag{5}$$

determines a local semiflow on triples (x_0, f, g). We shall discuss their last example further in Section X.

V. Stability and Other Asymptotic Properties

Given a flow π on $W \times \mathfrak{F}$, as described above, we now seek information about the asymptotic behavior of solutions as $t \to +\infty$. The LaSalle invariance principle (Theorem 4) does give some information by using the theory of Liapunov functions. If one cannot find a suitable Liapunov function, one can still get some information about the solutions by using the theory of topological dynamics. The most important concept used in this analysis is the ω-limit set. Specifically one is interested in determining the dynamical structure of the ω-limit set.

In order to show that the ω-limit set $\Omega(x, f)$ is nonempty, we will use Lemma 2 and determine conditions under which the motion $\pi(t, x, f) = (\varphi(t, x, f), f_t)$ is positively compact. Now $\pi(t, x, f)$ is positively compact when

(i) $\varphi(t, x, f)$ remains in a compact subset of W for all $t \geq 0$. [If $W = R^n$, then this is equivalent to demanding that $\varphi(t, x, f)$ be bounded on $t \geq 0$.]

(ii) $H^+(f)$ is compact in $\mathfrak{F}$.

In this case $\Omega(f)$ will be nonempty, as well as compact. The differential equations

$$x' = g(t, x), \qquad g \in \Omega(f), \tag{DE2}$$

are very important. They are called the *limiting equations* of f (cf. [62]).

The condition that "$H^+(f)$ be compact in $\mathfrak{F}$" naturally depends on the choice of $\mathfrak{F}$ and the topology on $\mathfrak{F}$. For example, if $f \in C = C(R \times W, R^n)$ (where C has the topology of uniform convergence on compact sets) then $H^+(f)$ is compact if and only if $f(t, x)$ is bounded and uniformly continuous on every set of the form $[0, \infty) \times K$, where K is compact in W (cf. [62]). [The uniqueness condition in the Carathéodory property would impose additional criteria on $H^+(f)$. These criteria can be dropped entirely if one uses the point of view of [66].]

Using the space described in (3), Wakeman [70] showed that $H^+(f)$ is compact if for each compact set $K \subset W$, there exists $M > 0$ such that

$$|f(t, x) - f(t, y)| \leq M|x - y| \qquad \text{for all} \quad t \geq 0 \quad \text{and all} \quad x, y \in K.$$

It is useful to recall that periodic motions and even almost periodic motions f_t are compact. Moreover, if f_t is positively compact and if $h_t \to 0$

in $\mathfrak{F}$ as $t \to +\infty$, then $(f + h)_t = f_t + h_t$ is positively compact, and f and $f + h$ have the same limiting equations. Thus we see that the semitrajectory $\gamma^+(f)$ can be precompact in a variety of spaces $\mathfrak{F}$. When $\mathfrak{F}$ itself is compact, the LaSalle invariance principle, together with the flow π given by (1), can be employed to good advantage. See [13, 15, 17, 27, 51, 70, 73] for examples.

In addition to the LaSalle invariance principle, Lemma 3 can sometimes be used to obtain asymptotic information. It is not difficult to show that boundedness and certain types of stability of bounded solutions (e.g., uniform stability and uniform asymptotic stability) transfer from solutions of

$$x' = f(t, x) \tag{DE1}$$

to solutions of the limiting equations (DE2) for f (cf. [62, 63]). The inverse problem of transferring information about solutions of the limiting equations (DE2) to solutions of (DE1) is of even more interest. This reverse problem is called the *inverse limit problem* [64]. Roughly speaking, the limiting equations are devoid of "transients" and presumably should be simpler than (DE1). Thus we wish to transfer information from the simpler limiting equation to some solution of (DE1). While this inverse limit problem has not been thoroughly studied (cf. [64]) the following result of Markus [25] is a good example of a theorem one might expect in this area:

Theorem 8. Let $f(t, x) = Ax + g(x) + h(t, x)$ be in $C(R \times R^n, R^n)$ with $h_t \to 0$ as $t \to +\infty$, $|g(x)| = o(|x|)$ as $|x| \to 0$, and A is a stable matrix. Then for all initial conditions x, with $|x|$ sufficiently small, the solution $\varphi(t, x, f) \to 0$ as $t \to +\infty$.

For related concepts of stability as applied to dynamical systems and differential processes, see [10–12, 71].

Similar asymptotic analysis questions can be asked for integral equations. Lemma 2, Lemma 3, and Theorem 4 can be extended to semiflows. These extensions can be used when π is determined by (4) or (5). Compactness of an orbit (4) depends on compactness of the corresponding orbits f_t and g_t and the boundedness and uniform continuity of $\varphi(t, f, g)$ on $[0, \infty)$ (cf. [36]). Similar remarks apply to (5).

As an example, consider

$$x'(t) + \int_0^t h(x(s))a(t - s) \, ds = k(t), \qquad x(0) = x_0, \tag{IE3}$$

where $k(+\infty)$ exists, $a \in L^1(0, \infty)$, and both h and k are continuous. In this case, $g(t, s, x) = -h(x)a(t - s)$ and $f(t, x) = k(t)$ in (5). Thus $g_t = g$ for all $t \geq 0$, $f_t \to k(+\infty)$, and so g_t and f_t are positively compact. If $x(t)$ is bounded, then the assumptions easily imply its uniform continuity on $[0, \infty)$. The limiting equation corresponding to (IE3) can be written in the form

$$x'(t) + \int_{-\infty}^{t} h(x(s))a(t - s)\, ds = k(+\infty) \qquad (-\infty < t < \infty). \qquad \text{(IE4)}$$

Analysis via limiting integral equations has been used in [36, Part III] for general equations, in [16, 29, 30, 33, 72] for asymptotically autonomous equations, and in [21, 32] for periodic and almost periodic limiting equations. The most systematic exploitation of limiting equations occurs in the work of Levin and Shea [18–20]. They always assume the existence of a bounded and uniformly continuous solution of the integral equation. Their conclusions are usually stated in terms of Ψ-sequences. The connection between Ψ-sequences and topological dynamics is studied in [65].

Levin and Shea consider equations of the form

$$X'(t) + \int_{-\infty}^{\infty} h(X(t - s))\, dA(s) = F(t), \qquad \text{(IE5)}$$

$$X(t) + \int_{-\infty}^{\infty} h(X(t - s))\, dA(s) = F(t), \qquad \text{(IE6)}$$

$$\int_{-\infty}^{\infty} h(X(t - s))\, dA(s) = F(t), \qquad \text{(IE7)}$$

on $-\infty < t < \infty$, where $A \in BV(-\infty, \infty)$, $A(-\infty) = 0$, $F \in L^\infty(-\infty, \infty)$, and $F(+\infty)$ exists. They study both linear and nonlinear cases. Note that (IE3) and (IE4) (for example) are special cases of (IE5). We need only put $A(s) = 0$ if $s \leq 0$, $A'(s) = a(s)$ on $s \geq 0$, define $F(t)$ in the obvious way in (IE3) or (IE4), and put $x(t) = 0$ for $t < 0$ in (IE3). A typical result is the following. Let $*$ denote the Fourier transform and let $S_a(\Lambda) = \{s: A^*(s) = -is, -\infty < s < \infty\}$.

Theorem 9. (See [19]).   With A and F as above and $h(x) = x$, suppose (IE5) has a bounded, uniformly continuous solution $X(t)$. Then there exists a function $\eta(t)$ with $\eta(\infty) = 0$ and

$$\lim_{t \to \infty} \left\{ \operatorname*{ess\ sup}_{t \leq s < \infty} |\eta'(s)| \right\} = 0$$

such that:

   (i) if $S_a(\Lambda) = \varnothing$, then $X(t) = F(\infty)/A(\infty) + \eta(t)$, on $-\infty < t < \infty$, and

   (ii) if $S_a(\Lambda) = \{\lambda_1, \lambda_2, \ldots, \lambda_n\}$ and $\lambda_k \neq 0$ for all k, then

$$X(t) = F(\infty)/A(\infty) + \sum_{k=1}^{n} C_k(t) \exp(i\lambda_k t) + \eta(t) \qquad (6)$$

on $-\infty < t < \infty$, where $C_k \in C^\infty \cap L^\infty(-\infty, \infty)$ and $C_k^{(i)}(+\infty) = 0$ for $1 \leq k \leq n$ and $i = 1, 2, 3, \ldots$.

   (iii) If $S_a(A) = \{\lambda_1, \lambda_2, \ldots, \lambda_n\}$ and some λ_k is zero, then (6) remains true but $F(\infty)$ must be zero.

Similar results can be proved for (IE6) and (IE7). Results can also be proved in nonlinear cases. See also [65].

VI. Periodic and Almost Periodic Solutions

Some of the first results that were discovered by exploiting the topological dynamical viewpoint for nonautonomous differential equations concern the existence of periodic and almost periodic solutions. The basic idea here is that one can describe the structure of an ω-limit set, usually in terms of certain stability concepts. The early results in this direction can be found in [7, 28, 61]. A different class of existence theorems for almost periodic solutions was started by Amerio [1] and later extended by Fink [8] and Seifert [59]. These theorems are based on various separatedness concepts.

Recently Sacker and Sell [53, 57], using the dynamical concept of a distal flow, have presented a theory in the context of skew-product flows that combines both of these approaches as special cases of the following

Theorem 10. Let π be the flow described by (1) on $W \times \mathfrak{F}$, where $\mathfrak{F}$ is the hull of an almost periodic function $f: R \times W \to R^n$. Let M be a compact invariant set in $W \times \mathfrak{F}$. Let $p: W \times \mathfrak{F} \to \mathfrak{F}$ be the projection mapping. Assume that

   (i) Card $(p^{-1}(g) \cap M) = N < \infty$ for some $g \in \mathfrak{F}$, and

   (ii) π has the distal property on M.

Then M is an N-fold covering space of $\mathfrak{F}$, and M can be written as the finite union of almost periodic minimal sets. In particular, for every $(x, f) \in M$ the solution $\varphi(t, x, f)$ is an almost periodic function of t.

In applications the set M is often chosen to be the ω-limit set $\Omega(x_0, f_0)$ of a positively compact motion $\pi(t, x_0, f_0)$. The *distal* property on M means that if (x_1, f), $(x_2, f) \in M$ and $x_1 \neq x_2$ then there is an $\alpha > 0$ such that $|\varphi(t, x_1, f) - \varphi(t, x_2, f)| \geq \alpha$ for all $t \in R$. This distal property and the finiteness property (i) are weaker than the uniform stability properties of [7, 28, 61] and the separatedness properties of [1, 8, 59]. For example, the following result of Miller [28] can be derived as a corollary of Theorem 10. Let $\mathfrak{F}$ be the hull of an almost periodic function, where $\mathfrak{F}$ satisfies conditions (i)–(iv) in Section III. If there exists a bounded uniformly asymptotically stable solution of $x' = f(t, x)$ for some $f \in \mathfrak{F}$, then there exists an almost periodic solution for every equation in $\mathfrak{F}$. (See [57] for details.)

VII. Generic Theory

The point of view underlying the "generic theories" of differential equations is a natural development in our study of the qualitative behavior of differential equations. Prior to the commencement of the study of the qualitative behavior of ordinary differential equations, which began with Poincaré nearly one hundred years ago, the emphasis in differential equations had been in the area of quadratures, which involves either the explicit solution of differential equations or expressing such solutions in terms of infinite series. If one recalls the application of the Poincaré–Bendixson theory to the Van der Pol equation

$$u'' + \varepsilon(u^2 - 1)u' + u = 0,$$

one gets an existence theorem for a nontrivial periodic solution, but not an explicit or even an implicit formula for this solution. As is well known, the Poincaré–Bendixson theory does exploit the special topological structure of the plane as expressed in the Jordan curve theorem. As one passes to differential equations defined on higher-dimensional spaces, one is often forced in the qualitative theory to accept weaker conclusions. One such class of conclusions is described in what is called the "generic theories."

The basic tenet of generic theories is to prove theorems of the following type:

A "typical" solution (function) has property "P."

Presumably in trying to establish a generic theorem as stated above, one is given a system (or a collection of solution functions) and one is investigating a certain property P. It remains then to decide on the meaning of the word "typical." Experience has shown that two concepts are commonly accepted, one measure-theoretic and the other metric-theoretic, viz.:

(i)   Let $(X, \mathfrak{M}, \mu)$ be a measure space. A measurable subset $A \subseteq X$ is "typical" if the complement $X - A$ has measure zero.

(ii)   Let X be a complete metric space. A subset $A \subseteq X$ is "typical" if it is residual, i.e., A is large in the sense of Baire—or A is the countable intersection of open dense sets.

The Fubini theorem for measures on product spaces would be an example of the generic theorems of a measure-theoretic nature, and Sard's theorem would be an example of the generic theorems of a metric-theoretic nature. In the theory of topological dynamics one finds several generic theorems. One is the Poincaré–Carathéodory recurrence theorem:

Theorem 11.   Let π be a flow on a compact space X and let μ be a positive invariant measure on X, i.e.,

$$\mu(\pi(t, A)) = \mu(A)$$

for all Borel sets $A \subseteq X$ and all $t \in R$. Then the set P of Poisson stable points in X is typical in X, i.e., $\mu(X - P) = 0$.

Recall that a point $p \in X$ is *Poisson stable* if for every neighborhood U of p the sets

$$\{t : t \geq 0 \text{ and } \pi(t, p) \in U\}, \qquad \{t : t \leq 0 \text{ and } \pi(t, p) \in U\},$$

are unbounded in R (cf. [50]). The Poincaré–Carathéodory recurrence theorem also has a metric-theoretic counterpart (cf. [52]). Another generic theorem is the following ergodic theorem of Birkhoff. (We shall use the fact that every a.p. minimal set is the space of a compact abelian group and therefore supports a Haar measure that is an invariant measure for the flow; cf. [49, 63].)

Theorem 12.   Let X be an a.p. minimal set for a flow π and let μ be the Haar measure on X with $\mu(X) = 1$. Then for every bounded measurable function $f: X \to R$ one has

$$\int_X f(x)\mu(dx) = \lim_{T \to \infty} \frac{1}{T} \int_0^T f(\pi(s, x)) \, ds$$

for almost all $x \in X$.

These are two examples of generic theorems that we shall use later. Other examples in the theory of topological dynamics and differential equations can be cited. One very important area is in the theory of structural stability, which is given in the chapter by Shub in this volume.

Let us conclude this section by considering one application of the Poincaré–Carathéodory recurrence theorem to a control-theoretic problem arising in celestial navigation [26]. The physical description of this space rescue problem has been treated elsewhere [26, 64] so we shall turn directly to the mathematical description.

Theorem 13. Let W be a compact Riemannian manifold and let $f(t, x, u)$ be a time-varying C^1-vector field depending on a control parameter $u \in \Omega$, where Ω is a compact neighborhood of the origin in R^m. Assume that f and its covariant spatial derivative are uniformly bounded for $(t, x, u) \in R \times W \times \Omega$. Let $\mathfrak{F}$ denote the hull of f, where u is treated as an independent parameter. Define the attainable set $A(t_0, x_0; \tau)$ as the collection of all $\varphi([u], t_0, x_0; t_0 + \tau)$, where $u(t) \in \Omega$ for $t_0 \leq t \leq t_0 + \tau$ and φ is the corresponding solution of

$$x' = f(t, x, u(t)), \qquad x(t_0) = x_0. \tag{DE3}$$

Assume that

$$\mathrm{div}_x f(t, x, 0) = 0 \qquad \text{on} \quad R \times W,$$

and that there is an $\eta = \eta(\tau) > 0$ such that for all $(t_0, x_0) \in R \times W$, the attainability set $A(t_0, x_0; \tau)$ contains a ball of radius η centered at $\varphi([0], t_0, x_0; t_0 + \tau)$.

Let μ be any invariant measure on $\mathfrak{F}$ with $\mu(\mathfrak{F}) = 1$. Then there is a measurable subset $\mathfrak{G} \subseteq \mathfrak{F}$ such that $\mu(\mathfrak{F} - \mathfrak{G}) = 0$ and if $f \in \mathfrak{G}$, then there is a $T > 0$ such that

$$A(t_0, x_0; s) = W \qquad \text{for all} \quad s \geq T. \tag{7}$$

The conclusion (7) says that one can steer from the initial point x_0 (at time t_0) to any point in the space W at precisely the time $t = s$. In the physical problem alluded to above, this means that one can effect the rescue of any derelict space ship, provided the equations of motion lie in the generic space $\mathfrak{G}$, i.e., with probability 1 rescue is possible. The proof of this is based on an application of the abstract Poincaré–Carathéodory recurrence theorem to the flow π on the product space $W \times \mathfrak{F}$ together with the Fubini theorem.

Theorem 13 then is another example of a generic theorem. It is also a good example of a theorem whose proof uses the construction described in Section III in an essential way. It does not appear that Theorem 13 is provable without using this construction (cf. [64]).

At this point one might object to the generic theories. The vector field $f(t, x, u)$ is determined completely by the gravitational field of the large celestial bodies and cannot be changed. Therefore one might ask, "What happens to the space rescue if my gravitational field f lies in the negligible set $\mathfrak{F} - \mathfrak{G}$?" The mathematical answer would be, "Well that would be unfortunate, but after all it is not very likely!" This answer, while irrefutable, is not very reassuring to the people living in the $(\mathfrak{F} - \mathfrak{G})$ worlds.

However, in the almost periodic case there is hope! One cannot prove directly that $(\mathfrak{F} - \mathfrak{G})$ is empty in this case, since the construction of G is irretrievably lost in the proof of the Poincaré–Carathéodory recurrence theorem. However, one can adapt a special control strategy, which works even for the $(\mathfrak{F} - \mathfrak{G})$ worlds, to show that conclusion (7) is valid for all $f \in \mathfrak{F}$. This strategy uses, in an essential way, Theorem 13 together with certain crucial properties of almost periodic functions (cf. [26]). So Theorem 13, with all its imperfections, actually saves the day for the almost periodic worlds!

Another class of generic theories, which we shall discuss in more detail shortly, concerns the behavior of solutions of linear differential equations (see also [40, 42, 43, 45, 47]).

VIII. Linear Theory: Dichotomies and Invariant Splittings

In the case of a linear differential system

$$x' = A(t)x, \tag{DE4}$$

where $x \in X$ (here $X = R^n$ or C^n) and $A \in \mathfrak{A}$, the corresponding flow

$$\pi(t, x, A) = (\varphi(t, x, A), A_t) \tag{8}$$

has the property that the mapping $\varphi(t, x, A)$ is linear in x. Therefore, one can write

$$\varphi(t, x, A) = \Phi(t, A)x,$$

where $\Phi(t, A)$ is a linear transformation on X. $\Phi(t, A)$ is referred to as the *fundamental solution* of (DE4). Because of the linearity, the flow π given by (8) is called a *linear skew-product flow*, LSPF for short.

As noted above, we will assume that the matrix-valued function $A(t)$, defined for $-\infty < t < \infty$, is a point in a function space $\mathfrak{A}$. In addition to conditions (i)–(iv) of Theorem 5, throughout this section we shall assume that $\mathfrak{A}$ *is a compact set*, for example, $\mathfrak{A}$ may be the hull of a bounded uniformly continuous function $A(t)$.

Next define the bounded set, the stable set, and the unstable set by:

$$\mathfrak{B} = \{(x, A) : |\varphi(t, x, A)| \text{ is uniformly bounded for } t \in R\},$$
$$\mathfrak{S} = \{(x, A) : |\varphi(t, x, A)| \to 0 \text{ as } t \to +\infty\},$$
$$\mathfrak{U} = \{(x, A) : |\varphi(t, x, A)| \to 0 \text{ as } t \to -\infty\}.$$

Also define the fibers

$$\mathfrak{B}(A) = \{x \in X : (x, A) \in \mathfrak{B}\},$$
$$\mathfrak{S}(A) = \{x \in X : (x, A) \in \mathfrak{S}\},$$
$$\mathfrak{U}(A) = \{x \in X : (x, A) \in \mathfrak{U}\}.$$

Because of the linearity of (DE4), the fibers $\mathfrak{B}(A)$, $\mathfrak{S}(A)$, and $\mathfrak{U}(A)$ are linear subspaces of X.

Definition 14. The LSPF π given by (8) is said to *admit an exponential dichotomy* at $A \in \mathfrak{A}$ if there is a projection $P = P(A)$ on X and positive constants α, K such that

$$|\Phi(t, A)P(A)\Phi^{-1}(s, A)| \le Ke^{-\alpha(t-s)}, \qquad s \le t,$$
$$|\Phi(t, A)[I - P(A)]\Phi^{-1}(s, A)| \le Ke^{-\alpha(s-t)}, \qquad t \le s.$$

It is shown in [54] that if π admits an exponential dichotomy at $A \in \mathfrak{A}$, then $\mathfrak{S}(A)$ is the range of $P(A)$, $\mathfrak{U}(A)$ is the null space of $P(A)$, and $\mathfrak{B}(A) = \{0\}$. Furthermore, in certain important cases, it is shown that if π admits an exponential dichotomy at $A \in \mathfrak{A}$ then $\mathfrak{B}$ is trivial, i.e., $\mathfrak{B} = \{0\} \times \mathfrak{A}$.

The converses of the above theorems have been studied recently by Sacker and Sell [54–56] and Selgrade [60]. The next result (see [55]) shows that the assumption that $\mathfrak{B}$ be trivial leads to some rather surprising conclusions about the flow $\sigma(A, \tau) = A_\tau$ on the base space $\mathfrak{A}$.

Theorem 15. Assume that $\mathfrak{B} = \{0\} \times \mathfrak{A}$, where $\mathfrak{A}$ is compact. Define

$$\mathfrak{A}_k = \{A \in \mathfrak{A} : \dim \mathfrak{S}(A) = k \text{ and } \dim \mathfrak{U}(A) = n - k\},$$

for $k = 0, 1, \ldots, n$, where $n = \dim X$. Then each $\mathfrak{A}_k$ is a compact invariant set in the flow σ on $\mathfrak{A}$. Furthermore, every α- and ω-limit set for the flow σ lies in precisely one $\mathfrak{A}_k$. Finally, either $\mathfrak{A} = \mathfrak{A}_k$ for some k, or there are at least two nonempty $\mathfrak{A}_k$'s. In the latter case, if

$$L = \min\{k : \mathfrak{A}_k \text{ is nonempty}\},$$

then $\mathfrak{A}_L$ is a stable attractor for the flow σ.

In the special case where $\mathfrak{A}$ is a compact minimal set one can say more (cf. [54, 60]).

Theorem 16.   Assume that $\mathfrak{B} = \{0\} \times \mathfrak{A}$, where $\mathfrak{A}$ is a compact minimal set. Then the following conclusions are valid:

(i)   There is an integer k such that

$$\dim \mathfrak{S}(A) = k \qquad \text{and} \qquad \dim \mathfrak{U}(A) = n - k$$

for all $A \in \mathfrak{A}$, where $n = \dim X$, and $X = \mathfrak{S}(A) + \mathfrak{U}(A)$ for all $A \in \mathfrak{A}$.

(ii) $\mathfrak{S}(A)$ and $\mathfrak{U}(A)$ vary continuously in $A \in \mathfrak{A}$, i.e., the mapping $P : X \times \mathfrak{A} \to X$ defined by

$$(x, a) \to P(A)x,$$

where $P(A)$ is the projection on X with range $= \mathfrak{S}(A)$ and null space $= \mathfrak{U}(A)$, is jointly continuous in (x, A).

(iii)   The rate of decay in $\mathfrak{S}$ and $\mathfrak{U}$ is exponential.

(iv) π admits an exponential dichotomy at every $A \in \mathfrak{A}$ where the associated projection $P(A)$ is given in (ii).

(v) $\mathfrak{S}$ and $\mathfrak{U}$ are subbundles of $X \times \mathfrak{A}$ and $X \times \mathfrak{A} = \mathfrak{S} + \mathfrak{U}$, a Whitney sum.

Conclusion (v) is simply a reformulation of conclusions (i) and (ii) (see [68] for more details).

The compactness of the base $\mathfrak{A}$ is crucial in the theory. Very little could be done without it. For example, compactness is used to prove the uniformity of various growth estimates. The stable and unstable sets $\mathfrak{S}$ and $\mathfrak{U}$ are defined with no reference to the rate of decay, but in conclusions (iii) and (iv) one concludes an exponential rate of decay, which is in fact uniform over $\mathfrak{A}$.

Several generalizations of Theorems 15 and 16 are possible. First, the theory does not really depend on the differential equation (DE4) but instead is a theory about a linear skew-product flow, that is, a flow π on a product space $X \times Y$

$$\pi(t, x, y) = (\varphi(t, x, y), \sigma(t, y)),$$

where the second coordinate σ depends only on y (hence a skew-product flow) and the first coordinate φ is linear in x (hence a linear skew-product flow) (cf. [14, 53, 54, 57]). Second, it is not crucial that the phase space of the LSPF π be a global product space; it is sufficient if the phase space is a vector bundle, which is locally a product space. In this way the theory described in Theorems 15 and 16 extends to the study of the linearized flow on a tangent bundle TM generated by a smooth vector field on a compact manifold M (cf. [54, 60, 68] for more details).

IX. Asymptotic Behavior of Linear Equations

Let us now turn our attention to a linear differential system with almost periodic coefficients:

$$x' = A(t)x, \tag{DE4}$$

where $x \in X$. Let $\mathfrak{A}$ denote the hull of $A = A(t)$ and let

$$\pi(t, x, A) = (\varphi(t, x, A), A_t) \tag{8}$$

denote the corresponding LSPF on $X \times \mathfrak{A}$.

The classical approach to the study of the asymptotic behavior of solutions of (DE4) has been in terms of the Liapunov-type numbers, or the Liapunov characteristic exponents. For each $(x, A) \in X \times \mathfrak{A}$ with $x \neq 0$, the four Liapunov-type numbers are defined by

$$\lambda_s^+(x, A) = \limsup_{T \to +\infty} \frac{1}{T} \log \|\varphi(T, x, A)\|,$$

$$\lambda_i^+(x, A) = \liminf_{T \to +\infty} \frac{1}{T} \log \|\varphi(T, x, A)\|,$$

$$\lambda_s^-(x, A) = \limsup_{T \to -\infty} \frac{1}{T} \log \|\varphi(T, x, A)\|,$$

$$\lambda_i^-(x, A) = \liminf_{T \to -\infty} \frac{1}{T} \log \|\varphi(T, x, A)\|.$$

Let $\Lambda_s^+(A) = \{\lambda_s^+(x, A): x \neq 0\}$ and similarly define $\Lambda_i^+(A)$, $\Lambda_s^-(A)$, and $\Lambda_i^-(A)$. It is not difficult to show that each of the $\Lambda(A)$-sets contains at most n distinct points, where $n = \dim X$.

One of the main problems arising in the study of the asymptotic behavior of solutions of (DE4) is to determine the stability or continuity of the

$\Lambda(A)$-sets. In general, one does not have continuity, but the theory is very complicated. We refer the reader to [22–24, 37–48] for more details.

Under certain circumstances though, one can prove a generic-type continuity for certain characteristic exponents. For example, Millionscikov [37, 40, 42, 43, 45, 47] has used somewhat different definitions for the characteristic exponents and shown the continuity of $\Lambda(A)$ for A belonging to a generic set in $\mathfrak{A}$. Let us illustrate this with Millionscikov's theory of the probable spectrum [47].

Let $n = \dim X$. Then for any nonsingular $n \times n$ matrix B, the matrix B^*B (where B^* denotes the adjoint matrix) is nonsingular and self-adjoint. Furthermore, B^*B has precisely n eigenvalues (counting multiplicity) all of which are positive real numbers. Let $d_1(B), \ldots, d_n(B)$ denote the positive square roots of these eigenvalues ordered by $d_1(B) \geq d_2(B) \geq \cdots \geq d_n(B)$.

Now consider $x' = A(t)x$, where $A \in \mathfrak{A}$. For $i = 1, 2, \ldots, n$, define

$$v_i(A) = \lim_{\tau \to +\infty} \left[\limsup_{S \to +\infty} \frac{1}{S\tau} \sum_{j=0}^{S-1} \log d_i(\Phi(\tau, A_{j\tau})) \right].$$

Since $\Phi(\tau, A)$ is continuous in A and the functions $d_i(B)$ are continuous, it is not difficult to see that for each i, the function $v_i(A)$ is a bounded measurable function of $A \in \mathfrak{A}$, where the measure of $\mathfrak{A}$ is the Haar measure μ. Since $\mathfrak{A}$ is an a.p. minimal set, the Birkhoff ergodic theorem 12 can be applied, and we conclude that

$$v_i(A) = \int_{\mathfrak{A}} v_i(A)\mu(dA) = \bar{v}_i \tag{9}$$

for almost every $A \in \mathfrak{A}$. The collection of real numbers $\{\bar{v}_i, \ldots, \bar{v}_n\}$ is the probable spectrum of A. We see then that the set $\{v_1(A), \ldots, v_n(A)\}$ is constant over the subset $\mathfrak{A}_0 \subseteq \mathfrak{A}$ for which (9) is valid.

More recently, a somewhat different theory has been developed by Sacker and Sell [58] (also see [60, 67]) to analyze the asymptotic behavior of solutions of (DE4). This theory, while still in its early stages of development, appears to give new insight into the structure of the Liapunov-type numbers. Let us now look at the details.

For $\lambda \in R$ define the LSPF π_λ on $X \times \mathfrak{A}$ by

$$\pi_\lambda(t, x, A) = (e^{-\lambda t}\varphi(t, x, A), A_t). \tag{10}$$

It is elementary to show that for each λ, π_λ is a flow on $X \times \mathfrak{A}$. For each

$\lambda \in R$, we define the corresponding bounded set $\mathfrak{B}_\lambda$, stable set $\mathfrak{S}_\lambda$, and unstable set $\mathfrak{U}_\lambda$ by

$$\mathfrak{B}_\lambda = \{(x, A) : |e^{-\lambda t}\varphi(t, x, A)| \text{ is uniformly bounded for } t \in R\},$$
$$\mathfrak{S}_\lambda = \{(x, A) : |e^{-\lambda t}\varphi(t, x, A)| \to 0 \text{ as } t \to +\infty\},$$
$$\mathfrak{U}_\lambda = \{(x, A) : |e^{-\lambda t}\varphi(t, x, A)| \to 0 \text{ as } t \to -\infty\}.$$

Since $\mathfrak{A}$ is an a.p. minimal set, it follows from Theorem 16 that if $\mathfrak{B}_\lambda = \{0\} \times \mathfrak{A}$, then $\mathfrak{S}_\lambda$ and $\mathfrak{U}_\lambda$ are invariant subbundles of $X \times \mathfrak{A}$ and

$$X \times \mathfrak{A} = \mathfrak{S}_\lambda + \mathfrak{U}_\lambda \qquad \text{(Whitney sum)}.$$

We define the resolvent of $\mathfrak{A}$ as

$$\rho(\mathfrak{A}) = \{\lambda \in R : \mathfrak{B}_\lambda = \{0\} \times \mathfrak{A}\}.$$

The complement of $\rho(\mathfrak{A})$ is the spectrum of $\mathfrak{A}$, i.e.,

$$\sigma(\mathfrak{A}) = R - \rho(\mathfrak{A}).$$

The next theorem describes the basic properties of the spectrum $\sigma(\mathfrak{A})$ and the corresponding LSPF π given by (8) (cf. [58]):

Theorem 17. Let π be the LSPF on $X \times \mathfrak{A}$ defined by (8), where $\mathfrak{A}$ is a compact minimal set and $\dim X \geq 1$. Then the spectrum $\sigma(\mathfrak{A})$ is a nonempty compact subset of R consisting of k nonoverlapping intervals $[a_i, b_i]$, where $k \leq \dim X$. Moreover, associated with each spectral interval $[a_i, b_i]$ there exists a nonempty invariant subset $\mathfrak{G}_i \subset X \times \mathfrak{A}$ with the following properties:

(i)   For each $A \in \mathfrak{A}$ the fiber

$$\mathfrak{G}_i(A) = \{x \in X : (x, A) \in \mathfrak{G}_i\}$$

is a linear subspace of X.

(ii)   There is an integer $n_i \geq 1$ such that

$$\dim \mathfrak{G}_i(A) = n_i$$

for all $A \in \mathfrak{A}$.

(iii) $\mathfrak{G}_i(A)$ varies continuously in $A \in \mathfrak{A}$ in the sense that there exists a continuous mapping $P_i : X \times \mathfrak{A} \to X$ given by

$$P_i(x, A) = P_i(A) \cdot x,$$

where $P_i(A)$ is a projection on X with range $= \mathfrak{G}_i(A)$. (This means that each $\mathfrak{G}_i$ is a nontrivial invariant subbundle of $X \times \mathfrak{A}$.) Furthermore one has $n_1 + \cdots + n_k = \dim X$ and

$$X = \mathfrak{G}_1(A) + \cdots + \mathfrak{G}_k(A)$$

for every $A \in \mathfrak{A}$. (This means that

$$X \times \mathfrak{A} = \mathfrak{G}_1 + \cdots + \mathfrak{G}_k$$

is a Whitney sum.)

What does this result tell us about the Liapunov-type numbers? Well if $(x, A) \in \mathfrak{G}_i$ and $x \neq 0$, then one can show that the four corresponding-type numbers $\lambda_s^+(x, A)$, $\lambda_i^+(x, A)$, $\lambda_s^-(x, A)$, and $\lambda_i^-(x, A)$ lie in the corresponding spectral intervals $[a_i, b_i]$. Also one can show that in the autonomous, or periodic, case the four $\Lambda(A)$-sets are the same and they agree with (i) the probable spectrum $\{\bar{v}_1, \ldots, \bar{v}_k\}$ as well as with (ii) the spectrum $\sigma(\mathfrak{A})$ defined above. The relationship between the probable spectrum and $\sigma(\mathfrak{A})$ in the almost periodic case is not known, although it seems reasonable to conjecture that

$$\text{probable spectrum} \subseteq \sigma(\mathfrak{A}),$$

when $\mathfrak{A}$ is an a.p. minimal set.

The study of the asymptotic behavior of solutions of (DE4) in the aperiodic case has over the last several years been viewed as an attempt to extend the Floquet theory to nonperiodic systems. An extensive discussion of the "Floquet problem for almost periodic systems" can be found elsewhere [67] so we will not go into details here. Suffice it to say that the problem reduces to asking:

$$\text{Do the spectral intervals } [a_i, b_i] \text{ reduce to points?} \qquad (11)$$

An affirmative answer to this would be particularly fortunate, because in this case, whenever (x, A) belongs to the associated subbundle $\mathfrak{G}_i$ (with $x \neq 0$), then the four Liapunov-type numbers must agree and be equal to $a_i \, (=b_i)$. In other words, the two limits

$$\lim_{T \to +\infty} \frac{1}{T} \log \|\varphi(T, x, A)\|, \qquad \lim_{T \to -\infty} \frac{1}{T} \log \|\varphi(T, x, A)\|,$$

exist and are equal.

In closing we should note that it is not at all clear at this time why (11) should always have an affirmative answer in the almost periodic case. R. McGehee has constructed a scalar equation

$$x' = a(t)x$$

(where $x \in R^1$, $a \in \mathfrak{A}$, and $\mathfrak{A}$ is a compact minimal set) with the property that the spectrum $\sigma(\mathfrak{A})$ consists of a nontrivial interval. In McGehee's example the function $a(t)$ is not (in fact, it cannot be) almost periodic in t.

X. Semigroups and Integrodifferential Equations

We wish to study the flows defined by (IE2) for the case where the coefficients of (IE2) are linear and of convolution type, i.e.,

$$x'(t) = Ax(t) + \int_0^t B(t-s)x(s)\,ds + F(t), \qquad x(0) = x_0, \quad t \geq 0, \quad \text{(IE8)}$$

where $B \in L^1(0, \infty)$. If $x(t) = \Psi(t)$ is given, $\Psi \in L^\infty(-\infty, 0)$, and if $x(t)$ solves the system

$$x'(t) = Ax(t) + \int_{-\infty}^t B(t-s)x(s)\,ds + f(t), \qquad t \geq 0, \qquad \text{(IE9)}$$

then $x(t)$ will solve (IE8) with

$$F(t) = f(t) + \int_{-\infty}^0 B(t-s)\Psi(s)\,ds, \qquad x_0 = \Psi(0).$$

Barbu and Grossman [3] study the asymptotic behavior of solutions of (IE9) by using that equation to define a C_0-semigroup and then applying the theory of semigroups. They put

$$X_1 = BC_1(-\infty, 0] = \{f \in C(-\infty, 0] : f(-\infty) \text{ exists and is finite}\}.$$

Given Ψ in X_1, let $x_0 = \Psi(0)$ and let $\varphi(t, x_0, \Psi)$ be the solution of (IE9) with $x(t) = \Psi(t)$ on $t \leq 0$. Define

$$T(t)\Psi \in X_1 \qquad \text{by} \qquad T(t)\Psi(s) = \varphi(t+s, x_0, \Psi) \quad \text{on} \quad -\infty < s \leq 0.$$

It is shown in [3] that $T(t)$ determines a C_0-semigroup on X_1. The same construction works in other spaces, e.g., $X_p = X_1 \cap L^p(-\infty, 0)$. They obtain exponential estimates on growth of solutions under various assumptions on the roots of the determinant $\det(sI - A - B^*(s)) = 0$, where s is a complex variable and $*$ denotes Laplace transform.

Miller [34] employs construction (5) of Section IV in order to construct a semigroup. Let

$$C_0(-\infty, 0] = \{f: (-\infty, 0] \to R^n : f \text{ is continuous and has compact support}\}.$$

Let

$$\varphi(f)(t) = \int_{-\infty}^{0} B(t - s)f(s)\,ds, \qquad t \geq 0, \quad f \in C_0(-\infty, 0].$$

Let

$$Y_0 = \{(f(0), \varphi(f)): f \in C_0(-\infty, 0]\}$$

be the linear space with norm

$$\|(f(0), \varphi(f))\| = |f(0)| + \sup\{|\varphi(f)(t)|: t \geq 0\},$$

and let Y be the completion of Y_0. For any pair (x_0, F) in Y, if $\varphi(t, x_0, f)$ solves (IE8), then (5) becomes

$$\pi(t, x_0, F) = \left(\varphi(t, x_0, F), F_t(\cdot) + \int_0^t B(s + \cdot)\varphi(t - s, x_0, F)\,ds\right).$$

It is shown that $\pi(t, x_0 F) = T(t)(x_0, F)$ determines a C_0-semigroup on Y. The main result in [34] is the following:

Theorem 18. Let $B \in L^1(0, \infty)$ and let Y, π, and $T(t)$ be as defined above. Then $\det(sI - A - B^*(s)) \neq 0$ in the half plane $\operatorname{Re} s \geq 0$ if and only if $T(t)$ is asymptotically stable in the sense that

(a) $\sup\{\|T(t)\|: t \geq 0\} < \infty$, and
(b) for any $y = (x, F)$ in Y, $T(t)y \to 0$ as $t \to \infty$.

Since the determinantal condition seems to be the standard asymptotic stability condition for (IE8) (see, for example, [35]), then its equivalence with (a) and (b) is a bit surprising. Normally for C_0-semigroups we expect and hope for exponential decay in the asymptotically stable case.

REFERENCES

[1] L. Amerio, Soluzioni quasi-periodiche o limitate, di systemi differenziali non lineari quasi-periodiche o limitati, *Ann. Math. Pura Appl.* **39** (1955), 97–119.
[2] Z. Artstein, Continuous dependence of solutions of Volterra integral equations, *SIAM J. Math. Anal.* **6** (1975), 446–456.
[3] V. Barbu and S. I. Grossman, Asymptotic behavior of linear integrodifferential systems, *Trans. Amer. Math. Soc.* **171** (1972), 277–288.
[4] C. C. Conley, The gradient structure of a flow I, *IBM Math. Sci. Dept. Tech. Rep.* (1972).

[5] C. M. Dafermos, An invariance principle for compact processes, *J. Differential Equations* **9** (1971), 239–252.

[6] C. M. Dafermos, Applications of the invariance principle for compact process II, *J. Differential Equations* **11** (1972), 416–424.

[7] L. Deysach and G. R. Sell, On the existence of almost periodic motions, *Michigan Math. J.* **12** (1965), 87–95.

[8] A. M. Fink, Semi-separated conditions for almost periodic solutions, *J. Differential Equations* **11** (1972), 245–251.

[9] A. M. Fink, Almost periodic differential equations, *Springer Lect. Notes Math.* **377** (1974).

[10] J. K. Hale, Dynamical systems and stability, *J. Math. Anal. Appl.* **26** (1969), 39–59.

[11] J. K. Hale and E. F. Infante, Extended dynamical systems and stability theory, *Proc. Nat. Acad. Sci. U.S.A.* **58** (1967), 405–409.

[12] J. K. Hale, J. P. LaSalle, and M. Slemrod, Theory of a general class of dissipative processes, *J. Math. Anal. Appl.* **39** (1972), 177–191.

[13] J. P. LaSalle, Asymptotic stability criterion, *Proc. Symp. Appl. Math.* **13**, 299–307. Amer. Math. Soc., Providence, Rhode Island, 1962.

[14] J. P. LaSalle, Invariance principles and stability theory for nonautonomous systems, *Proc. Greek Math. Soc. Carathéodory Symp.*, Athens, Sept. 3–7 (1973).

[15] J. J. Levin, On the global behavior of nonlinear systems of differential equations, *Arch. Rational Mech. Anal.* **6** (1960), 65–74.

[16] J. J. Levin, The qualitative behavior of a nonlinear Volterra equation, *Proc. Amer. Math. Soc.* **16** (1965), 711–718.

[17] J. J. Levin and J. A. Nohel, Global asymptotic stability for nonlinear systems of differential equations and applications to reactor dynamics, *Arch. Rational Mech. Anal.* **5** (1960), 194–211.

[18] J. J. Levin and D. F. Shea, Asymptotic behavior of bounded solutions of some functional equations, *in* "Contributions to Nonlinear Functional Analysis." Academic Press, New York, 1971.

[19] J. J. Levin and D. F. Shea, Tauberian theorems and functional equations, *in* "Ordinary Differential Equations." Academic Press, New York, 1972.

[20] J. J. Levin and D. F. Shea, On the asymptotic behavior of the bounded solutions of some integral equations, *J. Math. Anal. Appl.* **37** (1973), (I) 42–82, (II) 288–326, (III) 537–575.

[21] N. Levinson, A nonlinear Volterra integral equation arising in the theory of superfluidity, *J. Math. Anal. Appl.* **1** (1961), 1–11.

[22] J. C. Lillo, Perturbations of nonlinear systems, *Acta Math.* **103** (1960), 123–238.

[23] J. C. Lillo, Continuous matrices and the stability theory of differential systems, *Math. Z.* **73** (1960), 45–58.

[24] J. C. Lillo, A note on the continuity of characteristic exponents, *Proc. Nat. Acad. Sci. U.S.A.* **46** (1960), 247–250.

[25] L. Markus, Asymptotically autonomous differential systems. "Contributions to Nonlinear Oscillations" (S. Lefschetz, ed.), Vol. 3. Princeton Univ. Press, Princeton, New Jersey, 1956.

[26] L. Markus and G. R. Sell, Control in conservative dynamical systems: Recurrence and capture in aperiodic fields, *J. Differential Equations* **16** (1974), 472–505.

[27] R. K. Miller, Asymptotic behavior of solutions of nonlinear differential equations, *Trans. Amer. Math. Soc.* **115** (1965), 400–416.

[28] R. K. Miller, Almost periodic differential equations as dynamical systems with applications to the existence of a.p. solutions, *J. Differential Equations* **1** (1965), 337–345.

[29] R. K. Miller, Asymptotic behavior of solutions of nonlinear Volterra equations, *Bull. Amer. Math. Soc.* **72** (1966), 153–156.

[30] R. K. Miller, On Volterra's population equation, *SIAM J. Appl. Math.* **14** (1966), 446–452.
[31] R. K. Miller, The topological dynamics of Volterra integral equations, *in* "Studies in Applied Mathematics," Vol. 5. Academic Press, New York,
[32] R. K. Miller, Almost periodic behavior of solutions of a nonlinear Volterra system, *Quart. Appl. Math.* **28** (1971), 553–570.
[33] R. K. Miller, A system of Volterra integral equations arising in the theory of super-fluidity, *Anal. Stiintifice Ale Univ. Al. I Cuza din Iasi* **19** (1973), 349–364.
[34] R. K. Miller, Linear Volterra integrodifferential equations as semigroups, *Funkcial. Ekvac.* **17** (1974), 35–51.
[35] R. K. Miller, Structure of solutions of unstable linear Volterra integrodifferential equations, *J. Differential Equations* **15** (1974), 129–157.
[36] R. K. Miller and G. R. Sell, Volterra integral equations and topological dynamics, *Mem. Amer. Math. Soc.* **102** (1970).
[37] V. M. Millionscikov, On the stability of characteristic exponents of limit solutions of linear equations, *Dokl. Akad. Nauk SSSR* **166** (1966), 34–37.
[38] V. M. Millionscikov, Instability of the characteristic exponents of statistically regular systems. *Mat. Zametki* **2** (1967), 315–318.
[39] V. M. Millionscikov, The connection between the stability of characteristic exponents and almost reducibility of systems with almost periodic coefficients, *Differencial'nye Uravnenija* **3** (1967), 2127–2134.
[40] V. M. Millionscikov, A metric theory of linear systems of differential equations, *Dokl. Akad. Nauk SSSR* **179** (1968), 20–23.
[41] V. M. Millionscikov, On the spectral theory of nonautonomous linear systems of differential equations, *Trudy Moskov. Mat. Obsc.* **18** (1968), 147–186.
[42] V. M. Millionscikov, Statistically regular systems, *Mat. Sb.* **75** (1968), 140–151.
[43] V. M. Millionscikov, A criterion for the stability of the probable spectrum of linear systems of differential equations with recurrent coefficients and a criterion for the almost reducibility of systems with almost periodic coefficients, *Mat. Sb.* **78** (1969), 179–201.
[44] V. M. Millionscikov, The instability of the singular exponents and the nonsymmetry of the relation of reducibility of linear systems of differential equations, *Differencial'nye Uravnenija* **5** (1969), 749–750.
[45] V. M. Millionscikov, Systems with integral division which are everywhere dense in the set of all linear systems of differential equations, *Differencial'nye Uravnenija* **5** (1969), 1167–1170.
[46] V. M. Millionscikov, Structurally stable properties of linear systems of differential equations, *Differencial'nye Uravnenija* **5** (1969), 1775–1784.
[47] V. M. Millionscikov, On the theory of characteristic Lyapunov exponents, *Mat. Zametki* **7** (1970), 503–513.
[48] V. M. Millionscikov, Linear systems of ordinary differential equations. *Actes Congr. Internat. Math., Nice, 1970* Tome 2, pp. 915–919. Gauthier-Villars, Paris, 1971. Also in *Differencial'nye Uravnenija* **7** (1971), 387–390.
[49] L. Nachbin, "The Haar Integral." Van Nostrand-Reinhold, Princeton, New Jersey, 1965.
[50] V. V. Nemytskii and V. V. Stepanov, "Qualitative Theory of Differential Equations." Princeton Univ. Press, Princeton, New Jersey, 1960.
[51] N. Onuchic, Invariance properties in the theory of ordinary differential equations with applications to stability theory, *SIAM J. Control* **5** (1971), 97–104.
[52] J. C. Oxtoby, "Measure and Category." Springer, New York, 1971.
[53] R. J. Sacker and G. R. Sell, Skew-product flows, finite extensions of minimal transformation groups and almost periodic differential equations. *Bull. Amer. Math. Soc.* **79** (1973), 802–805.

[54] R. J. Sacker and G. R. Sell, Existence of dichotomies and invariant splittings for linear differential systems I, *J. Differential Equations* **15** (1974), 429–458.

[55] R. J. Sacker and G. R. Sell, Existence of dichotomies and invariant splittings for linear differential systems II (to appear).

[56] R. J. Sacker and G. R. Sell, Existence of dichotomies and invariant splittings for linear differential systems III (to appear).

[57] R. J. Sacker and G. R. Sell, Lifting properties in skew-product flows with applications to differential equations (to appear).

[58] R. J. Sacker and G. R. Sell, A spectral theory for linear almost periodic differential equations. Preliminary report, *Proc. Internat. Conf. Differential Equations, Los Angeles, California* (September, 1974) (to appear).

[59] G. Seifert, A condition for almost periodicity with some applications to functional–differential equations, *J. Differential Equations* **1** (1965), 393–408.

[60] J. F. Selgrade, Isolated invariant sets for flows on vector bundles (to appear).

[61] G. R. Sell, Periodic solutions and asymptotic stability, *J. Differential Equations* **2** (1966), 143–157.

[62] G. R. Sell, Nonautonomous differential equations and topological dynamics I and II, *Trans. Amer. Math. Soc.* **127** (1967), 241–283.

[63] G. R. Sell, "Lectures on Topological Dynamics and Differential Equations." Van Nostrand-Reinhold, Princeton, New Jersey, 1971.

[64] G. R. Sell, Topological dynamical techniques for differential and integral equations, *in* "Ordinary Differential Equations," pp. 287–304. Academic Press, New York, 1972.

[65] G. R. Sell, A Tauberian condition and skew product flows with applications to integral equations, *J. Math. Anal. Appl.* **43** (1973), 388–396.

[66] G. R. Sell, Differential equations without uniqueness and classical topological dynamics, *J. Differential Equations* **14** (1973), 42–56.

[67] G. R. Sell, The Floquet problem for almost periodic linear differential equations, *Proc. Internat. Conf. Differential Equations, Dundee, Scotland, March, 1974* (to appear).

[68] G. R. Sell, Linear differential systems, Lecture Notes, Univ. of Minnesota (1974).

[69] T. Ura, Sur le Courant extérieur a une région invariante, prolongements d'une caractéristique et parde de stabilité, *Funkcial. Ekvac.* **2** (1959), 143–200.

[70] D. R. Wakeman, Invariance and stability properties for nonautonomous differential equations, *J. Differential Equations* **17** (1975), 259–295.

[71] J. A. Walker and E. F. Infante, Some results on the precompactness of orbits of dynamical systems, *CDS Tech. Rep.* **74-2**, Brown Univ. (1974).

[72] D. G. Weis, The asymptotic behavior of some Volterra integral equations arising from heat radiation problems, *J. Math. Anal. Appl.* (to appear).

[73] T. Yoshizawa, Asymptotic behavior of a perturbed system, *Internat. Symp. Nonlinear Differential Equations Nonlinear Mech.* pp. 80–85. Academic Press, New York, 1963.

Chapter 6: PARTIAL DIFFERENTIAL EQUATIONS

Nonlinear Oscillations under Hyperbolic Systems*

LAMBERTO CESARI
Department of Mathematics
University of Michigan, Ann Arbor, Michigan

Here we consider hyperbolic systems of first-order partial differential equations in the canonic forms of Courant–Lax and of Schauder. For these systems in a slab $D_a = [0 \leq x \leq a, y \in E^r]$ of the xy-space E^{r+1}, we formulate certain boundary value problems, and we state corresponding theorems of existence of the solutions, and of their uniqueness and continuous dependence on the data (Theorems A and B in Sections 1 and 2) [2, 3]. In particular, if all the data are periodic of given periods in the y-coordinates, then the solutions are also periodic of the same periods. Theorems A and B contain a specific hypothesis, which can be briefly stated by saying that the relevant matrices have "dominant main diagonal." In Section 3 we show by an example that the conclusions of the same theorems may not hold if such specific condition is not satisfied. In Section 5 we describe a boundary value problem of the above type, which has been proposed as a simple preliminary model for an otherwise extremely complex resonance phenomenon of nonlinear optics. By a well-known algebraic process (Section 4) we then show in Section 5 that the problem can be reduced to a Schauder system with relative boundary conditions, to which Theorem B applies.

1. A Boundary Value Problem for Courant–Lax Hyperbolic Systems

Quasi-linear hyperbolic systems of first-order partial differential equations are often written in the Courant–Lax canonic form

$$\partial z_i/\partial x + \sum_{k=1}^{r} \rho_{ik}(x, y, z)(\partial z_i/\partial y_k) = f_i(x, y, z), \qquad (x, y) \in D, \quad i = 1, \ldots, m,$$

(1)

* This research was partially supported by AFOSR Research Project 71-2122 at the University of Michigan.

where x, $y = (y_1, \ldots, y_r)$ are the $r + 1$ independent variables, $r \geq 1$, $z = z(x, y) = (z_1, \ldots, z_m)$, $(x, y) \in D$, are the m unknown functions, and ρ_{ik}, f_i are given functions defined in $D \times E^m$.

Here we assume that D is an unbounded region, namely, a (thin) slab $D = I_a \times E^r, I_a = [x \,|\, 0 \leq x \leq a]$. Instead of the usual Cauchy problem (with data at $x = 0$), we consider here the following more general boundary value problems, with data on m, in general distinct hyperplanes $x = a_i$, $0 \leq a_i \leq a$, $i = 1, \ldots, m$ (though we do not exclude that some or all of the a_i coincide).

I. Let a_i, $0 \leq a_i \leq a$, $i = 1, \ldots, m$, be arbitrary numbers, let $\psi_i(y)$, $y \in E^r$, $i = 1, \ldots, m$, be given functions, and let us determine a solution $z(x, y) = (z_1, \ldots, z_m)$, $(x, y) \in I_a \times E^r$, of (1) satisfying

$$z_i(a_i, y) = \psi_i(y), \qquad y \in E^r, \quad i = 1, \ldots, m. \tag{2}$$

In particular, if $a_i = 0$, $i = 1, \ldots, m'$, $a_i = a$, $i = m' + 1, \ldots, m$, $0 \leq m' \leq m$, problem (2) reduces to the problem of assigning m' of the functions z_i on the hyperplane $x = 0$ and the remaining functions z_i on the hyperplane $x = a$. If $a_i = 0$, $i = 1, \ldots, m$, problem (2) reduces to the Cauchy problem.

We shall also consider the following more general problem.

II. Let a_i, $0 \leq a_i \leq a$, $i = 1, \ldots, m$, be arbitrary numbers, let $\psi_i(y)$, $c_{ij}(y)$, $y \in E^r$, $i, j = 1, \ldots, m$, be given functions, and let us determine a solution $z(x, y) = (z_1, \ldots, z_m)$, $(x, y) \in D = I_a \times E^r$, of (1) satisfying

$$\sum_{j=1}^{m} c_{ij}(y) z_j(a_i, y) = \psi_i(y), \qquad i = 1, \ldots, m. \tag{3}$$

In other words, we assign m distinct linear combinations of the z_j with coefficients depending on y, on m different hyperplanes $x = a_i$.

Problem II reduces to problem I when $[c_{ij}(y)]$ is the identity matrix

$$[c_{ij}(y)] = [\delta_{ij}], \qquad \delta_{ii} = 1, \quad \delta_{ij} = 0, \quad \text{for} \quad i \neq j, \quad i, j = 1, \ldots, m.$$

We shall consider here problem I in all its generality, and problem II when $[c_{ij}(y)]$ is a matrix with "dominant main diagonal." Since each equation in (3) can be multiplied by arbitrary nonzero factors, it is convenient, for the sake of simplicity, to express this condition in the following form:

There is a number σ_0, $0 \leq \sigma_0 < 1$, such that for $c_{ij}(y) = \delta_{ij} + \tilde{c}_{ij}(y)$, we have

$$\sum_{j=1}^{m} |\tilde{c}_{ij}(y)| \leq \sigma_0 < 1, \qquad y \in E^r, \quad i = 1, \ldots, m. \tag{4}$$

If $[c_{ij}]$ is the identity matrix, that is, for problem I, we have $\tilde{c}_{ij}(y) = 0$ and we can take $\sigma_0 = 0$.

The problem of prescribing the values of the functions z_i on m distinct hyperplanes $x = a_i$ (problem I), or of m linear combinations of them on the same hyperplanes (problem II), for hyperbolic systems at the level of generality that we shall take into consideration is rather new. Problem I for very particular systems was considered by Niccoletti in 1897, and more recently was mentioned by Hukuhara. Aspects of these problems were discussed later by different authors (see references in [3]).

Let $a_0 > 0$, $\Omega > 0$ be given constants, and let I_{a_0}, Ω denote the intervals $I_{a_0} = [x \,|\, 0 \leq x \leq a_0] \subset E^1$, $\Omega = [-\Omega, \Omega]^m \subset E^m$. We denote by $|y|$, $|z|$ the norms in E^r and E^m defined by $|y| = \max [|y_k|, k = 1, \ldots, r]$, $|z| = \max [|z_i|, i = 1, \ldots, m]$.

By repeated use of Banach's fixed-point theorem we have proved in [2] the following theorem of existence, uniqueness, and continuous dependence on the data for system (1) with boundary conditions II.

Theorem A. Let $\rho_{ik}(x, y, z)$, $f_i(x, y, z)$, $i = 1, \ldots, m$, $k = 1, \ldots, r$, be functions defined in $I_{a_0} \times E^r \times \Omega$, measurable in x for every (y, z), continuous in (y, z) for every x, and let us assume that there are nonnegative functions $m(x), l(x), n(x), l_1(x), 0 \leq x \leq a_0, m, l, n, l_1 \in L_1[0, a_0]$, such that, for all (x, y, z), $(x, \bar{y}, \bar{z}) \in I_{a_0} \times E^r \times \Omega$, $i = 1, \ldots, m$, $k = 1, \ldots, r$, we have

$$|\rho_{ik}(x, y, z)| \leq m(x), \qquad |f_i(x, y, z)| \leq n(x),$$

$$|\rho_{ik}(x, y, z) - \rho_{ik}(x, \bar{y}, \bar{z})| \leq l(x)[|y - \bar{y}| + |z - \bar{z}|],$$

$$|f_i(x, y, z) - f_i(x, \bar{y}, \bar{z})| \leq l_1(x)[|y - \bar{y}| + |z - \bar{z}|].$$

Let $c_{ij}(y)$, $y \in E^r$, $i, j = 1, \ldots, m$, be given functions, and let us assume that there are constants $0 \leq \sigma_0 < 1$, Γ, $\Lambda \geq 0$, such that, for all y, $\bar{y} \in E^r$ and $i, j = 1, \ldots, m$, we have

$$|c_{ij}(y)| \leq \Gamma, \qquad |c_{ij}(y) - c_{ij}(\bar{y})| \leq \Lambda |y - \bar{y}|,$$

and (4) holds. Let $\psi_i(y)$, $y \in E^r$, $i = 1, \ldots, m$, be given functions continuous in E^r, and let us assume that there are constants ω_0, $\Lambda_0 \geq 0$, such that for all y, $\bar{y} \in E^r$ and $i = 1, \ldots, m$, we have

$$\omega_0 < (1 - \sigma_0)\Omega, \qquad |\psi_i(y)| \leq \omega_0, \qquad |\psi_i(y) - \psi_i(\bar{y})| \leq \Lambda_0 |y - \bar{y}|.$$

Then there is a constant $Q > 0$ and a number a, $0 < a \leq a_0$, such that for any system of numbers a_i, $0 \leq a_i \leq a$, $i = 1, \ldots, m$, there are functions $z(x, y) = (z_1, \ldots, z_m)$, $(x, y) = (x, y_1, \ldots, y_r) \in I_a \times E^r$, continuous in $I_a \times E^r$, absolutely continuous in x for every y, such that for all (x, y), $(x, \bar{y}) \in I_a \times E^r$ and $i = 1, \ldots, m$, we have

$$|z_i(x, y)| \leq \Omega, \qquad |z_i(x, y) - z_i(x, \bar{y})| \leq Q|y - \bar{y}|,$$

and the functions z_i satisfy (3) everywhere in E^r, and equations (1) almost everywhere in $I_a \times E^r$.

A proof of this existence theorem is given in [2] together with estimates of the value of a, $0 < a \leq a_0$, which depends only on the constants ω_0, Ω, σ_0, Λ, Γ, Λ_0, and given functions m, l, n, l_1. We also prove in [1] that the solution $z(x, y) = (z_1, \ldots, z_m)$ of the problem (1), (3) is unique in the class designated in [2] and depends continuously on $\psi = (\psi_1, \ldots, \psi_m)$ in the uniform topologies.

Whenever all functions $\rho_{ik}(x, y, z)$, $f_i(x, y, z)$, $\psi_i(y)$, $c_{ij}(y)$, are periodic of period $T_s > 0$ in each variable y_s, $s = 1, \ldots, r$, then the unique solution $z(x, y) = (z_1, \ldots, z_m)$ is also periodic of the same period T_s in each variable y_s, $s = 1, \ldots, r$.

2. Extension of Theorem A to Schauder's Systems

A canonic form more general than (1) for quasilinear hyperbolic systems was proposed by Schauder:

$$\sum_{j=1}^{m} A_{ij}(x, y, z)\left[\partial z_j/\partial x + \sum_{k=1}^{r} \rho_{ik}(x, y, z)(\partial z_j/\partial y_k)\right] = f_i(x, y, z),$$

$$(x, y) \in D, \quad i = 1, \ldots, m, \qquad (5)$$

where x, $y = (y_1, \ldots, y_r)$ are the independent variables, $z(x, y) = (z_1, \ldots, z_m)$ the m unknown functions, where $\rho_{ik}(x, y, z)$, $f_i(x, y, z)$ are as in Section 1, and $A_{ij}(x, y, z)$ are given bounded functions defined in $D \times \Omega$ with $\det [A_{ij}] \geq \mu > 0$, μ a constant. For $[A_{ij}(x, y, z)] = [\delta_{ij}]$, the identity matrix, system (5) reduces to system (1).

We shall assume as before that D is a (thin) slab $D = I_a \times E^r$, and we consider for (5) the same boundary value problems I and II of Section 1.

In order to present a general theorem of existence, uniqueness, and continuous dependence on the data, we assume that both matrices $[A_{ij}(x, y, z)]$ and $[c_{ij}(y)]$ have "dominant main diagonal" in a sense that will be explained below. First, as in Section 1, let $c_{ij}(y) = \delta_{ij} + \tilde{c}_{ij}(y), i, j = 1, \ldots, m$, and

$$\sigma_0 = \underset{i=1,\ldots,m}{\text{Max}} \underset{y \in E^r}{\text{Sup}} \sum_{j=1}^{m} |\tilde{c}_{ij}(y)|, \tag{6}$$

so that $\sigma_0 = 0$ when $[c_{ij}]$ is the identity matrix. Concerning the matrix $[A_{ij}(x, y, z)]$ first we denote by α_{ij} the cofactor of A_{ij} divided by det $[A_{ij}]$, that is, $\alpha_{ij} = (A^{-1})_{ji}$ and let

$$A_{ij}(x, y, z) = \delta_{ij} + \tilde{A}_{ij}(x, y, z), \qquad \alpha_{ij}(x, y, z) = \delta_{ij} + \tilde{\alpha}_{ij}(x, y, z),$$

$$\sigma_1 = \underset{i=1,\ldots,m}{\text{Max}} \underset{(x, y, z) \in D \times \Omega}{\text{Sup}} \sum_{h=1}^{m} |\tilde{A}_{ih}(x, y, z)|,$$

$$\sigma_2 = \underset{i=1,\ldots,m}{\text{Max}} \underset{(x, y, z) \in D \times \Omega}{\text{Sup}} \sum_{h=1}^{m} |\tilde{\alpha}_{hi}(x, y, z)|, \tag{7}$$

$$\sigma_3 = \underset{i=1,\ldots,m}{\text{Max}} \underset{(x, y, z) \in D \times \Omega}{\text{Sup}} \sum_{s=1}^{m} \sum_{h=1}^{m} |\tilde{\alpha}_{si}(x, y, z)| \, |\tilde{A}_{sh}(x, y, z)|,$$

$$\sigma = \sigma_1 + \sigma_2 + \sigma_3.$$

Then $\sigma \geq 0$, and $\sigma = 0$ when A is the identity matrix. In the existence theorem below, we shall assume

$$\sigma + \sigma_0 + \sigma\sigma_0 < 1. \tag{8}$$

Thus, if $[A_{ij}]$ is the identity matrix [system (1)], then $\sigma = 0$, and all we need is that $\sigma_0 < 1$. If $[c_{ij}]$ is the identity matrix (problem I), then $\sigma_0 = 0$, and all we need is that $\sigma < 1$.

By repeated use of Banach's fixed-point theorem we have proved in [3] the following theorem of existence, uniqueness, and continuous dependence on the data for system (5) with boundary conditions II.

Theorem B.   Let $A_{ij}(x, y, z), i, j = 1, \ldots, m$, be continuous functions on $I_{a_0} \times E^r \times \Omega$ with det $[A_{ij}] \geq \mu > 0$, μ a constant, and let us assume that there are constants H, $C > 0$ and a function $\mathring{m}(x) \geq 0$, $0 \leq x \leq a_0$, $\mathring{m} \in L_1[0, a_0]$, such that, for all $(x, y, z), (x, \bar{y}, \bar{z}), (\bar{x}, y, z) \in I_{a_0} \times E^r \times \Omega$ and all $i, j = 1, \ldots, m$, we have

$$|A_{ij}(x, y, z)| \leq H,$$

$$|A_{ij}(x, y, z) - A_{ij}(x, \bar{y}, \bar{z})| \leq C[|y - \bar{y}| + |z - \bar{z}|],$$

$$|A_{ij}(x, y, z) - A_{ij}(\bar{x}, y, z)| \leq \left| \int_{x}^{\bar{x}} \mathring{m}(\alpha) \, d\alpha \right|.$$

If $\alpha_{ij}(x, y, z)$ denotes the cofactor of A_{ij} in the matrix $[A_{ij}]$ divided by det $[A_{ij}]$, that is, $\alpha_{ij} = (A^{-1})_{ji}$, then certainly there are constants H', $C' > 0$, and a function $\overset{\circ}{m}'(x) \geq 0$, $0 \leq x \leq a_0$, $\overset{\circ}{m}' \in L_1[0, a_0]$ such that, as above,

$$|\alpha_{ij}(x, y, z)| \leq H',$$

$$|\alpha_{ij}(x, y, z) - \alpha_{ij}(x, \bar{y}, \bar{z})| \leq C'[|y - \bar{y}| + |z - \bar{z}|],$$

$$|\alpha_{ij}(x, y, z) - \alpha_{ij}(\bar{x}, y, z)| \leq \left| \int_x^{\bar{x}} \overset{\circ}{m}'(\alpha) \, d\alpha \right|.$$

Let $\rho_{ik}(x, y, z), f_i(x, y, z), i = 1, \ldots, m, k = 1, \ldots, r$, be functions defined in $I_{a_0} \times E^r \times \Omega$, measurable in x for every (y, z), continuous in (y, z) for every x, and let us assume that there are nonnegative functions $m(x)$, $l(x)$, $n(x)$, $l_1(x)$, $0 \leq x \leq a_0$, $m, l, n, l_1 \in L_1[0, a_0]$, such that for all (x, y, z), $(x, \bar{y}, \bar{z}) \in I_{a_0} \times E^r \times \Omega$, $i = 1, \ldots, m, k = 1, \ldots, r$, we have

$$|\rho_{ik}(x, y, z)| \leq m(x), \qquad |f_i(x, y, z)| \leq n(x),$$
$$|\rho_{ik}(x, y, z) - \rho_{ik}(x, \bar{y}, \bar{z})| \leq l(x)[|y - \bar{y}| + |z - \bar{z}|],$$
$$|f_i(x, y, z) - f_i(x, \bar{y}, \bar{z})| \leq l_1(x)[|y - \bar{y}| + |z - \bar{z}|].$$

Let $\psi_i(y)$, $c_{ij}(y)$, $y \in E^r$, $i, j = 1, \ldots, m$, be given bounded continuous functions in E^r, and let us assume that there are constants $\omega_0, \Gamma_0, \Lambda_0, \tau_0$, $0 < \omega_0 < \Omega, \Lambda_0, \tau_0 \geq 0$, such that for all $y, \bar{y} \in E^r$ and $i = 1, \ldots, m$, we have

$$|\psi_i(y)| \leq \omega_0, \qquad |\psi_i(y) - \psi_i(\bar{y})| \leq \Lambda_0|y - \bar{y}|,$$

$$|c_{ij}(y)| \leq \Gamma_0, \qquad \sum_{j=1}^m |c_{ij}(y) - c_{ij}(\bar{y})| \leq \tau_0|y - \bar{y}|.$$

With the notations (6)–(8) (with $a = a_0$) we also assume that $\sigma + \sigma_0 + \sigma\sigma_0 < 1$, and then $H \leq 1 + \sigma$, $H' \leq 1 + \sigma$, $\Gamma_0 \leq 1 + \sigma_0$.

Then, for $a, \omega_0, \tau_0, C, C'$ sufficiently small, $0 < a \leq a_0, \omega_0, \tau_0, C, C' > 0$, and for every system of numbers a_i, $0 \leq a_i \leq a$, $i = 1, \ldots, m$, there are a constant $Q > 0$, a function $\chi(x) \geq 0$, $0 \leq x \leq a$, $\chi \in L_1[0, a]$, and a vector function $z(x, y) = (z_1, \ldots, z_m)$, $(x, y) \in I_a \times E^r$, continuous in $I_a \times E^r$, satisfying (3) everywhere in E^r, satisfying (5) almost everywhere in $I_a \times E^r$, and such that, for all (x, y), $(x, \bar{y})$, $(\bar{x}, y) \in I_a \times E^r$ and $i = 1, \ldots, m$, we have

$$|z_i(x, y)| \leq \Omega, \qquad |z_i(x, y) - z_i(x, \bar{y})| \leq Q|y - \bar{y}|,$$

$$|z_i(x, y) - z_i(\bar{x}, y)| \leq \left| \int_x^{\bar{x}} \chi(\alpha) \, d\alpha \right|.$$

The function $z(x, y)$ is unique and depends continuously on $\psi(y) = (\psi_1, \ldots, \psi_m)$ for z and ψ in classes described in [3]. Also, computable estimates of a, ω_0, C, C', τ_0 are given in [3], which depend only on the constants Ω, Λ_0, σ, σ_0, and on the functions $\overset{\circ}{m}$, $\overset{\circ}{m}'$, m, n, l, l_1, but not on the numbers a_i, $0 \leq a_i \leq a$, $i = 1, \ldots, m$.

Again, as in Section 1, whenever all functions $A_{ij}(x, y, z)$, $\rho_{ik}(x, y, z)$, $f_i(x, y, z)$, $\psi_i(y)$, $c_{ij}(y)$ are periodic of period $T_s > 0$ in each variable y_s, $s = 1, \ldots, r$, then the unique solution $z(x, y) = (z_1, \ldots, z_m)$ is also periodic of the same period T_s in each variable y_s, $s = 1, \ldots, r$.

3. A Counterexample

We show here by an example that the uniqueness parts of Theorems A and B may not hold when the "dominant main diagonal" conditions are not satisfied.

Let $m = 2, r = 1, \rho_1, \rho_2$ constants, $\rho_1 \neq \rho_2, [c_{ij}] = [1, 1; 1, -1], f_1 = f_2 = 0$, $\psi_1 = \psi_2 = 0$, so that problem (1), (II) reduces to the linear homogeneous problem in the strip $0 \leq x \leq a$, $-\infty < y < +\infty$:

$$\partial z_1/\partial x + \rho_1 \, \partial z_1/\partial y = 0, \qquad \partial z_2/\partial x + \rho_2 \, \partial z_2/\partial y = 0,$$
$$z_1(0, y) + z_2(0, y) = 0, \qquad z_1(a, y) - z_2(a, y) = 0,$$

This system has ∞-many solutions. In particular, the following family of trigonometrical polynomial solutions is trivially singled out:

$$z_1(x, y) = \sum_{s=0}^{N} c_s \sin\left[(2s + 1)\pi b^{-1}(-\rho_1 x + y)\right],$$

$$z_2(x, y) = -\sum_{s=0}^{N} c_s \sin\left[(2s + 1)\pi b^{-1}(-\rho_2 x + y)\right],$$

where $b = (\rho_1 - \rho_2)a$, c_s arbitrary constants. Note that here we can take $a > 0$ as small as we want.

4. Reduction of Hyperbolic Systems in Two Independent Variables to Schauder's Canonic Form

Quasilinear hyperbolic systems in $m \geq 1$ unknowns $W = (W_1, \ldots, W_m)$ and two independent variables x, t, of the form

$$\sum_{j=1}^{m} [a_{ij}(x, t, W)(\partial W_j/\partial x) + b_{ij}(x, t, W)(\partial W_j/\partial t)] = c_i(x, t, W),$$

$$i = 1, \ldots, m, \qquad (9)$$

can be reduced solely by algebraic operations, as is well known, to Schauder's systems (5).

Indeed, let us denote by $A(\rho)$ the $m \times m$ matrix $A(\rho) = [a_{ij}\rho - b_{ij}, i, j = 1, \ldots, m]$, and by $\rho_1, \ldots, \rho_m$ the m roots, distinct or coincident, of the algebraic equation det $A(\rho) = 0$ (the characteristic equation). For every root ρ of this equation, we denote by M the usual algebraic multiplicity, and by μ, $1 \le \mu \le M$, the geometric multiplicity, that is, the dimension of the linear space of all vectors $h = \text{col } (h_1, \ldots, h_m)$ satisfying $(A(\rho))_{-1}h = 0$, or

$$\rho \sum_{i=1}^{m} a_{ij} h_i - \sum_{i=1}^{m} b_{ij} h_i = 0, \qquad i = 1, \ldots, m.$$

We assume here that the roots ρ of the equation det $A(\rho) = 0$ are all real, not necessarily distinct, but for each of them geometric and algebraic multiplicities coincide, or $\mu = M$. Then we can locally determine systems $h_r = \text{col } (h_{r1}, \ldots, h_{rm})$, $r = 1, \ldots, m$, of independent vectors such that $(A(\rho_r))_{-1}h_r = 0$, $r = 1, \ldots, m$. The last hypothesis we need is that we can choose vectors $h_r, r = 1, \ldots, m$, as above in such a way that det $(h_1, \ldots, h_m) \ne 0$ in the entire domain covered by (x, t, z). Then system (9) is equivalent to the one we obtain by taking linear combinations of them with factors $h_{r1}, \ldots, h_{rm}$, and then letting r run from 1 to m. In other words, we have the system

$$\sum_{i=1}^{m} h_{ri} \sum_{j=1}^{m} [a_{ij}(\partial W_j/\partial x) + b_{ij}(\partial W_j/\partial t)] = \sum_{i=1}^{m} h_{ri} c_i, \qquad r = 1, \ldots, m,$$

or

$$\sum_{j=1}^{m} \left[\left(\sum_{i=1}^{m} h_{ri} a_{ij} \right)(\partial W_j/\partial x) + \left(\sum_{i=1}^{m} h_{ri} b_{ij} \right)(\partial W_j/\partial t) \right] = \sum_{i=1}^{m} h_{ri} c_i, \qquad r = 1, \ldots, m,$$

where now $\rho_r \sum_{i=1}^{m} a_{ij} h_{ri} = \sum_{i=1}^{m} b_{ij} h_{ri}$. We have now the Schauder system

$$\sum_{j=1}^{m} \left(\sum_{i=1}^{m} a_{ij} h_{ri} \right)(\partial W_j/\partial x + \rho_r \, \partial W_j/\partial t) = \sum_{i=1}^{m} h_{ri} c_i,$$

or

$$\sum_{j=1}^{m} A_{rj}(\partial W_j/\partial x + \rho_r \, \partial W_j/\partial t) = F_r, \qquad r = 1, \ldots, m.$$

5. A Hyperbolic Problem in Nonlinear Optics

For a problem of nonlinear optics (see Remark 1), the following system of nonlinear Maxwell equations with boundary conditions has been proposed:

$$-\partial H/\partial x = k^2 \, \partial E/\partial t + \varepsilon E \, \partial E/\partial t, \qquad -\partial E/\partial x = \partial H/\partial t,$$
$$0 \leq x \leq a, \quad -\infty < t < \infty, \qquad (10)$$
$$E(0, t) + H(0, t) = 2\omega(t), \qquad E(a, t) - H(a, t) = 0, \qquad -\infty < t < \infty,$$

where $k > 1$ is a known constant, $\varepsilon > 0$ and $a > 0$ are "small" constants, and $\omega(t)$, $-\infty < t < +\infty$, is a known smooth periodic function of given period T. Relations (10) have been made dimensionless by choosing suitable units. This is a simple reduction of problem (10) to a Schauder system (5) with boundary conditions II, satisfying the "dominant main diagonal condition." By changing E, H into $k^{-1}u$ and v, and by changing t into $t = k\bar{t}$, problem (10) becomes

$$u_x + v_t = 0, \qquad v_x + (1 + \varepsilon'u)u_t = 0,$$
$$k^{-1}u(0, t) + v(0, t) = 2\overline{\omega}(t), \qquad k^{-1}u(a, t) - v(a, t) = 0, \qquad (11)$$

where $\varepsilon' = \varepsilon k^{-2}$ and we have simply written t for $\bar{t}$. By introducing new unknowns $U = 2^{-1}(u + v)$, $V = 2^{-1}(u - v)$, so that $u = U + V$, $v = U - V$, problem (11) becomes

$$U_x + V_x + U_t - V_t = 0,$$
$$U_x - V_x + (1 - \varepsilon'u)U_t + (1 + \varepsilon'u)V_t = 0,$$
$$2^{-1}(k^{-1} + 1)U(0, t) + 2^{-1}(k^{-1} - 1)V(0, t) = \overline{\omega}(t), \qquad (12)$$
$$2^{-1}(k^{-1} - 1)U(a, t) + 2^{-1}(k^{-1} + 1)V(a, t) = 0,$$

where $u = U + V$. System (12) is of the form (9) with $a_{11} = 1, a_{12} = 1, b_{11} = 1,$ $b_{12} = -1, a_{21} = 1, a_{22} = -1, b_{21} = 1 + \varepsilon'u, b_{22} = 1 + \varepsilon'u.$
The characteristic equation is now det $A(\rho) = 0$, or

$$\det [\rho - 1, \rho + 1; \rho - (1 + \varepsilon'u), -\rho - (1 + \varepsilon'u)] = 0,$$

or $\rho^2 = 1 + \varepsilon'u, u = U + V$. For $|U|, |V| \leq M$ and $0 \leq \varepsilon' < 1/2M$, the roots are real and distinct, $\rho_1 = (1 + \varepsilon'u)^{1/2}$, $\rho_2 = -(1 + \varepsilon'u)^{1/2}$. We shall now choose nonzero real vectors $h_1 = \text{col } (h_{11}, h_{12})$, $h_2 = \text{col } (h_{21}, h_{22})$ such that $(A(\rho_r))_{-1}h_r = 0$, $r = 1, 2$, that is,

$$(\rho_1 - 1)h_{11} + (\rho_1 - (1 + \varepsilon'u))h_{12} = 0,$$
$$(\rho_1 + 1)h_{11} + (-\rho_1 - (1 + \varepsilon'u))h_{12} = 0,$$

and analogous equations hold for h_{21}, h_{22} with ρ_2 replacing ρ_1. We take here

$$h_{11} = (\varepsilon' u)^{-1}(1 + \varepsilon' u - \rho_1), \qquad\qquad h_{12} = (\varepsilon' u)^{-1}(\rho_1 - 1),$$

$$h_{21} = (2 + \varepsilon' u - 2\rho_2)^{-1}(1 + \varepsilon' u - \rho_2), \qquad h_{22} = (2 + \varepsilon' u - 2\rho_2)^{-1}(\rho_2 - 1),$$

$$\Delta = \det [h_{11}, h_{12} ; h_{21}, h_{22}] = -2(1 + \varepsilon' u)^{1/2}((1 + \varepsilon' u)^{1/2} + 1)^{-2}.$$

Thus, for $\varepsilon' = 0, \Delta = -2^{-1} \neq 0$, and for $|U|, |V| \leq M$ and $\varepsilon' \geq 0$ sufficiently small, we also have $\Delta \neq 0$. System (21) is now equivalent to the system

$$h_{r1}(U_x + V_x + U_t - V_t) + h_{r2}(U_x - V_x + (1 + \varepsilon' u)U_t + (1 + \varepsilon' u)V_t) = 0,$$
$$r = 1, 2,$$

or, after manipulations, to the Schauder system,

$$(U_x + \rho_1 V_t) + \theta(V_x + \rho_1 V_t) = 0, \qquad (U_x + \rho_2 U_t) + (V_x + \rho_2 V_t) = 0,$$
$$2^{-1}(k^{-1} + 1)U(0, t) + 2^{-1}(k^{-1} - 1)V(0, t) = \overline{\omega}(t),$$
$$2^{-1}(k^{-1} - 1)U(a, t) - 2^{-1}(k^{-1} + 1)V(a, t) = 0,$$

where $\theta = [(1 + \varepsilon' u)^{1/2} - 1]/[(1 + \varepsilon' u)^{1/2} + 1]$. The boundary data are of type II with $c_{11} = c_{22} = 2^{-1}(k^{-1} + 1)$, $c_{12} = c_{21} = 2^{-1}(k^{-1} - 1)$; hence $\tilde{c}_{ij} = c_{ij} - \delta_{ij} = 2^{-1}(k^{-1} - 1)$, $\sigma_0 = 1 - k^{-1} < 1$. The coefficients of the system are $[A_{ij}] = [1, \theta; \theta, 1]$, and for $\varepsilon' = 0$, $\theta = 0$ and $[A_{ij}] = [\delta_{ij}]$, $\sigma = 0$. Thus, for $|U|, |V| \leq M$ and $\varepsilon \geq 0$ sufficiently small, we certainly have $\sigma + \sigma_0 + \sigma\sigma_0 < 1$.

Remark 1. If a monochromatic strong laser beam is focused on a thin crystal, then after emerging the light shows a measurable component of the second harmonic radiation. For instance, a ruby laser, 6940 Å, gives rise to the 3470 Å component. Franken and Ward [4] discuss at great length the experimental and theoretical basis of this phenomenon. (See also Bloembergen [1].) In his lecture [5], Graffi considered the main nonlinear problems of electromagnetism, and particularly those of nonlinear optics. From Graffi's exposition, as well as from those of the authors mentioned above, it appears that Maxwell equations (10) represent the simplest model for a preliminary study of phenomena of nonlinear optics, otherwise an extremely complex field. The boundary conditions for the plane waves in (10) can be justified as follows (from correspondence between Graffi and the author). If the crystal in the xyz-space is the slab $0 \leq x \leq a$, then in the empty space $-\infty < x < 0$ the linear Maxwell equations $-H_x = E_t$, $-E_x = H_t$ have solutions $E(x, t) = \varphi(x + t) + \psi(x - t)$, $H(x, t) = -\varphi(x + t) + \psi(x - t)$, ψ a known progressive wave, φ a regressive wave, and thus $E(0, t) + H(0, t) = 2\psi(-t)$. For $a < x < +\infty$, again $E(x, t) = \overline{\varphi}(x + t) + \overline{\psi}(x - t)$, $H(x, t) =$

$-\overline{\varphi}(x + t) + \overline{\psi}(x - t)$, with $\overline{\varphi} = 0$ (no regressive wave), and then $E(a, t)$ $- H(a, t) = 0$. If we assume that E and H are continuous across the surface of the crystal, then we have the boundary conditions $E(0, t) + H(0, t) = 2\omega(t)$, $E(a, t) - H(a, t) = 0$ for the nonlinear Maxwell equations (10) at $x = 0$ and $x = a$, with $\omega(t) = \psi(-t)$. The study of problem (10), with boundary conditions assigning the values of linear combinations of E and H on the surface of the slab, was the motivation of our work in [2, 3], some results of which we have presented in Sections 1–3.

Remark 2.   The same equations (10) in usual units are $-H_x = \varepsilon_2(E_t + \varepsilon E E_t)$, $-E_x = \mu_2 H_t$ in the crystal, that is, for $0 \leq x \leq a$ (and $-H_x = \varepsilon_1 E_t$, $-E_x = \mu H_t$ in the empty space, and thus for $x < 0$, and for $x > a$). Here ε_1, ε_2, μ_1, μ_2 are the dielectric constant and the magnetic permeability, in the crystal and in the empty space, respectively, $\varepsilon_1 \leq \varepsilon_2$, $\mu_1 = \mu_2$. Corresponding boundary conditions (10), and reduction to a Schauder system (5), with boundary conditions II, satisfying (8), valid for all values of the constants $\varepsilon_1 \leq \varepsilon_2$, $\mu_1 = \mu_2$, and $\varepsilon > 0$ sufficiently small, will be discussed elsewhere in papers by the author and Bassanini.

For boundary conditions of the Leontovich–Schelkunoff type in other problems of electromagnetism, compare Schelkunoff [8] and Graffi [6]. For other studies of nonlinear Maxwell equations, compare, e.g., Rivlin and Venkataraman [7].

REFERENCES

[1] N. Bloembergen, "Nonlinear Optics." Benjamin, New York, 1965.

[2] L. Cesari, A boundary value problem for quasilinear systems. *Rend. Mat. Univ. Parma* (to appear).

[3] L. Cesari, A boundary value problem for quasilinear hyperbolic systems in Schauder's canonic form. *Ann. Scuola Norm. Sup. Pisa* (4) **1** (1974), 311–358.

[4] P. F. Franken and J. F. Ward, Optical harmonics and nonlinear phenomena. *Rev. Mod. Phys.* **35** (1963), 23–39.

[5] D. Graffi, Problemi nonlineari nella teoria del campo elettromagnetico. *Mem. Accad. Sci. Modena* **11** (1967), 172–196.

[6] D. Graffi, Sulle condizioni al contorno approssimate nell'elettromagnetismo. *Rend. Accad. Sci. Bologna* **5** (1958), 88–94.

[7] R. S. Rivlin and R. Venkataraman, Harmonic generation in an electromagnetic wave. *J. Appl. Math. Phys.* **24** (1973), 661–675.

[8] S. A. Schelkunoff, "Electromagnetic Waves." Van Nostrand, New York, 1943.

Liapunov Methods for a One-Dimensional Parabolic Partial Differential Equation

NATHANIEL CHAFEE

School of Mathematics
Georgia Institute of Technology, Atlanta, Georgia

In recent years several authors have used techniques based on Liapunov's second method to investigate stability phenomena for partial differential equations [1–4, 6–9].

The first step in such an investigation is to interpret the given partial differential equation as a dynamical system on some appropriate phase space. The remaining part of the investigation is to use the already existing Liapunov theory for general dynamical systems [10–12] to help determine the stability properties of the particular dynamical system at hand. This Liapunov theory was developed mainly in the context of ordinary and functional differential equations. Applying it to the study of partial differential equations seems to involve serious difficulties, but the articles already cited [1–4, 6–9] indicate that success is possible.

The subject of this chapter is some of the author's recent work involving the procedure just indicated. This work concerns the following equations:

$$u_t(x, t) = u_{xx}(x, t) + f(x, u(x, t)), \qquad 0 < x < \pi, \quad 0 < t < s, \qquad \text{(1a)}$$

$$u_x(0, t) = u_x(\pi, t) = 0, \qquad 0 \le t < s, \qquad \text{(1b)}$$

$$u(x, 0) = \phi(x), \qquad 0 \le x \le \pi. \qquad \text{(1c)}$$

Here, f is a given C^2-smooth function mapping $[0, \pi] \times R$ into R, where R denotes the real number system. The initial datum ϕ is any C^1-smooth function mapping $[0, \pi]$ into R such that $\phi'(0) = \phi'(\pi) = 0$. Equations (1) are to be solved for a function u continuously mapping a domain of the form $[0, \pi] \times [0, s), 0 < s \le +\infty$, into R such that

(i) u_x exists and is continuous on $[0, \pi] \times [0, s)$;
(ii) u_t and u_{xx} exist and are continuous on $(0, \pi) \times (0, s)$;
(iii) u satisfies Eqs. (1) on the domains indicated.

We are interested in the asymptotic behavior of solutions of Eqs. (1).

In a separate article [5] we have given a detailed report of our investigations for Eqs. (1). Here we shall only describe some of our more salient results.

Let X be the space of all C^1-smooth functions $\phi: [0, \pi] \to R$ such that $\phi'(0) = \phi'(\pi) = 0$. For each $\phi \in X$ we let $\|\phi\|^{(0)} = \sup\{|\phi(x)|: 0 \le x \le \pi\}$ and $\|\phi\|^{(1)} = \|\phi\|^{(0)} + \sup\{|\phi'(x)|: 0 \le x \le \pi\}$. The space X as normed by $\|\ \|^{(1)}$ will be our phase space for Eqs. (1). The norm $\|\ \|^{(0)}$ will only play an auxiliary role.

For any $\phi \in X$, Eqs. (1) have a unique noncontinuable solution $u(\phi)$ defined on a domain $[0, \pi] \times [0, s(\phi))$, where $0 < s(\phi) \le +\infty$. For each $x \in [0, \pi]$, $t \in [0, s(\phi))$, we let $u(x, t; \phi)$ denote the value of $u(\phi)$ at (x, t). Thus, we obtain a strongly continuous semigroup $\{U(t)\}$ on X by setting $U(t)\phi = u(\cdot, t; \phi)$ for all $\phi \in X$, $t \in [0, s(\phi))$.

For any $\phi \in X$ we let $\gamma(\phi)$ denote the *orbit* of $u(\phi)$, that is, $\gamma(\phi) = \{U(t)\phi: 0 \le t < s(\phi)\}$.

By an *equilibrium solution* of (1) we mean an element $\psi \in X$ such that $s(\psi) = +\infty$ and $\gamma(\psi) = \{\psi\}$.

Theorem 1. Let $\phi \in X$ and suppose that $\gamma(\phi)$ is bounded with respect to $\|\ \|^{(0)}$. Then $s(\phi) = +\infty$ and, with respect to $\|\ \|^{(1)}$, the solution $u(\phi)$ has a nonempty compact connected invariant ω-limit set $\omega(\phi)$. Moreover, $U(t)\phi \to \omega(\phi)$ as $t \to +\infty$, the convergence here being relative to $\|\ \|^{(1)}$.

The proof occurs in two main steps. First, one shows that $\gamma(\phi)$ must be precompact relative to $\|\ \|^{(1)}$. This part of the proof very much involves the smoothness properties of solutions of (1). Second, one uses standard arguments from dynamical systems theory to obtain the existence of $\omega(\phi)$ and its required properties.

Now we introduce a Liapunov functional $V: X \to R$ by setting

$$V(\phi) = \int_0^\pi \left[\tfrac{1}{2}\phi'(x)^2 - \int_0^{\phi(x)} f(x, \eta)\, d\eta \right] dx, \qquad \phi \in X.$$

One can show that for any $\phi \in X$,

$$\dot{V}(U(t)\phi) = -\int_0^\pi u_t(x, t; \phi)^2\, dx, \qquad 0 < t < s(\phi).$$

This result together with the LaSalle invariance principle [12] leads us to the following theorem.

Theorem 2. If ϕ and $\omega(\phi)$ are as in Theorem 1, then each element $\psi \in \omega(\phi)$ is an equilibrium solution of Eqs. (1).

Theorem 2 can be used to obtain stability properties for equilibrium solutions of Eqs. (1). Suppose that $\psi \in X$ is such an equilibrium solution. Without loss of generality we can assume that ψ is the origin in X. Thus,

$$f(x, 0) = 0, \qquad 0 \le x \le \pi. \tag{2}$$

For any number $b > 0$ we let $Q[0, b]$ be the set of all $\phi \in X$ such that $0 \le \phi(x) \le b$ for every $x \in [0, \pi]$.

Theorem 3. Given Eq. (2), suppose that there exists a number $b > 0$ such that

(i) $f(x, \xi) \le 0$ for all $x \in [0, \pi]$ and $\xi \in (0, b]$; and
(ii) for any $\xi \in [0, b]$ there exists an $\bar{x} \in [0, \pi]$ such that $f(\bar{x}, \xi) < 0$.

Then, for each $\phi \in Q[0, b]$ we have $s(\phi) \le +\infty$, $\gamma(\phi) \subseteq Q[0, b]$, and $\| U(t)\phi \|^{(1)} \to 0$ as $t \to +\infty$.

On the basis of Theorem 3 one can readily formulate criteria for instability and asymptotic stability of the equilibrium solution $\psi = 0$. In the case of asymptotic stability one also obtains a region of attraction for $\psi = 0$ (see [5, Section 5]).

By a nonconstant equilibrium solution of Eqs. (1) we mean an equilibrium solution ψ that as a function on $[0, \pi]$ is nonconstant.

We close this chapter with an interesting instability theorem for which we must assume that the function f in Eq. (1a) does not depend on x. That is, the domain of f is R and Eq. (1a) has the form

$$u_t(x, t) = u_{xx}(x, t) + f(u(x, t)), \qquad 0 < x < \pi, \quad 0 < t < s.$$

Theorem 4. If, in the sense just indicated, f does not depend on x and if ψ is an isolated nonconstant equilibrium solution of Eqs. (1), then ψ is unstable.

REFERENCES

[1] J. F. G. Auchmuty, Lyapunov methods and equations of parabolic type, *Lecture Notes in Mathematics* **322**, Springer-Verlag, New York, 1973.
[2] J. M. Ball, Stability theory for an extensible beam, *J. Differential Equations* **14** (1973), 399–418.

[3] N. Chafee and E. F. Infante, A bifurcation problem for a nonlinear partial differential equation of parabolic type, *Appl. Anal.* **4** (1974), 17–37.

[4] N. Chafee, A stability analysis for a semilinear parabolic partial differential equation, *J. Differential Equations* **17** (1974), 522–540.

[5] N. Chafee, Asymptotic behavior for solutions of a one-dimensional parabolic equation with homogeneous Neumann boundary conditions, *J. Differential Equations* **18** (1975), 111–134.

[6] C. M. Dafermos, An invariance principle for compact processes, *J. Differential Equations* **9** (1971), 239–252.

[7] C. M. Dafermos, Applications of the invariance principle for compact processes, I. Asymptotically dynamical systems, *J. Differential Equations* **9** (1971), 291–299.

[8] C. M. Dafermos, Uniform processes and semicontinuous Liapunov functionals, *J. Differential Equations* **11** (1972), 401–415.

[9] C. M. Dafermos, Applications of the invariance principle for compact processes. II. Asymptotic behavior of solutions of a hyperbolic conservation law, *J. Differential Equations* **11** (1972), 416–424.

[10] J. K. Hale and E. F. Infante, Extended dynamical systems and stability theory, *Proc. Nat. Acad. Sci. U.S.A.* **58** (1967), 405–409.

[11] J. K. Hale, Dynamical systems and stability, *J. Math. Anal. Appl.* **26** (1969), 39–59.

[12] J. P. LaSalle, An invariance principle in the theory of stability, *Int. Symp. Differential Equations Dynam. Syst.* (J. K. Hale and J. P. LaSalle, eds.), pp. 277–286. Academic Press, New York, 1967.

Discontinuous Periodic Solutions of an Autonomous Wave Equation

J. P. FINK, WILLIAM S. HALL, and A. R. HAUSRATH
Department of Mathematics
University of Pittsburgh, Pittsburgh, Pennsylvania

In this chapter we indicate how a convergent two-variable procedure developed by the authors [2] can be used to analyze the behavior of the autonomous nonlinear wave equation

$$y_{1t} = y_{2x} + \varepsilon(y_1 - y_1^3), \qquad y_{2t} = y_{1x}. \tag{1}$$

As we shall see, the interesting qualitative information about (1), such as its asymptotic behavior and the existence and stability of periodic solutions, is conveniently analyzed in terms of a certain ordinary differential equation in the "slow" time $\tau = \varepsilon t$. The analysis produces quantitative data as well, so we can also present explicit formulas approximating the transient behavior for long but finite times and the form of the infinitely many periodic solutions to (1).

The main result for (1) is that a solution having a 2π-periodic initial condition converges with increasing time to a time-periodic limit function of the same period. The limiting value is approximated by combinations of left- and right-traveling waves. When viewed in the moving frame, each appears as a bounded measurable (L_∞) discontinuous function that is positive when its initial value is positive, negative when it is negative, and zero at the zeros of the initial point. The positive and negative amplitudes of the wave are the same at each point where it is nonzero. Its magnitude depends only on the measure of the set of zeros of the initial function, but not on the initial amplitude. An initial point having a set of zeros of measure zero has a limit solution that is exponentially asymptotically stable and that acts as a limit cycle since it attracts to it any other initial value with the same general shape.

Equation (1) arises in a variety of applications. For example, suppose electromagnetic plane waves are traveling between parallel conducting surfaces in a region where the conductivity varies with the electric field as $\sigma_0(\alpha|E|^2 - 1)$. An application of Maxwell's equations and a suitable scaling of dependent and independent variables gives (1) along with the boundary conditions

$$y_1(t, 0) = y_1(t, \pi) = 0. \tag{2}$$

Another example is a resistance-loaded, battery-driven transmission line in which the series resistance per unit length is zero but the shunt leakage has a cubic characteristic approximating that of an Esaki diode. After a suitable analysis it is found that the line voltage and current obey (1) and (2). Finally, there is an equation related to (1),

$$z_{tt} - z_{xx} = \varepsilon(z_t - z_t^3), \qquad z(t, 0) = z(t, \pi) = 0, \tag{3}$$

which arises in the study of galloping oscillations of a power line in the wind. The standard transformation $y_1 = z_t$ and $y_2 = z_x$ reduces (3) to (1) and (2).

A few authors have discussed (1), (3), and related equations. Most of these studies are formal, such as the papers of Chikwendu and Kevorkian [1], Keller and Kogelman [4], and Myerscough [6]. However, Kurzweil [5] has given a rigorous analysis of the van der Pol string

$$z_{tt} - z_{xx} = \varepsilon(1 - z^2)z_t, \qquad z(t, 0) = z(t, 1) = 0, \tag{4}$$

and the authors have studied (1) in [3]. Recently we have obtained many new results about this equation and some appear here for the first time.

The analysis of (1) begins after it is transformed to a more convenient form. We note that (1) is an evolution equation

$$\dot{y} = Ay + \varepsilon F(y), \tag{5}$$

where all the solutions to the unperturbed part are 2π-periodic in t when the initial values are 2π-periodic in x (the D'Alembert representation). Let $\{E(t)\}$ be the group of operators solving $\dot{y} = Ay$. Put $y(t) = E(t)u(t)$ to obtain the equation for u,

$$\dot{u} = \varepsilon f(t, u) = \varepsilon E(-t)F(E(t)u), \tag{6}$$

where f is a 2π-periodic Lipschitzian vector field.

Generally we anticipate that a solution to (6) will contain oscillations at the rate of the period of the vector field with an amplitude varying slowly with time. To pick out the periodic part we use a device due to Cesari and Hale and replace (6) by the relation

$$dw/d\sigma = \varepsilon\{f(\sigma, w) - Mf(w)\}, \tag{7}$$

where M is the mean value operator

$$Mf(w) = \frac{1}{2\pi} \int_0^{2\pi} f(s, w(s)) \, ds. \tag{8}$$

Since the vector field in (7) has no mean value, every solution $w(\sigma, v, \varepsilon)$ with initial value v is 2π-periodic in σ. Of course, it does not satisfy (6). However, we can show that for each v in L_∞, $w(\sigma, v, \varepsilon)$ and its Frechet derivative in v, $w_v(\sigma, v, \varepsilon)$, form σ-continuous Lipschitzian vector fields themselves. This is true even though the right-hand side of (7) is not continuous in σ on L_∞.

The initial value v is still at our disposal. Let us vary it so that

$$u(t, \varepsilon) = w(\sigma, v(\tau), \varepsilon), \qquad \sigma = t, \quad \tau = \varepsilon t, \tag{9}$$

will satisfy (6). An application of the chain rule to (9) along with (7) and (6) shows that v must satisfy

$$w_v(\tau/\varepsilon, v, \varepsilon) \, dv/d\tau = Mf(w(\cdot, v, \varepsilon)). \tag{10}$$

From the representation of w as an integral equation (valid when ε is small)

$$w(\sigma, v, \varepsilon) = v + \varepsilon \int_0^\sigma \{f(s, w(s, v, \varepsilon)) - Mf(w(\cdot, v, \varepsilon))\} \, ds, \tag{11}$$

it can be deduced that w_v^{-1} exists and is bounded and Lipschitz continuous in v when ε is small. Thus, (10) can be solved by Picard's theorem on some interval $0 \leq \tau \leq a$.

Let $v(\tau, \varepsilon)$ satisfy (10) on $[0, a]$. Because the integrand in (11) has no mean value, $u(t, \varepsilon) - v(\varepsilon t, \varepsilon)$ is $O(\varepsilon)$ as $\varepsilon \to 0$ on $0 \leq t \leq a/\varepsilon$. Furthermore, if $v(\varepsilon)$ is a (stable) equilibrium point to (10), then $w(t, v(\varepsilon), \varepsilon)$ is a (stable) 2π-periodic solution to (6). Using the continuity of $v(\tau, \varepsilon)$ in ε, it follows that the solutions $v(\tau)$ to (10) when $\varepsilon = 0$,

$$\frac{dv}{d\tau} = \frac{1}{2\pi} \int_0^{2\pi} f(s, v) \, ds, \tag{12}$$

are also asymptotic to the solutions to (6). Finally, isolated equilibrium points v_0 to (12) correspond to periodic solutions of (6) and these are exponentially asymptotically stable if the same is true for v_0.

Equation (12) is, of course, the classical averaged equation for (6). Although the results described above are simply the usual conclusions of the method of averaging, our techniques are neither generalizations nor applications of that theory.

After a tedious calculation we obtain the specific form of Eq. (12):

$$v_1' = \tfrac{1}{2}v_1 - \tfrac{1}{8}\{v_1{}^3 + 3v_1v_2{}^2 + 3\alpha v_1\},$$
$$v_2' = \tfrac{1}{2}v_2 - \tfrac{1}{8}\{v_2{}^3 + 3v_1{}^2v_2 + 3\alpha v_2\},$$

(13)

where the functional α is given by

$$\alpha(\tau) = \frac{1}{2\pi} \int_0^{2\pi} \{v_1{}^2(\tau, x) + v_2{}^2(\tau, x)\}\, dx.$$

(14)

These can be almost uncoupled by letting $v_1 = w_1 + w_2$ and $v_2 = w_1 - w_2$ to get

$$w_i' = (\tfrac{1}{2} - \tfrac{3}{4}\beta)w_i - \tfrac{1}{2}w_i{}^3,$$

for each $i = 1, 2$. Here β has the same form as α but with w_1 and w_2 replacing v_1 and v_2.

If the values of $w_i(0, x) = a_i(x)$ are specified, then the solution to (15) can be reduced to quadrature. These formulas, which are too long to give here, describe the approximate transient behavior for times of order ε^{-1} for any given initial value.

By letting $\tau \to +\infty$ in $w_i(\tau, x)$, the equilibrium points for (13) can be found. These are simply sums and differences of w_1 and w_2, where

$$w_i(x) = \{2/(2 + 3A)\}^{1/2} \operatorname{sgn} a_i(x).$$

(16)

Here $A = (A_1 + A_2)/2A$ and A_i is the measure of the set where $a_i(x) \neq 0$. A direct application of the implicit function theorem shows that each $w_i(x)$ is isolated and so corresponds to a periodic solution.

The variational equation for (13) is quite interesting. Its form varies with x. At those x's where $a_i(x) = 0$, it has an unstable node. Even where $a_i(x) \neq 0$, the solution is unstable if $A < 2$. However, if $A = 2$ and $a_i(x) \neq 0$, then the solution decays as $e^{-t/4}$. Thus, the initial points whose sets of zeros have measure zero generate exponentially asymptotically stable periodic solutions. Furthermore, each of these attracts any initial function having the same signum value as the limit function and thus acts as a limit cycle by drawing to it large classes of initial points.

REFERENCES

[1] S. C. Chikwendu and J. Kevorkian, A perturbation method for hyperbolic equations with small nonlinearities, *SIAM J. Appl. Math.* **22** (1972), 235–258.

[2] J. P. Fink, W. S. Hall, and A. R. Hausrath, A convergent two-time method for periodic differential equations, *J. Differential Equations* **15** (1974), 459–498.

[3] J. P. Fink, W. S. Hall, and A. R. Hausrath, Discontinuous periodic solutions for a nonlinear, autonomous wave equation, *Proc. Royal Irish Acad.* (to appear).

[4] J. B. Keller and S. Kogelman, Asymptotic solutions of initial value problems for nonlinear partial differential equations, *SIAM J. Appl. Math.* **18** (1970), 748–758.

[5] J. Kurzweil, van der Pol perturbations of the equation for a vibrating string, *Czechoslovak Math. J.* **17** (1967), 588–608.

[6] C. J. Myerscough, The growth of full span galloping oscillations, Central Electricity Research Laboratories, Leatherhead, Surrey, England, Lab. notes No.'s RD/L/N271/73 and RD/L/N51/72.

Continuous Dependence of Forced Oscillations for $u_t = \nabla \cdot \gamma(|\nabla u|)\nabla u$

THOMAS I. SEIDMAN

Department of Mathematics
University of Maryland, Baltimore County, Baltimore, Maryland

We consider the autonomous nonlinear parabolic equation of the title with forcing provided by t-periodic boundary data. Under appropriate conditions on $\gamma(\cdot)$ there is a unique t-periodic weak solution and we are concerned to show that this depends continuously on the data and on the function γ. The equation arises, for example, in the analysis of induced currents in a nonlinearly ferromagnetic material (∇u is the field, γ the reluctivity).

Let Ω be a bounded region in $\mathfrak{R}^n$ with smooth boundary $\partial\Omega$. With $T := \mathfrak{R}/Z$ let $Q := \Omega \times T$ and $\Sigma := \partial\Omega \times T$. Considering functions and data as defined on Q or Σ "builds in" the periodicity. Thus, rather than seeking periodicity among solutions (e.g., by applying a fixed-point theorem to the period map for the initial value problem), we look for solutions in a space of t-periodic functions.

It is convenient to impose the conditions on $\gamma(\cdot)$ in terms of $g(r) := r\gamma(r)$; we assume g is positive and differentiable for $r > 0$ and that, for some $p > 1$,

(i) $g(r) = \mathcal{O}(r^{p-1})$ as $r \to \infty$;
(ii) $m(r) > 0$ for $r > 0$, where $m(r) := \inf\{\rho^{2-p}g'(\rho) : \rho \geq r\}$.

With this p, let

$$Y := \{u \text{ defined on } Q : \nabla u \in L_p(Q \to \mathfrak{R}^n)\},$$
$$Y_0 := \{v \in Y : v \text{ satisfies } homogeneous \text{ b.c.}\}.$$

We assume that the b.c. are such that Y_0 is a Banach space with

$$\|u\| := \left[\int_Q |\nabla u|^p\right]^{1/p}$$

(this follows for Dirichlet b.c. by the Poincaré inequality; for Neumann b.c., as in the application, it is necessary to impose an auxiliary condition on Y_0 such as $\int_\Omega u \equiv 0$). Let

$$X := \left\{ \varphi \in Y : \left[v \mapsto \int_Q \varphi_t v \right] \in Y_0^* \right\}$$

and assume the boundary data $\tilde{\varphi}$ are taken from a space F (of functions on Σ) for which the extension map $\tilde{\varphi} \mapsto \varphi$ (φ satisfying the b.c. with data $\tilde{\varphi}$) is continuous from F to X. Thus, $v := u - \varphi$ is to satisfy

$$v_t = \nabla \cdot \gamma(|\nabla v + \nabla \varphi|)[\nabla v + \nabla \varphi] - \varphi_t$$

in the sense that $v_t = -A(v + \varphi) - \varphi_t$, with A given by

$$A(u)w := \int_Q \gamma(|\nabla u|)\nabla u \cdot \nabla w, \qquad u \in Y, \qquad w \in Y_0.$$

From the conditions on $\gamma(\cdot)$ it follows that $A: Y \to Y_0^*$ is well defined and continuous.

Lemma. For $u_1, u_2 \in Y$ such that $(u_1 - u_2) \in Y_0$,

$$[A(u_1) - A(u_2)](u_1 - u_2) \geq \Psi(\|u_1 - u_2\|; \|u_2\|),$$

with

$$\Psi(r, R) := r^p m(cr) K_p(r/R).$$

Here $c := \frac{1}{8}[4 \text{ meas } (\Omega)]^{1/p}$ and $K_p(s)$ is independent of s for $p \geq 2$ and $\mathcal{O}(s^{2-p})$ as $s \to 0$ for $1 < p < 2$; $p, m(\cdot)$ provide the only dependence of Ψ on $\gamma(\cdot)$.

Suppose now that $v^0 = u^0 - \varphi^0$ (with $v^0 \in Y_0$ and $\varphi^0 \in X$) satisfies $v_t^0 = -A(v^0 + \varphi^0) - \varphi_t^0$ and let $v \in Y_0$ satisfy $v_t = -A(v + \varphi) - \varphi_t$ for some $\varphi \in X$. Since $\int_Q v_t v = 0$ for differentiable $v \in Y$, one has

$$[A(v + \varphi) - A(v^0 + \varphi)](v - v^0)$$

$$= [A(v^0 + \varphi^0) - A(v^0 + \varphi)](v - v^0) + \int_Q (\varphi - \varphi^0)_t(v - v^0).$$

The right-hand side goes to 0 as $\varphi \to \varphi^0$ in X whence, by the lemma, $v \to v^0$ in Y_0, so $u = v + \varphi \to u^0$ in Y.

Similarly, we consider $\gamma \to \gamma^0$ with γ, γ^0 satisfying the conditions uniformly [i.e., fixed p and $\mathcal{O}(r^{p-1})$, $m(\cdot)$ uniform in (i), (ii), respectively] and let u, u^0 satisfy the corresponding equations with the same data. Then

$$[A(u) - A(u^0)](u - u^0) = [A(u^0) - A^0(u^0)](u - u^0) \le C\rho\|u - u^0\|,$$

with C depending on $\|u^0\|$ (but not on γ or otherwise on γ^0, φ^0) and

$$\rho := \sup \{(1 + r)^{1-p}|g(r) - g^0(r)| : r > 0\}.$$

Thus, by the lemma, $u \to u^0$ in Y as $\rho \to 0$ $(\gamma \to \gamma^0)$.

REFERENCE

T. I. Seidman, "Periodic Solutions of a Non-Linear Parabolic Equation," *J. Differential Equations* (to appear).

Partial Differential Equations and Nonlinear Hydrodynamic Stability

J. T. STUART

Mathematics Department
Imperial College, London, England

1. Introduction

Hydrodynamic stability is concerned with the evolution of laminar, or streamline, flow to the turbulent, or eddying, form. Often transition to turbulence entails an initial bifurcation from a given simple laminar flow to a complicated laminar flow of wavelike structure. The greatest, albeit modest, success of applied mathematics in the subject lies in the description of this relatively simple part of transition. Even after many decades of effort, we have no complete mathematical or physical theory of transition to turbulence. For this reason, this chapter is concerned with transitions or bifurcations from a basic form of laminar flow to a more complicated, wavelike form.

Many technical and detailed mathematical problems have been studied. For the purposes of this chapter, we shall refer to three such examples, chosen from areas that we have found to be of special interest. We should emphasize that we are concerned with formal applied mathematics: apart from just a few papers, which have appeared in the last few years, the subject has not been developed rigorously in the truly mathematical sense. The three examples of this chapter, the study of which would benefit from greater rigor, concern (i) wavy Taylor vortices in the concentric case, (ii) Taylor vortices in the eccentric case, and (iii) wave systems in plane Poiseuille flow. These topics are treated briefly in Sections 2–4, in a deliberately simplified manner.

2. Wavy Taylor Vortices

We consider two long concentric circular cylinders of radii a (inner) and b (outer), the speed of the inner being q_1 with the outer at rest.

Moreover we let $d = b - a$, and denote the kinematic viscosity by v. It is known that, if the parameter $Ta = (q_1 a/v)^2 (d/a)^3$ exceeds a critical value of order 1700, the first bifurcation point, the simple laminar circumferential flow is replaced by Taylor-vortex flow, which is periodic along the axis and is composed of a steady swirling in the form of toroidal vortices, superimposed on a circumferential flow. In fact, the simple laminar flow becomes unstable and bifurcates to the Taylor-vortex flow, the axial wavelength being given experimentally at the critical value of the Taylor number.

At a still higher Taylor number, constituting a second bifurcation point, the Taylor-vortex flow itself becomes unstable and bifurcates to a stable wavy vortex flow, in which the Taylor vortices develop a periodicity and propagate with time in the circumferential direction.

A quantitative measure of the flow field is given by the torque, which is required to be exerted on the inner cylinder in order to maintain the flow. In simple laminar flow the torque rises with Taylor number. Above the first bifurcation point, moreover, the stable Taylor vortex flow requires a greater torque than the simple laminar flow, which is now unstable; in addition, the torque continues to rise with the Taylor number. Above the second bifurcation point, however, the stable wavy vortex flow has a *lower* torque than the unstable Taylor-vortex flow at the same Taylor number.

The study of the instability of simple laminar flow and of the development of Taylor-vortex flow above the first bifurcation point has made considerable strides, both experimentally and theoretically, and these have been summarized by Stuart [1]. Agreement between theory and experiment is good. Correspondingly, the problem of instability of Taylor-vortex flow has been studied, with reasonable agreement with experiment as to the location of the second bifurcation Taylor number. The work is also described in [1]. More recently, Eagles [2] has calculated the torque of wavy vortices and shown that, subject to certain assumptions, the calculated wavy-vortex torque is lower than that of Taylor vortices at the same Taylor number, in agreement with experiment.

Problems remain to be solved: for example, the second bifurcation point, as it has been described here, is in truth a cluster of bifurcation points, with different circumferential wave numbers; the mechanism of choice of the circumferential wave number is not known. Moreover, to our knowledge, there is no mathematically rigorous treatment of the second bifurcation, which involves time and angular dependence. The field seems ripe for mathematical treatment.

3. Eccentric Taylor Vortices

In Section 2, the simple laminar flow, whose instability is first treated, is very simple indeed. In practice, however, more complex basic laminar flows arise, for example, in the aerodynamical and geophysical sciences. Usually the basic flow varies significantly in the flow direction, in contrast to the simple circumferential flow between concentric circular cylinders discussed previously. An example of this type occurs in the journal bearing of lubrication technology. This problem may be idealized as the flow between two circular cylinders, whose axes are displaced by an amount $a\varepsilon\delta$, where a, b are the cylinder radii and $\delta = (b - a)/a$. The quantity δ, which is often small in practice, is known as the clearance ratio, while ε $(0 \leq \varepsilon < 1)$ is the eccentricity. The fact of this eccentric-cylinder configuration having a load-bearing capability is dependent on ε being nonzero.

There is one main way in which the instability problem of the basic flow, or first bifurcation, differs from that for the concentric case: with the assumption of sinusoidal periodicity along the axis and of exponential dependence on time, the "concentric" differential equations, which yield the critical Taylor number through an eigenvalue problem, are ordinary. In contrast, because the basic flow in the eccentric case varies in the circumferential (or flow) direction and, furthermore, depends on that circumferential angle, the differential equations, which yield the critical Taylor number, are partial with radial and circumferential derivatives.

From an experimental point of view, there are several ways in which the eccentricity manifests itself in the development of the Taylor vortices. In many experiments, but not all, the initial Taylor number rises monotonically with eccentricity, as illustrated in the data summarized by DiPrima and Stuart [3]. Second, there is a particular experiment, due to Vohr [4], which indicates that the position of maximum vortex strength occurs some 50° downstream from the maximum gap. Third, Vohr (4) shows that the angle at which the torque bifurcates at the critical Taylor number shows significant dependence on ε.

A formal perturbation theory, which is assumed valid for small ε and utilizes the specified relation $\delta^{1/2} = k\varepsilon$, where k is a positive constant, has been developed in [3]. The authors find that the critical Taylor number does rise monotonically with ε, and agreement with experiment is quite good. This arises from a linear calculation. As a byproduct it is found that maximum vortex strength lies at 90° downstream from maximum gap. When the perturbation scheme is extended to include nonlinear effects, as DiPrima and Stuart [5] have done, the choice of parametric values appropriate to experiment does yield a position of maximum Taylor-vortex strength more nearly at 50°. This effect of nonlinearity is made evident through the structure of the first term in the expansion of the solution of the nonlinear partial differential equation for the Taylor-vortex flow, in the form of a function of a radial variable times a function $B(\phi)$, where ϕ is a circumferential angle. This function $B(\phi)$ is found to satisfy

$$k \, dB(\phi)/d\phi = \gamma B(\phi) \cos \phi + T_1 \gamma_1 B(\phi) - \gamma_2 B^3(\phi),$$

and arises essentially from the method of multiple scales; γ, γ_1, and γ_2 are positive numbers; k has been defined earlier, while the positive number T_1 dictates the amount by which the Taylor number exceeds its critical value. A choice of k and T_1 appropriate to experiment yields the 50° mentioned. Caution is necessary, however, because of the experimental magnitude of ε, which may be too large for application of the theory. Theoretical results for torque are not yet available.

4. Wave Systems in Shear Flows

In the problems of Taylor vortices in Sections 2 and 3, attention has been focused on bifurcations associated with the development of a new stable solution associated with a particular wave number or pair of wave numbers. In many boundary-layer situations, however, it is found that streamwise-propagating waves, arising from instability of a basic laminar flow, have energy spread over a spectrum of wave numbers. This wave system evolves, in a way that is imperfectly understood, to produce turbulence. A theoretical example of this situation, but one that has experimental difficulties, occurs in plane Poiseuille flow, which occurs when fluid is driven under pressure between two parallel planes. We now concentrate briefly on that case.

An essential parameter is the Reynolds number $R = U_0 h/\nu$; here U_0 is the maximum value of the velocity distribution, which is parabolic in the coordinate between the two parallel planes, $2h$ the distance between them, and ν the kinematic viscosity. A velocity perturbation mode of the form $\exp[i\alpha(x - ct)]$, where x is downstream distance, t time, and α and c constants, is exponentially amplified in x and/or t if R is greater than a critical value R_c. This occurs at a particular value of α, namely, α_c. For larger Reynolds numbers there is a band of values of α for which instability takes place. We wish to discuss how a wave system can evolve with energy spread over this band of wave numbers. A number of authors have independently studied problems of this type with application to traveling waves in shear flows, including DiPrima et al. [6], Newell and Whitehead [7], and Stewartson and Stuart [8]. The latter especially treated plane Poiseuille flow directly by formal expansion in terms of a parameter ε proportional to $(R - R_c)$. By the method of multiple scales, the first term in the assumed series for the perturbation velocity is found to be proportional to $\varepsilon^{1/2}A(\xi, \tau) \exp[i\alpha_c(x - ct)]$, where $\xi = \varepsilon^{1/2}(x - C_g t)$, $\tau = \varepsilon t$, C_g is the group velocity, and $A(\xi, \tau)$ satisfies

$$\frac{\partial A}{\partial t} - a_2 \frac{\partial^2 A}{\partial \xi^2} = (1 + i\omega)A + kA|A|^2,$$

a_2 and k being complex numbers and ω real.

There are two features of the above amplitude equation to which we draw attention: (i) near R_c and α_c, the real part of k is positive, so that the bifurcation is subcritical, leading to unstable modes with $R < R_c$ provided their amplitude is large enough, as described in [1] and by Chen and Joseph [9]; (ii) a phenomenon of a self-focusing "burst" to larger amplitudes can occur, for which reference is made to Davey et al. [10] and papers cited there.

ACKNOWLEDGMENT

The author is pleased to thank NATO for a research grant for travel, enabling him to attend this meeting.

REFERENCES

[1] J. T. Stuart, Nonlinear stability theory, *Ann. Rev. Fluid Mech.* **3** (1971), 347–370.

[2] P. M. Eagles, On the torque of wavy vortices, *J. Fluid Mech.* **62** (1974), 1–9.

[3] R. C. DiPrima and J. T. Stuart, Nonlocal effects in the stability of flow between eccentric rotating cylinders, *J. Fluid Mech.* **54** (1972), 393–415.

[4] J. H. Vohr, An experimental study of Taylor vortices and turbulence in flow between eccentric rotating cylinders, *J. Lub. Tech.* **90** (1968), 285–296.

[5] R. C. DiPrima and J. T. Stuart, Development and effects of super-critical Taylor-vortex flow in a lightly loaded journal bearing, *J. Lub. Tech.* **96** (1974), 28–35.

[6] R. C. DiPrima, W. Eckhaus, and L. A. Segel, Nonlinear wave-number interactions in near critical two-dimensional flows, *J. Fluid Mech.* **49** (1971), 705–744.

[7] A. C. Newell and J. A. Whitehead, Review of the finite bandwidth concept, *in* "Instability of Continuous Systems" (*IUTAM Symp., Herrenalb., Germany*, 1969) (H. Leipholz, ed.), pp. 284–289. Springer, Berlin, 1971.

[8] K. Stewartson and J. T. Stuart, A nonlinear instability theory for a wave system in plane Poiseuille flow, *J. Fluid Mech.* **48** (1971), 529–545.

[9] T. S. Chen and D. D. Joseph, Subcritical bifurcation of plane Poiseuille flow, *J. Fluid Mech.* **58** (1973), 337–351.

[10] A. Davey, L. M. Hocking, and K. Stewartson, On the nonlinear evolution of three-dimensional disturbances in plane Poiseuille flow, *J. Fluid Mech.* **63** (1974), 529–536.

Chapter 7 : CONTROL THEORY

On Normal Control Processes

ROBERTO CONTI
Istituto Matematico U. Dini
Università di Firenze, Florence, Italy

1

The concept of normality for a control process represented by an ordinary differential equation

$$dx/dt - A(t)x = B(t)u(t) \qquad \text{(U)}$$

was introduced by J. P. LaSalle in order to ensure uniqueness of time optimal controls. The meaning and the applicability of such notions are analyzed here in a broader context than that of LaSalle.

The symbols used in (U), as usual, represent t, a real variable in a compact interval $I = [\tau, T]$; $x = x(t)$, a real n-vector; $u(t)$, a real m-vector; $A(t)$ and $B(t)$, real matrices of types $n \times n$, $n \times m$, respectively. The functions $t \to A(t)$ and $t \to B(t)u(t)$ are (Lebesgue) measurable and integrable in I. We shall further denote by $H(t)$ the $n \times m$ matrix

$$H(t) = X(T)X^{-1}(t)B(t),$$

where X is any fundamental matrix solution of the linear equation

$$dx/dt - A(t)x = 0.$$

According to LaSalle [1, p. 65] the control process (U) is said to be a normal one iff

$$y \neq 0, \quad j = 1, \ldots, m \Rightarrow y^*h_j(t) \neq 0, \qquad \text{a.e.} \quad t \in I, \qquad (1.1)$$

where $y \in E^n$, y^* is the transpose of y, and $h_j(t)$ is the jth column of $H(t)$.

Relation (1.1) is equivalent to the following: for each $y \neq 0$ there is a unique u such that

$$y^* \int_I H(t)u(t)\, dt = \sup_{w \in B_{\infty, \infty}} y^* \int_I H(t)w(t)\, dt, \qquad (1.2)$$

$$u \in B_{\infty, \infty}, \qquad (1.3)$$

where $B_{\infty, \infty}$ denotes the unit ball of the Banach space $L^\infty(I, l_m^\infty)$.

The normality condition ensures that for every $v \in E^n$ belonging to the domain of zero controllability (i.e., to the set of initial states v that can be transferred to zero in a finite time), corresponding to the set of controls $C = B_{\infty, \infty}$, there is a unique time optimal control steering the state to zero.

2

The concept of normality can be defined for a class of control sets C more general than $B_{\infty, \infty}$. In [2] we have considered the case $C = B_{p', r'}$, the unit ball of the Banach space $L^{p'}(I, l_m^{r'})$, with $1 < p' \leq \infty$, $1 \leq r' \leq \infty$. In that case (U) is said to be normal for $B_{p', r'}$ iff there is a unique solution u of (1.2) and (1.3) with $B_{p', r'}$ replacing $B_{\infty, \infty}$. Then it can be proved that (if $m \geq 2$) there are six different types of normal processes, which can be characterized in terms of A (via X) and B.

If we denote by $N_{p', r'}$ the property of (U) being normal for $B_{p', r'}$ we have the following implications:

$$
\begin{array}{ccccc}
N_{p', \infty} & \Rightarrow & N_{p', r'} & \Leftarrow & N_{p', 1} \\
\Uparrow & & \Uparrow & & \Uparrow \\
N_{\infty, \infty} & \Rightarrow & N_{\infty, r'} & \Leftarrow & N_{\infty, 1}
\end{array}
$$

Here $N_{\infty, \infty}$ corresponds to LaSalle's original definition and $N_{p', r'}$, $1 < p' < \infty$, $1 < r' < \infty$, to the complete controllability of (U).

When A and B are independent of t, then there are only three different types, since $N_{p', r'} = N_{\infty, r'}$ for $1 \leq r' \leq \infty$, and they can be characterized directly in terms of A and B.

3

Let us now consider normality in a more abstract setting, as an attempt to extend it to infinite-dimensional control processes.

We are given two Banach spaces U and X, a linear continuous map $L: U \to X$, and a convex set $C \subset U$, not reduced to a single point. For each $x' \in X'$ (the normed dual of X), the set of u such that

$$x'(Lu) = \sup_{w \in C} x'(Lw), \qquad (3.1)$$

$$u \in C, \qquad (3.2)$$

is convex (possibly empty). Then we say that L is normal for C iff for each $x' \in X'$, $x' \neq 0$, there is at most one u satisfying (3.1) and (3.2).

A comparison with Section 2 gives $U = L^{p'}(I, l_m^{r'})$, $X = E^n$, L defined by $Lu = \int_I H(t)u(t)\, dt$, and $C = B_{p', r'}$.

With reference to [1, p.65] we also say that L is essentially normal for C iff the set LC is strictly convex. This means that every closed support plane to LC meets LC at just one point.

It is easy to prove

Theorem 1. If L is normal for C and LC is closed then L is also essentially normal for C.

L can be said to be completely controllable by means of U iff

$$\overline{LU} = X,$$

i.e., iff the image LU of U is a dense subspace of X. Then we have

Theorem 2. If L is normal for some $C \subset U$ then L is completely controllable by means of U. Conversely if L is completely controllable by means of U, then L is normal for any $C \subset U$ that is strictly convex.

4

As we observed at the beginning, normality was introduced as a condition to ensure the uniqueness of time optimal controls. In fact, if $u \in B_{\infty, \infty}$ is a time optimal control, then Lu belongs to the boundary of the attainable set $LB_{\infty, \infty}$; hence by the supporting property of closed convex sets in E^n there must exist some $y \neq 0$ satisfying (1.2).

The same holds for the case $C = B_{p', r'}$.

Unfortunately this is no longer true for (3.1) in the general case, since a convex set LC of an infinite-dimensional space X does not necessarily have the supporting property. In other words, the notion of normal control process seems to be confined to the case of processes for which a maximum principle holds [3].

REFERENCES

[1] H. Hermes and J. P. LaSalle, "Functional Analysis and Time Optimal Control." Academic Press, New York, 1969.
[2] R. Conti, On normal control processes, *J. Optimization Theory Appl.* **14** (1974), 497–503.
[3] L. S. Pontryagin *et al.*, "The Mathematical Theory of Optimal Processes." Wiley (Interscience), New York, 1962.

Projection Methods for Hereditary Systems

H. T. BANKS* and JOHN A. BURNS†‡
Lefschetz Center for Dynamical Systems
Division of Applied Mathematics
Brown University, Providence, Rhode Island

1. Introduction

Here we consider a general projection method for obtaining approximations to a hereditary system. We shall define the system and present the approximating systems in this section. The main results are stated in Section 2.

The space of R^n-valued square integrable functions defined on an interval $[a, b]$ will be denoted by $L_2(a, b)$. Let $r > 0$ be a fixed real number. If $x: [-r, +\infty) \to R^n$ is given, then $x_t: [-r, 0] \to R^n$ is defined by $x_t(s) = x(t + s)$, $s \in [-r, 0]$, for each $t \geq 0$.

We assume that L is a linear transformation with domain in the linear space of R^n-valued Lebesgue measurable functions defined on $[-r, 0]$, such that L restricted to the space of continuous functions is a bounded operator. As in [2], L is required to satisfy the following conditions:

(C) If $t_1 > 0$ and $x \in L_2(-r, t_1)$, then the function $g(t) = L(x_t)$ is defined almost everywhere on $[0, t_1]$, and depends only on the equivalence class of the function x. Also, g belongs to $L_1(0, t_1)$ and there is a continuous function Γ such that

$$\int_0^t |L(x_s)|\, ds \leq \Gamma(t)\left(\int_{-r}^t |x(s)|^2\, ds\right)^{1/2}, \qquad \text{for all} \quad t \in [0, t_1].$$

The system is defined by the linear retarded functional differential equation

$$\dot{x}(t) = L(x_t) + f(t), \qquad t \geq 0, \tag{1.1}$$

---

* This research was supported in part by the U.S. Army Research Office under DA-ARO-D-31-124-73-G130, in part by the National Science Foundation, GP 28931X2, and in part by the Air Force Office of Scientific Reseach under AFOSR 71-2078C.

† This research was supported in part by the National Science Foundation under GP 28931X2 and in part by Air Force Office of Scientific Research under AFOSR 71-2078C.

‡ Present Address: Department of Mathematics, Virginia Polytechnic Institute and State University, Blacksburg, Virginia.

and the initial data

$$x(0) = \eta, \qquad x_0 = \phi, \tag{1.2}$$

where $\eta \in R^n$, $\phi \in L_2(-r, 0)$, and $f \in L_2(0, t_1)$ for every $t_1 > 0$.

A solution is a function $x \in L_2(-r, t_1)$, such that x is absolutely continuous (a.c.) for $t \geq 0$, x satisfies (1.1) a.e. on $[0, t_1]$, $x(0) = \eta$, and $x(s) = \phi(s)$ a.e. on $[-r, 0]$. In [2] it is shown that there is a unique solution to (1.1) and (1.2) defined on $[-r, \infty)$, and this solution depends continuously on initial data $(\eta, \phi) \in R^n \times L_2(-r, 0)$. Let Z denote the product space $R^n \times L_2(-r, 0)$ and for $t \geq 0$ define $S(t): Z \to Z$ by $S(t)(\eta, \phi) = (x(t), x_t)$, where x is the unique solution to system (1.1) and (1.2) with $f \equiv 0$. The following lemma may be found in [2].

Lemma 1.1. The semigroup $S(t)$ is strongly continuous and has a closed, densely defined infinitesimal generator. If $\mathscr{A}$ denotes this generator, then:

(i)   The domain of $\mathscr{A}$ is given by $\mathscr{D}(\mathscr{A}) = \{(\eta, \phi): \phi \text{ is a.c.}, \dot{\phi} \in L_2(-r, 0),$ and $\eta = \phi(0)\}$.

(ii)   If $(\eta, \phi) \in \mathscr{D}(\mathscr{A})$, then $\mathscr{A}(\eta, \phi) = (L(\phi), \dot{\phi})$.

(iii)   There exist constants γ and M such that the spectrum of $\mathscr{A}$ lies to the left of the line $\operatorname{Re} \lambda < \gamma$, and $|S(t)(\eta, \phi)| \leq M e^{(\gamma + \varepsilon)t} |(\eta, \phi)|$ for each $\varepsilon > 0$.

In order to define the approximating systems we make the following assumption:

(A)   There exists a sequence of finite-dimensional subspaces of Z, denoted by Z^N, a sequence of projections $P^N: Z \to Z^N$, and a sequence of linear operators $\mathscr{A}^N: Z^N \to Z^N$ such that:

(i) $|\exp \mathscr{A}^N t|$ is uniformly exponentially bounded,

(ii) $|P^N(\xi, \psi) - (\xi, \psi)| \to 0$ for all $(\xi, \psi) \in Z$,

(iii) $|\mathscr{A}^N P^N(\xi, \psi) - \mathscr{A}(\xi, \psi)| \to 0$ for all $(\xi, \psi) \in \mathscr{D}(\mathscr{A})$.

Note that condition (A) implies that $\mathscr{A}^N$ "converges" to $\mathscr{A}$ in the sense of Trotter (see [3, 6].) Therefore, we introduce the approximating system defined by the ordinary vector differential equation

$$\dot{w}(t) = \mathscr{A}^N w(t) + P^N(f(t), \theta), \qquad t \geq 0, \tag{1.3}$$

and the initial value

$$w(0) = P^N(\eta, \phi), \tag{1.4}$$

where θ denotes the zero function in $L_2(-r, 0)$.

Let

$$z^N(t;f) = \exp(\mathscr{A}^N t)P^N(\eta, \phi) + \int_0^t \exp[\mathscr{A}^N(t - \sigma)]P^N(f(\sigma), \theta)\,d\sigma$$

denote the solution to system (1.3) and (1.4), and define $z(t; f)$ by

$$z(t;f) = S(t)(\eta, \phi) + \int_0^t S(t - \sigma)(f(\sigma), \theta)\,d\sigma.$$

2. Statement of Results

We now present our main results concerning the representation of solutions of (1.1) and (1.2), and convergence of approximating solutions. Proofs of these statements, along with examples and an application to optimal control theory, will appear in [1].

The following theorem extends the results stated in [2], and provides an important representation theorem.

Theorem 2.1.   Let $t_1 > 0$, $(\eta, \phi) \in \mathscr{D}(\mathscr{A})$, and $f \in L_2(0, t_1)$. If $x(\cdot; f)$ denotes the solution to (1.1) and (1.2), then $z(t;f) = (x(t;f), x_t(\cdot;f))$.

If assumption (A) is made, then one may conclude from Trotter's theorem [3, 6] that

$$|\exp(\mathscr{A}^N t)P^N(\xi, \psi) - S(t)(\xi, \psi)| \to 0 \qquad \text{for all} \quad (\xi, \psi) \in Z, \quad t > 0.$$

Using this result, we can establish the following.

Theorem 2.2.   Let $t_1 > 0$ and $t \in [0, t_1]$. If assumption (A) holds, then:

(i)   The operator $F(t): L_2(0, t_1) \to Z$ defined by

$$F(t)f = \int_0^t S(t - \sigma)(f(\sigma), \theta)\,d\sigma$$

is a compact linear operator;

(ii) $|z^N(t; f) - z(t; f)| \to 0$ as $N \to \infty$, uniformly in f for f in bounded subsets of $L_2(0, t_1)$.

Corollary 2.1.   If assumption (A) holds and f^K converges weakly to f in $L_2(0, t_1)$, then for each t, $z^N(t; f^K) \to z(t; f)$ strongly as $N, K \to \infty$.

Corollary 2.2. If assumption (A) holds and $\mathcal{F}: L_2(0, t_1) \to \mathcal{C}([0, t_1], Z)$ is defined by $\mathcal{F}[f](t) = z(t; f)$, then $\mathcal{F}$ is a compact affine operator. Here $\mathcal{C}$ is the space of Z-valued functions, continuous on $[0, t_1]$, taken with the supremum norm.

We emphasize that the approximating system is largely determined by the selection of the spaces Z^N and the projections P^N. For example, by an appropriate choice of Z^N and P^N we can obtain a widely used finite-difference scheme for approximating solutions to (1.1) and (1.2). Consequently, results such as those found in [4] become special cases of the above theorems.

The importance of Corollary 2.1 in establishing approximation results for optimal control problems should be obvious. A more detailed discussion of this aspect of these results will be given in [1]. An application of the Trotter theorem to the problem of finding approximate optimal controls for another type of infinite-dimensional system may be found in [5].

REFERENCES

[1] H. T. Banks and J. A. Burns, An abstract framework for approximate solutions to optimal control problems governed by hereditary systems, *Proc. Internat. Conf. Differential Equations* pp. 10–25. Academic Press, New York, 1975.

[2] Ju. G. Borisovic and A. S. Turbabin, On the Cauchy problem for linear nonhomogeneous differential equations with retarded argument, *Soviet Math. Dokl.* **10** (1969), 401–405.

[3] T. Kato, "Perturbation Theory for Linear Operators." Springer-Verlag, Berlin and New York, 1966.

[4] Iu. M. Repin, On the approximate replacement of systems with lag by ordinary dynamical systems, *Appl. Math. Mech.* **29** (1965), 254–264.

[5] H. Sasai and E. Shimemura, On the convergence of approximating solutions for linear distributed parameter optimal control problems, *SIAM J. Control* **9** (1971), 263–273.

[6] H. F. Trotter, Approximation of semi-groups of operators, *Pacific J. Math.* **8** (1958), 887–919.

Lower Bounds for the Extreme Value of a Parabolic Control Problem

WERNER KRABS

Fachbereich Mathematik der Technischen
Hochschule Darmstadt, West Germany

1. Statement of the Problem and Basic Results

In connection with optimal heating of solids, the following parabolic control problem arises that has been previously treated in [1, 8]. We are looking for a measurable function u on a given finite interval $[0, T]$, $T > 0$, such that under the conditions

$$|u(t)| \leq 1 \qquad \text{for almost every} \quad t \in [0, T], \tag{1}$$

$$y_t(x, t) = y_{xx}(x, t) \qquad \text{for} \quad 0 < t \leq T, \quad 0 \leq x < 1, \tag{2}$$

$$y_x(0, t) = 0, \tag{3}$$
$$\qquad \text{for} \quad t \in (0, T],$$
$$by(1, t) + y_x(1, t) = bu(t) \tag{4}$$

$$y(x, 0) = 0 \qquad \text{for} \quad x \in [0, 1], \tag{5}$$

the quantity

$$\|y(\cdot, T) - y_T\| = \sup_{x \in [0, 1]} |y(x, T) - y_T(x)| \tag{6}$$

is to be minimized, where y_T is a given function in a suitable function space $F[0, 1]$ and b is a given positive constant.

Yegorov [8] has shown that for each $u \in L^\infty[0, T]$ there is exactly one generalized solution $y = y(u, x, t)$ of (2)–(5) such that for $y(u, \cdot, T)$ the representation formula

$$y(u, x, T) = \sum_{k=1}^{\infty} A_k \mu_k^2 \cos \mu_k x \int_0^T \exp\left[-\mu_k^2(T - t)\right] u(t) \, dt \tag{7}$$

holds, where the μ_k's form the sequence of positive solutions of the transcendental equation $\mu \tan \mu = b$ and the A_k's are given by

$$A_k = \frac{2 \sin \mu_k}{\mu_k + \sin \mu_k \cos \mu_k}, \qquad k = 1, 2, \ldots.$$

291

Using (7) it can be shown that by defining

$$(Su)(x) = y(u, x, T), \qquad x \in [0, 1], \tag{8}$$

we obtain a linear mapping $S: L^\infty[0, T] \to C[0, 1]$ that is continuous and compact. Therefore, it is sensible to choose $F[0, 1] = C[0, 1]$.

If we put

$$Q = \{u \in L^\infty[0, T] : |u| \le 1 \text{ a.e.}\}, \tag{9}$$

we can define the extreme value of the above control problem by

$$\rho(y_T, Q) = \inf_{u \in Q} \|Su - y_T\|, \tag{10}$$

where $\|\cdot\|$ denotes the maximum norm on $C[0, 1]$.

An optimal control $\hat{u} \in Q$ is then defined by

$$\|S\hat{u} - y_T\| = \rho(y_T, Q). \tag{11}$$

Yegorov [8] has further shown that the existence of an optimal control $\hat{u} \in Q$ is ensured. By recent results of Weck [5] the additional condition $\rho(y_T, Q) > 0$ implies that there is exactly one optimal control $\hat{u} \in Q$ that, furthermore, satisfies the "bang–bang" condition $|\hat{u}| = 1$ a.e. on $[0, T]$.

The purpose of this chapter is to indicate how lower bounds for the extreme value (10) of the control problem can be obtained that yield arbitrarily close estimates. The results are given without proofs, which can be found in [2, 4].

Similar approaches for obtaining lower bounds of the extreme value of control problems with distributed parameters have been given by Yavin [6, 7].

2. Lower Bounds for the Extreme Value via Dualizing the Problem

2.1. THE DIRECT APPROACH

As in [4] we start with the well-known duality formula

$$\rho(y_T, Q) = \max_{y^* \in B^*} \left\{ y^*(y_T) - \sup_{u \in Q} y^*(Su) \right\} \tag{12}$$

(see, e.g., [3, p. 62]), where $S: L^\infty[0, T] \to Y = C[0, 1]$ is the linear mapping (8) and B^* denotes the unit ball of the dual space Y^* of Y. On using (7) and introducing the adjoint mapping $S^*: Y^* \to L^\infty[0, T]^*$ one can verify for every $u \in L^\infty[0, T]$ and $y^* \in Y^*$ that

$$(S^*y^*)(u) = \int_0^T (\hat{S}^*y^*)(t)u(t)\, dt,$$

where

$$(\hat{S}^*y^*)(t) = \sum_{k=1}^\infty A_k \mu_k^2 y^*(\cos \mu_k \cdot) \exp\left[-\mu_k^2(T - t)\right] \tag{13}$$

is in $L^1[0, T]$. This implies

$$\sup_{u \in Q} y^*(Su) = \|\hat{S}^*y^*\|_1, \tag{14}$$

and hence

$$\rho(y_T, Q) = \max_{y^* \in B^*} \{y^*(y_T) - \|\hat{S}^*y^*\|_1\}, \tag{15}$$

where $\|\cdot\|_1$ denotes the L^1-norm.

Equation (14) also holds if $L^\infty[0, T]$ is replaced by $C[0, T]$, which means that only continuous control functions are admitted. The extreme value (10) then is not changed; however, the existence of optimal controls is in general no more ensured.

If $y^* \in B^*$ is a positive linear functional, then due to the monotonicity of the mapping (8), it follows that $\hat{S}^*y^*$ given by (13) is a nonnegative function in $L^1[0, T]$ thus yielding the estimation

$$y^*(y_T) - \sum_{k=1}^\infty A_k y^*(\cos \mu_k \cdot)[1 - \exp(-\mu_k^2 T)] \le \rho(y_T, Q). \tag{16}$$

2.2. THE INDIRECT APPROACH

Let I be the set of all the continuous linear functionals on $Y = C[0, 1]$ of the form

$$y^*(y) = \int_0^1 y_*(x)y(x)\, dx, \qquad y \in Y,$$

with $\hat{y}_* \in Y$ and $\int_0^1 |y_*(x)|\, dx \le 1$.

Then by [4], we have for the extreme value (10)

$$\rho(y_T, Q) = \sup_{y \in I} \{y^*(y_T) - \|\hat{S}^*y^*\|_1\}, \tag{17}$$

where again $\hat{S}^*y^*$ is given by (13). For each $y^* \in I$ the function $\hat{S}^*y^*$ can be computed in the following way: Let $z = z(x, t)$ be the "classical solution" of the "backward initial boundary value problem"

$$z_t(x, t) = -z_{xx}(x, t), \qquad (x, t) \in (0, 1) \times [0, T], \tag{18}$$

$$z_x(0, t) = 0$$
$$bz(1, t) + z_x(1, t) = 0 \qquad \text{for} \quad t \in [0, T), \tag{19}$$

$$z(x, T) = y_*(x), \qquad x \in [0, 1]. \tag{20}$$

By applying partial integration to

$$\int_0^T \int_0^1 [z_t(x, t) + z_{xx}(x, t)]y(u, x, t) \, dx \, dt = 0$$

for each $u \in C[0, T]$, where $y(u, \cdot, \cdot)$ is the corresponding "classical" solution of (2)–(5), it can be easily shown that

$$(S^*y^*)(u) = \int_0^T (\hat{S}^*y^*)(t)u(t) \, dt = b \int_0^T z(1, t)u(t) \, dt. \tag{21}$$

Defining Z as the set of all the solutions of (18)–(20) such that $z \in C([0, 1] \times [0, T])$, $z_t z_{xx} \in C((0, 1) \times [0, T))$,

$$z_x(1, t) = \lim_{x \to 1-0} z_x(x, t), \quad t \in [0, T); \quad \text{and} \quad \int_0^1 |z(x, T)| \, dx \le 1,$$

we conclude, by virtue of (17) and (21) and the remark following Eq. (15), that

$$\rho(y_T, Q) = \sup_{z \in Z} \left\{ \int_0^1 z(x, T)y_T(x) \, dx - b \int_0^T |z(1, t)| \, dt \right\}. \tag{22}$$

Taking, for instance,

$$z(x, t) = \sum_{k=1}^N C_k \exp(\mu_k^2 t) \cos \mu_k x,$$

we have that $z \in Z$ if the C_k's are chosen such that

$$\int_0^1 \left| \sum_{k=1}^N C_k \exp(\mu_k^2 T) \cos \mu_k x \right| \, dx \le 1. \tag{23}$$

Hence

$$\sum_{k=1}^{N} C_k \exp\left(\mu_k^2 T\right) \int_0^1 y_T(x) \cos \mu_k x \, dx$$

$$- b \int_0^T \left| \sum_{k=1}^{N} C_k \exp\left(\mu_k^2 t\right) \cos \mu_k \right| dt \le \rho(y_T, Q). \tag{24}$$

Furthermore, it can be shown that by varying N and $C_1, \ldots, C_N$ such that (23) holds, the supremum of the left-hand side of (24) equals $\rho(y_T, Q)$.

2.3. CONCLUDING REMARK

By using the duality formula (12) it can be shown that if the extreme value (10) of the control problem is positive, then the unique optimal control $\hat{u}$ can be characterized in terms of a more refined bang–bang principle as follows: For each $\delta \in (0, T)$ we have $|u| = 1$ on $[0, T - \delta]$ with the exception of a finite number of points, i.e., the jumps of $\hat{u}$ are enumerable and can only accumulate at $t = T$ (for details, see [2]).

REFERENCES

[1] A. G. Butkovskiy, "Distributed Control Systems." Elsevier, New York, 1969.
[2] K. Glashoff and W. Krabs, Dualität und Bang-Bang-Prinzip bei einem parabolischen Rand-Kontrollproblem, *Bonner Math. Schriften* 77 (1975), 1–8.
[3] R. B. Holmes, A course on optimization and best approximation, *Springer Lecture Notes Math.* 257. Springer-Verlag, Berlin and New York, 1972.
[4] W. Krabs, Zur Berechnung des Extremalwertes bei einem parabolischen Rand-Kontrollproblem, *Beitr. Numer. Math.* 3, (1975), 129–139.
[5] N. Weck, Über Existenz, Eindeutigkeit und das "Bang-Bang-Prinzip" bei Kontrollproblemen aus der Wärmeleitung, *Bonner Math. Schriften* 77 (1975), 9–19.
[6] Y. Yavin, Lower bounds on the cost functional for a class of distributed systems, *in* Proc. IFAC Symp. Control Distributed Parameter Syst. Vol. I. Banff, Canada, 1971.
[7] Y. Yavin, Lower bounds on the cost functional for systems governed by partial differential equations, *J. Optimization Theory Appl.* 11 (1973), 605–612.
[8] J. V. Yegorov, Some problems in the theory of optimal control, *USSR Comput. Math. Phys.* 3 (1963), 1209–1232.

Controllability for Neutral Systems of Linear Autonomous Differential–Difference Equations

MARC Q. JACOBS*
Department of Mathematics
University of Missouri, Columbia, Missouri

C. E. LANGENHOP
Department of Mathematics
Southern Illinois University, Carbondale, Illinois

We consider systems typified by

$$\dot{x}(t) = A_{-1}\dot{x}(t-h) + A_0 x(t) + A_1 x(t-h) + Bu(t), \qquad (1)$$

where $h > 0$, the A_i are $n \times n$ constant real matrices, and B is an $n \times m$ constant real matrix. The control vector u takes values in R^m and is assumed to be in L_2 on compact intervals. The natural state space of x for such systems is the Sobolev space $W_2^{(1)}([-h, 0], R^n)$ of absolutely continuous functions $\phi: [-h, 0] \to R^n$ with derivative ϕ' in L_2.

Considering the initial time t as 0, we denote the solution x of (1) on $t \geq 0$ such that $x_0 = \phi$ by $x(\cdot, \phi, u)$. Here x_τ is defined by $x_\tau(\theta) = x(\tau + \theta)$, $\theta \in [-h, 0]$. We say that system (1) is *controllable on* $[0, \tau]$ iff for all $\phi, \psi \in W_2^{(1)}([-h, 0], R^n)$ there exists $u \in L_2([0, \tau], R^m)$ such that $x_\tau(\cdot, \phi, u) = \psi$. Of particular interest is the *attainable set from 0* defined by

$$\mathscr{A}(\tau) = \{\psi = x_\tau(\cdot, 0, u) \,|\, u \in L_2([0, \tau], R^m)\}.$$

Obviously, $\mathscr{A}(\tau) \subset W_2^{(1)}([-h, 0], R^n)$ and it is easy to see that (1) is controllable on $[0, \tau]$ iff $\mathscr{A}(\tau) = W_2^{(1)}([-h, 0], R^n)$. Moreover, controllability of (1) on $[0, \tau]$ implies $\tau > h$.

For any integer $v \geq 1$, define the controllability matrix $C_v = [B, A_{-1}B, \ldots, A_{-1}^{v-1}B]$. One can prove the following:

Proposition. If (1) is controllable on $[0, \tau]$ and $p \geq 1$ is the integer such that $ph \leq \tau < (p+1)h$, then rank $C_p = n$; in particular, if $\tau > nh$, then rank $C_n = n$.

* This research was supported by the National Science Foundation under grants GP 33882 and GF 37298.

The solutions $x(\cdot, \phi, u)$ of (1) are given by the variation of constants formula

$$x(t, \phi, u) = x(t, \phi, 0) + \int_0^t \Phi(t - s)Bu(s)\, ds, \qquad t \ge 0,$$

where $\Phi(t)$ is the $n \times n$ transition matrix corresponding to (1). The system (1) is *Euclidean controllable* on an interval $[0, \sigma]$ if $\{\psi(-h) | \psi \in \mathscr{A}(\sigma)\} = R^n$. A necessary and sufficient condition for (1) to be *Euclidean controllable on* $[0, \sigma]$ is that the controllability Gramian

$$G(\sigma) = \int_0^\sigma \Phi(s)BB^*\Phi^*(s)\, ds$$

have rank n. One may show that rank $C_n = n$ implies rank $G(\tau - h) = n$ if $\tau > nh$, but not if $\tau \le nh$. Examples show that rank $C_n = n$ is not by itself a sufficient condition for (1) to be controllable on $[0, \tau]$, $\tau > nh$ (cf. Example 3 below). However, the condition is sufficient to establish a result that can often be substituted for controllability (cf. [2]). That is, rank $C_n = n$ implies that $\mathscr{A}(\tau)^\perp$ has finite dimension and $\mathscr{A}(\tau)$ is closed in $W_2^{(1)}([-h, 0], R^n)$, or what is the same, the solution operator $u \mapsto x_\tau(\cdot, 0, u)$ has closed range and finite deficiency in $W_2^{(1)}([-h, 0], R^n)$ (cf. [4]). Indeed, this property persists for quite general neutral functional differential equations of the form

$$\frac{d}{dt} \mathscr{D}(x_t) = L(x_t) + Bu(t), \qquad (2)$$

considered by Hale and Meyer [3]. In Eq. (2), $\mathscr{D}$ and L are linear operators of the form

$$L(\phi) = \int_{-h}^0 d\eta(\theta)\phi(\theta), \qquad \mathscr{D}(\phi) = \phi(0) - \int_{-h}^0 d\mu(\theta)\phi(\theta),$$

$\phi \in W_2^{(1)}([-h, 0], R^n)$, where η, μ are $n \times n$ matrix functions of bounded variation on $[-h, 0]$ and $\mu(0) = 0$ with μ continuous at 0. Let $\phi(0) - \mathscr{D}(\phi) = g(\phi)$ and let $g_0(\phi) = A_{-1}\phi(-h)$, $\phi \in W_2^{(1)}([-h, 0], R^n)$. Define the difference operator $\mathscr{D}_0$ by $\mathscr{D}_0(\phi) = \phi(0) - g_0(\phi)$.

Theorem 1. Let $\tau > h$ and let p be the integer such that $ph \le \tau < (p + 1)h$. If rank $C_p = n$, then the solution operator $u \mapsto x_\tau(\cdot, 0, u)$, $u \in L_2([t_0, t_1], R^n)$ corresponding to

$$\frac{d}{dt} \mathscr{D}_0 x_t = L(x_t) + Bu(t) \qquad (3)$$

has closed range and finite deficiency in $W_2^{(1)}([-h, 0], R^n)$. Moreover, if $g - g_0$ is sufficiently small or if $g = g_0 + g_1$, where $g_1(\phi) = \int_{-h}^{0} d\lambda(\theta)\phi(\theta)$ and λ is an $n \times n$ matrix function whose entries are in $W_2^{(2)}([-h, 0], R)$, then the solution operator $u \mapsto x_\tau(\cdot, 0, u)$, $u \in L_2([0, \tau], R^n)$ corresponding to (2) has closed range and finite deficiency in $W_2^{(1)}([-h, 0], R^n)$.

An easy corollary to Theorem 1 is that the solution operator having closed range and finite deficiency is a *generic property* of systems of the form (2).

Returning now to the controllability problem for system (1) we note that in [1] it was shown that $\mathscr{A}(\tau)$ [for system (1)] is a nondecreasing set-valued function of τ that is constant after nh. In view of this we confine our attention to the situations $\tau > nh$ and rank $C_n = n$. Using operational techniques developed in [1], we can obtain explicit criteria for controllability of (1) under these circumstances. The nature of the criteria is effectively demonstrated in the case $m = 1$, that is, when B is a single column and u a scalar function.

In applying operational techniques it is convenient to introduce certain function spaces:

$$X_0 = \{x: (-\infty, \tau] \to R^n \,|\, x(t) = 0 \text{ for } t \leq 0 \text{ and } x|_{[0, \tau]} \in W_2^{(1)}([0, \tau], R^n)\},$$
$$U_0 = \{u: (-\infty, \tau] \to R \,|\, u(t) = 0 \text{ for } t \leq 0 \text{ and } u|_{[0, \tau]} \in L_2([0, \tau], R)\}.$$

For $f \in X_0$ or $f \in U_0$ the shift operator S is defined by $(Sf)(t) = f(t - h)$, $t \leq \tau$. For $f \in X_0$ the derivative operator is denoted by D, so that $(Df)(t) = \dot{f}(t)$, $t \leq \tau$ (a.e.). We identify S^0 and D^0 with the identity operator.

The extension of $x(\cdot, 0, u)$ to a function $x \in X_0$ satisfies

$$(I_n D - A_{-1} SD - A_0 - A_1 S)x = Bu, \qquad (1_0)$$

if u here denotes the extension to U_0 of the control u defined on $[0, \tau]$. Treating D and S formally as scalars, we define

$$P(D, S) = \text{adj}(I_n D - A_{-1} SD - A_0 - A_1 S) = \sum_{j=0}^{n-1} P_j(D)S^j.$$

Here adj denotes the transposed matrix of cofactors so the $P_j(D)$ are $n \times n$ matric polynomials in D. With these we define the operator matrix

$$K(D) = [P_0(D)B, P_1(D)B, \ldots, P_{n-1}(D)B],$$

the submatrices $P_j(D)B$ being formal products rather than the results of the operators $P_j(D)$ acting on the constant function B.

Using some ideas from [5], we have proven the following in [1]:

Lemma 1. If $\tau > nh$ and B is $n \times 1$, then for every $\psi \in W_2^{(1)}([-h, 0], R^n)$ there exists $u \in U_0$ such that the solution $x \in X_0$ of (1_0) satisfies $x_\tau = \psi$ iff there exists $\omega \in W_2^{(n)}([\tau - h, \tau], R^n)$ such that

$$K(D)\omega(t) = \psi(t - \tau), \qquad t \in [\tau - h, \tau], \tag{4a}$$

$$D^i\omega_j(\tau - h) = D^i\omega_{j+1}(\tau), \qquad i = 0, \ldots, n-1, \quad j = 1, \ldots, n-1, \tag{4b}$$

where ω_j denotes the jth component of ω.

Using this, we are able to prove

Theorem 2. If $\tau > nh$, B is $n \times 1$, and rank $C_n = n$, then system (1) is controllable on $[0, \tau]$ iff the problem (4a) and (4b) has a solution for every $\psi \in W_2^{(1)}([-h, 0], R^n)$; moreover, this holds iff the homogeneous problem, (4a) with $\psi = 0$ and (4b), has only the trivial solution.

Under the assumption that rank $C_n = n$ it is the case that $\Delta(D) = \det K(D)$ has degree $n(n-1)$ in D. The solutions of $K(D)\omega = 0$ thus are of the form

$$\omega(t) = \sum_{i=1}^{n(n-1)} \gamma^i v_i(t)$$

for some $\gamma^i \in R^n$, $i = 1, \ldots, n(n-1)$, if the v_i form a basis for the scalar solutions of $\Delta(D)v = 0$. Using a basis consisting of functions of the form $t^\alpha e^{\lambda t}$, we obtain

Theorem 3. If $\tau > nh$, B is $n \times 1$, and rank $C_n = n$, then (1) is controllable on $[0, \tau]$ implies that for all complex λ either $\Delta(\lambda) \neq 0$ or

$$w(\lambda) \equiv K(\lambda) \begin{bmatrix} 1 \\ e^{-\lambda h} \\ \vdots \\ e^{-(n-1)\lambda h} \end{bmatrix} \neq 0. \tag{5}$$

When B is $n \times 1$ and rank $C_n = n$, there exists a nonsingular $n \times n$ constant matrix T such that with $y = Tx$ the controllability of (1) is equivalent to the controllability of

$$(I_n D - G_{-1} SD - G_0 - G_1 S)y = B_0 u, \tag{6}$$

where

$$B_0 = \begin{bmatrix} 0 \\ 0 \\ \vdots \\ 0 \\ 1 \end{bmatrix} \quad \text{and} \quad G_{-1} = \begin{bmatrix} 0 & 1 & 0 & \cdots & 0 \\ 0 & 0 & 1 & & 0 \\ \vdots & & & & \vdots \\ 0 & 0 & 0 & & 1 \\ g_1 & g_2 & g_3 & \cdots & g_n \end{bmatrix}$$

G_{-1} being the companion matrix for the characteristic polynomial of A_{-1}. For any matrix M with n rows, let $\tilde{M}$ denote the submatrix consisting of the first $n - 1$ rows. For the canonical form (6) of Eq. (1) the corresponding matrix $K(\lambda)$ is determined by $\tilde{G}_0$ and $\tilde{G}_1$. In fact, one can easily show that the necessary condition in Theorem 3 is equivalent to

$$\text{rank}[\tilde{I}_n \lambda - \tilde{G}_{-1}\lambda e^{-\lambda h} - \tilde{G}_0 - \tilde{G}_1 e^{-\lambda h}] = n - 1$$

for all complex λ. One may readily apply the results of Theorems 2 and 3 to specific cases. The following examples illustrate the methods.

Example 1. Let $\tilde{G}_1 = 0$ and $\tilde{G}_0 = [\text{diag}(a_1, \ldots, a_{n-1}), 0]$. If $\tau > nh$, then (6) is controllable on $[0, \tau]$ iff none of the a_i, $i = 1, \ldots, n - 1$, is zero.

Example 2. $(n = 3)$. Let

$$\tilde{G}_0 = \begin{bmatrix} 0 & 0 & 0 \\ -1 & a & 0 \end{bmatrix} \quad \text{and} \quad \tilde{G}_1 = \begin{bmatrix} 0 & 0 & b^2 \\ 0 & 4 & 0 \end{bmatrix}$$

If $\tau > 3h$, then (6) is controllable on $[0, \tau]$ iff $b \neq 0$ and $a \neq \pm b - 3e^{\mp bh}$.

Example 3. $(n = 2)$. Let $\tilde{G}_0 = [\alpha_0 \, \beta_0]$, $\tilde{G}_1 = [\alpha_1 \beta_1]$. If $\tau > 2h$, then (6) is controllable on $[0, \tau]$ iff $\beta_0 = 0$, $\beta_1 + \alpha_0 + \alpha_1 e^{\beta_1 h} \neq 0$, or $\beta_0 \neq 0$, $\beta_0 + (\lambda + \beta_1)e^{-\lambda h} \neq 0$ for λ such that $(\lambda - \alpha_0)(\lambda + \beta_1) + \beta_0 \alpha_1 = 0$.

REFERENCES

[1] H. T. Banks, M. Q. Jacobs, and C. E. Langenhop, Characterization of the controlled states in $W_2^{(1)}$ of linear hereditary systems, *SIAM J. Control* **13** (1975), 611–649.
[2] H. T. Banks and M. Q. Jacobs, An attainable sets approach to optimal control of functional differential equations with function space side conditions, *J. Differential Equations* **13** (1973), 127–149.

[3] J. K. Hale and K. R. Meyer, A class of functional differential equations of neutral type, *Mem. Amer. Math. Soc.* **76** (1967).

[4] T. Kato, "Perturbation Theory of Linear Operators." Springer-Verlag, Berlin and New York, 1966.

[5] S. A. Minjuk, On complete controllability of linear systems with delay, *Differential Urav.* **8** (1972), 254–259.

Local Controllability of a Hyperbolic Partial Differential Equation

WILLIAM C. CHEWNING

Department of Mathematics and Computer Science
University of South Carolina, Columbia, South Carolina

Introduction

Let D be a parallelepipedon in R^N with boundary B. By $H^r(X)$, we mean the (possibly fractional) Sobolev space of order r on X. We let T exceed the diameter of D in what follows. The chief result we report is

Theorem 1. There is an open ball V about $0 \in H \equiv H^2(D) \times H^1(D)$ such that for $(g_1, g_2) \in V$ there exists a unique control $\hat{h} \in H^{3/2}(B \times [0, T])$ for which the equation

$$u_{tt} = \Delta u + f(u, u_t); \qquad u \Big|_{B \times [0, T]} = \hat{h}; \tag{1}$$

$$u(\mathbf{x}, 0) = 0, \qquad u_t(\mathbf{x}, 0) = 0,$$

has a unique solution in $C([0, T]; H)$ and $u(\mathbf{x}, T) = g_1$, $u_t(\mathbf{x}, T) = g_2$. The nonlinear function f and its Fréchet derivative are each assumed to be Lipschitz continuous at $0 \in H$.

For $f \equiv 0$, Russell [2] has established Theorem 1 for any $(g_1, g_2) \in H$ and any bounded domain D with piecewise smooth boundary. Our approach is to use Russell's result for the linear case together with the inverse function theorem to show that the nonlinear solution operator at time T is a local homeomorphism at $0 \in H$.

After outlining the method of attack that leads to Theorem 1, we discuss some numerical work that illustrates it.

Development of Theory

It is convenient to rewrite (1) as

$$\dot{w} = Aw + F(w); \qquad w\Big|_{B \times [0,\, T]} = h; \qquad w(0) = 0, \tag{2}$$

in which $w(t) = (u(\circ, t), (u_t(\circ, t)) \in H$, $F = \binom{0}{f}: H \to H$, and $A = \left(\begin{smallmatrix} 0 & I \\ \Delta & 0 \end{smallmatrix}\right)$ is an unbounded linear operator on H, and $h = (\hat{h},\ \hat{h}_t) \in H^{3/2}(B + [0,\ T]) \times H^{1/2}(B \times [0,\ T])$.

In order to prove that (2) has a solution in $C([0,\ T];\ H)$ we split (2) into two problems:

$$\dot{v} = Av; \qquad v\Big|_{B \times [0,\, T]} = h; \qquad v(0) = 0, \tag{3}$$

and

$$\dot{z} = Az + F(v + z); \qquad z\Big|_{B \times [0,\, T]} = 0; \qquad z(0) = 0. \tag{4}$$

It is clear that the solution w of (2) is obtained by $w(t) = v(t) + z(t)$.

Our first step is to establish

Theorem A. For any $u \in H$, there is a control

$$h_u \in H^{3/2}(B \times [0,\ T]) \times H^{1/2}(B \times [0,\ T]),$$

such that (3) has a unique solution $w_u(t)$ in $C([0,\ T];\ H)$ with $w_u(T) = u$.

One then can use standard iterative techniques to prove

Theorem B. If v_u denotes the solution to (3) as noted in Theorem A, and u is sufficiently small, then

$$\dot{z} = Az + F(z + v_u); \qquad z\Big|_{B \times [0,\, T]} = 0; \qquad z(0) = 0. \tag{4'}$$

has a unique solution $z_u \in C([0,\ T];\ H)$.

With these theorems established, we note that

$$w_u(t) \equiv v_u(t) + z_u(t)$$

is a solution to (2) with boundary control $h = h_u$, where h_u is the control that steers the linear problem from 0 to u. Thus $w_u(T)$ is not exactly u, but "misses" u due to the presence of F.

We define $G: U \subset H \to H$ by $G(u) = w_u(T)$, with U some open set about 0 for which Theorem B holds. Note that $G(u) = u + z_u(T)$. We then prove

Theorem C. G is continuously Fréchet differentiable at $0 \in H$, $G(0) = 0$, and $G'(0)$ is a linear homeomorphism.

Once Theorem C is established, an application of the inverse function theorem finishes the proof of Theorem 1.

Algorithm. Suppose we are given the system (1) with desired terminal state $(g_1, g_2) \in H$. To find the control $\hat{h}(g_1, g_2)$ we must

(1) Find $G^{-1}(g_1, g_2) \equiv (f_1, f_2)$.
(2) Compute the control that would steer $(0, 0)$ to (f_1, f_2) for the linear problem.

In practice, we can estimate $G^{-1}(u)$ by $[G'(0)]^{-1}u$ for small u in H. $[G'(0)]^{-1}$ can be approximated by solving a linear integral equation on H. Finding the controls for a linear problem is a relatively easy matter of solving some related Cauchy problems (see [1] for more explanation).

Numerical results. As an example, the following system was solved:

$$D = [0, 1] \times [0, 1] \subset R^2, \qquad T = 1.96,$$
$$u_{tt} = u_{xx} + u_{yy} + 0.1u_t, \qquad (g_1, g_2) = (0.1 \sin^2 \pi x \sin^2 \pi y, 0).$$

Of course, the term $0.1u_t$ is not a nonlinearity, but it was treated as a nonlinear perturbation to test the algorithm. Using grid sizes of $\Delta x = \Delta y = 0.1$, $\Delta T = 0.07$ and $\Delta x = \Delta y = 0.05$, $\Delta T = 0.035$, the problem was solved twice and Richardson extrapolation applied to the two approximate answers. The resulting approximate control steered $(0, 0)$ to a state (u_1, u_2) at time T, and if we define the relative error of the method

$$E^2 \equiv \frac{\|u_1 - g_1\|_2^2 + \|u_2 - g_2\|_1^2}{\|g_1\|_1^2 + \|g_2\|_1^2},$$

where $\|f\|_r$ is the Sobolev norm of order r of f, then our program produced $E^2 = 0.036$. [The crude grid sizes necessitated by limited computer storage capacity make it difficult to work with accuracy in a norm so exacting as $H^2(D)$.]

REFERENCES

[1] Chewning, W. C., Controllability of the non-linear wave equation in several space variables (to appear in *SIAM J. Control*).
[2] Russell, David L., A unified boundary controllability theory for hyperbolic and parabolic partial differential equations, *Studia Appl. Math.* **LII**, No. 3 (1973).

A Connection between Optimal Control and Disconjugacy

E. N. CHUKWU*
Department of Mathematics
The Cleveland State University, Cleveland, Ohio

O. HÁJEK
Department of Mathematics
Case Western Reserve University, Cleveland, Ohio

Consider the control system

$$\dot{x}(t) = A(t)x(t) + B(t)u(t), \qquad \text{(S)}$$

with fixed initial data $x(t_0) = x_0$.

The vector $x \in R^n$, A and B are $n \times n$- and $n \times m$-matrix-valued functions with $n - 2$ and $n - 1$ continuous derivatives, respectively, and $u \in R^m$ is a vector-valued function with values constrained to lie in the unit cube U of R^m. If $x(t, u)$ denotes the absolutely continuous solution of (S) satisfying $x(t_0, u) = x_0$, the problem is to determine a control u^*, subject to its constraints, in such a way that the solution $x = x(t, u^*)$ of (S) satisfies a termination condition $Z(x, t) = 0$ in minimum time $t^* \geq 0$. Such a control will be termed time optimal.

Let $X(t, t_0)$ be the principal matrix solution (at t_0) of the matrix differential equation

$$\dot{X} = A(t)X.$$

Definition. Each m-vector-valued function y,

$$y(t) = c'X^{-1}(t, t_0)B(t), \qquad c \in R^n,$$

will be called an index of the control system (S).

We have isolated and named a concept that appears implicitly in control theory. The connection is that under assumptions of normality each optimal control u must have the form

$$u = \operatorname{sgn} y' \qquad \text{for some index} \quad y.$$

---

* Supported in part by 1974 Cleveland State University Research Initiation Award.

Proposition 1. If (S) is controllable, then its indices (or rather their transposes y) form an n-dimensional subspace of the solution space of the nth order indicial equation in m-space

$$y^{(n)} = \sum_{k=0}^{n-1} A_k y^{(k)}, \tag{T}$$

with continuous m-square matrices A_k.

If (S) is normal, then each A_k can be taken diagonal, whereupon the indices determine (T) uniquely.

(T) will be called the indicial equation of the control system (S). For obvious reasons we will call (T) uncoupled if the matrices A_k are diagonal.

Examples of indicial equation. The control system with scalar equation

$$\ddot{x} + a(t)\dot{x} + b(t)x = u, \qquad -1 \le u \le 1, \tag{1}$$

has the indicial equation

$$\ddot{y} - a\dot{y} + (b - \dot{a})y = 0. \tag{2}$$

Analogously,

$$\dddot{x} + a\ddot{x} + b\dot{x} + cx = u(t) \tag{3}$$

has the indicial equation

$$\dddot{y} - a\ddot{y} + (b - 2\dot{a})\dot{y} + (\dot{b} - c - \ddot{a})y = 0. \tag{4}$$

Incidentally, (2) is the adjoint equation of (1) [more precisely the scalar equivalent of the adjoint equation to the first-order two-dimensional version of (1)]. (4) is the adjoint of (3).

Definition 1. A collection Y of solutions y of (T), all defined on a common interval I, will be termed disconjugate on I, if every member $y \not\equiv 0$, and the sum of the number of roots (on I, counting multiplicities) of its m coordinates is at most $n - 1$.

Definition 2. In the same situation Y is disconjugate at a point $t_0 \in I$ if it is disconjugate in some neighborhood of t_0. Also Y is geometrically disconjugate in I if it is such relative to any coordinate basis for R^m (i.e., if $MY = \{My: y \in Y\}$ is disconjugate for any nonsingular constant matrix M).

Our notion of disconjugacy generalizes that employed in the study of scalar nth order equations. Ours seems more applicable to control theory than the concepts and results known to the authors for $m > 1$ (Coppel [1] for $n = 2$ and Nehari [3] for $n = 1$ and general m).

The set of solutions of (1) has a natural structure of a linear space over the reals, with dimension nm. We will be concerned with its linear subspaces, and primarily with subspaces of dimension n. Again the collection of these subspaces has a natural structure of the Grassmann manifold $M_n(R^{nm})$, an analytic compact manifold of dimension $n^2m(m - 1)$ (Warner [4]).

An n-dimensional subspace L of the solution space of (T) can be described by specifying an initial time t_0 and n constant matrices $N_0, \ldots, N_{n-1}$ of type $m \times n$ as follows: $y \in L$ if and only if there exists $c \in R^n$ such that

$$y^{(k)}(t_0) = N_k c, \quad \text{for} \quad k = 0, \ldots, n - 1.$$

With this description we have the following:

Theorem. L is disconjugate at t_0 if and only if the following holds: for any integers $r_j \geq 0$ with $\sum_1^m r_j = n$, the n vectors

$$e_j' N_k, \quad 1 \leq j \leq m, \quad 0 \leq k \leq r_j - 1,$$

are linearly independent (e_j is the jth basic unit vector in R^m).

Theorem. The collection Y of all n-dimensional subspaces of the solution space of (T) that has each member disconjugate at t_0 is open and dense in the Grassmann manifold $M_n(R^{nm})$.

No $n + 1$-dimensional subspace is disconjugate at any point.

Definition 3. The system (S) is strictly normal at t_0 if

$$\text{rank}[\Gamma^i b_j(t_0) : 1 \leq j \leq m, 0 \leq i \leq r_j - 1] = n,$$

for all choices of integers r_j with $\sum_{j=1}^m r_j = n$. The operator is

$$\Gamma = -A(t) + d/dt$$

(with $\Gamma^{i+1} = \Gamma\Gamma^i$ for $i = 0, 1, \ldots$) [2].

Corollary. The collection of indices of (S) is disconjugate at t_0 if and only if (S) is strictly normal at t_0.

As a consequence of this it is now possible to construct optimal feedback controls for the system (S).

REFERENCES

[1] W. A. Coppel, "Disconjugacy." Springer-Verlag, Berlin and New York, 1971.
[2] H. Hermes and J. P. LaSalle, "Functional Analysis and Time Optimal Control." Academic Press, New York, 1969.
[3] Z. Nehari, Non-oscillation and disconjugacy of systems of linear differential equations, *J. Math. Anal. Appl.* **42** (1973), 237–254.
[4] F. W. Warner, "Foundations of Differential Manifolds and Lie Groups." Scott Foresman, Glenview, Illinois, 1971.

Control for Linear Volterra Systems without Convexity

THOMAS S. ANGELL

Department of Mathematics
University of Delaware, Newark, Delaware

We report here results concerning the existence of optimal and optimal bang–bang controls for systems whose dynamics are described by Volterra integral equations that are linear in the state variables. Using techniques similar to those developed by Cesari (see, for example, [2, 4]) for control systems monitored by ordinary and partial differential equations, we are able to establish a Neustadt-type existence theorem for usual solutions in the absence of convexity assumptions (see Neustadt [6]). We also state a result concerning the existence of an optimal bang–bang controller when the control domain is fixed and compact. Similar techniques have been used by Suryanaryana to discuss these questions for linear control problems with total differential equations [7]. Existence theorems for related (nonlinear) systems are discussed in [1] and, in a more general setting, by Warga [8]. Here, the particular form of our equations allows us to introduce the notion of generalized solution in a simple manner, and to avoid convexity assumptions. Unlike [8], we also consider the possibility of unbounded controls.

Specifically, we consider the Volterra-type integral equation

$$x(t) = x(t_1) + \int_{t_1}^{t} [h(t, s)x(s) + g(t, s)f(s, u(s))] \, ds, \qquad t_1 \leq t \leq t_2,$$

where the state variable $x(t) = (x^1, \ldots, x^n) \in E^n$ and the control variable $u(t) = (u^1, \ldots, u^m) \in E^m$. Let $A = [t_1{}^*, \, t_2{}^*] \times Z \subset E^{n+1}$ and write $I_A = [t_1{}^*, \, t_2{}^*]$. For each $t \in A$, let $U(t) \subset E^m$ be closed and let $M_0 = \{(t, u) | t \in I_A, u \in U(t)\}$. We require that the function f be continuous on M_0 and that the kernels g and h be continuous on $\Delta = \{(t, \, s) : t_1{}^* \leq s \leq t \leq t_2{}^*\}$ and zero for $t_1{}^* \leq t < s \leq t_2{}^*$.

We consider admissible pairs of functions $\{x, u\}$, each pair consisting of a measurable function $u : [t_1, \, t_2] \to E^m$ and a corresponding function $x : [t_1, \, t_2] \to E^n$, which satisfies the Volterra integral equation

$$x(t) = x(t_1) + \int_{t_1}^{t} [h(t, s)x(s) + g(t, s)f(s, u(s))] \, ds,$$
$$t_1{}^* \leq t_1 \leq t \leq t_2 \leq t_2{}^*, \tag{1a}$$

311

subject to boundary conditions

$$\eta(x) = (t_1, x(t_1), t_2, x(t_2)) \in B = B_1 \times B_2, \qquad (1b)$$

where B is closed in E^{2n+2} with $B_1 \subset E^{n+1}$ compact, as well as constraints

$$(t, x(t)) \in A \qquad \text{for all} \quad t \in [t_1, t_2], \qquad (1c)$$

$$u(t) \in U(t) \qquad \text{for almost all} \quad t \in [t_1, t_2]. \qquad (1d)$$

We assume that the class of all admissible pairs Ω is nonempty, and consider the Mayer problem $\min_\Omega I[x, u] = \min_\Omega e(t_1, x(t_1), t_2, x(t_2))$, where e is a real-valued function defined and continuous on $B \subset E^{2n+2}$.

If $\{x, u\} \in \Omega$ we refer to u as an admissible control and to x as an admissible trajectory. The necessary closure and compactness properties of the set of admissible trajectories can be ensured by requesting that the function f satisfy a growth condition with respect to a "comparison function" H that is defined and continuous on M_0. More precisely, the function f is said to satisfy the growth condition (γ_H) if, for any $\varepsilon > 0$, there exists a nonnegative locally L-integrable function ψ_ε such that $|f(t, u)| \le \psi_\varepsilon(t) + \varepsilon H(t, u)$ for all $(t, u) \in M_0$.

Let $K \ge 0$ be a fixed constant and denote by Ω_K the set consisting of all $\{x, u\} \in \Omega$ such that $H(t, u(t))$ is L-integrable and $\int_{t_1}^{t_2} H(t, u(t)) \, dt \le K$. Let $R = \{y(t) = f(t, u(t)) | \{x, u\} \in \Omega_K\}$ and let $\hat{R}$ be the set of extensions of elements of R to all of I_A by the value zero. Let

$$C^0 = \left\{ z(t) = \int_{t_1}^{t} [h(t, s)x(s) + g(t, s)f(s, u(s))] \, ds, \{x, u\} \in \Omega_K \right\}$$

and let $\hat{C}^0$ be the set of extensions of elements of C^0 to all of I_A by continuity and constancy. Then, as a special case of the results of [1] we have the following

Theorem. If the function f satisfies (γ_H), the set $\hat{R}$ is weakly relatively compact in $L_1^n(I_A)$ and $\hat{C}^0$ is an equicontinuous equibounded set.

Associated with the present problem is the orientor field $Q_H(t)$ defined by the functions f and H, namely,

$$Q_H(t) = \{(z^0, z) | z^0 \ge H(t, u), z = f(t, u), u \in U(t)\}$$

Under the hypothesis that the set $Q_H(t)$ is convex for all $t \in I_A$, one may apply variants of closure theorems proved by Cesari [3] to prove the existence of an optimal solution. The present situation is particularly simple since the orientor fields do not depend on the state variable and no additional seminormality conditions [e.g., the upper semicontinuity condition property (Q)] are needed.

However, the linear character of the control system (1a)–(1d) allows one to establish existence without assuming convexity of the $Q_H(t)$. We introduce, for nonconvex $Q_H(t)$, the concept of weak solution in a manner completely analogous to the familiar case of ordinary differential control systems (see, e.g., [2] or [5]). That is, we consider triples $\{x(t), p(t), u^*(t)\}$, where $x(t) = (x^1, \ldots, x^n)$, $p(t) = (p^1, \ldots, p^{n+1})$ is measurable, $u^*(t) = (u^{(1)}, \ldots, u^{(n+1)})$, $t_1 \le t \le t_2$, and for $i = 1, \ldots, n + 1$, $p^i \ge 0$, $\sum_{i=1}^{n+1} p^i = 1$, $u^{(i)}(t) \in U(t)$. Let $\tilde{f}(t, p, u^*) = (\tilde{f}_1, \ldots, \tilde{f}_n)$, where

$$\tilde{f}_j(t, p, u^*) = \sum_{i=1}^{n+1} p^{(i)} f_j(t, u^{(i)}),$$

and let

$$H^*(t, p, u^*) = \sum_{i=1}^{n+1} p^i H(t, u^{(i)}).$$

Then the triple $\{x, p, u^*\}$ is called a generalized solution to (1a)–(1d) provided

$$x(t) = x(t_1) + \int_{t_1}^{t} [h(t, s)x(s) + g(t, s)\tilde{f}(s, p(s), u^*(s))] \, ds, \qquad t_1 \le t \le t_2. \quad (2)$$

Let Ω^* denote the class of all generalized solutions such that $\eta(x) \in B$ with $H^*(t, p(t), u^*(t))$ L-integrable in $[t_1, t_2]$ and

$$\int_{t_1}^{t_2} H^*(t, p(t), u^*(t)) \, dt \le M.$$

We write $J[x, p, u^*] = e(\eta(x))$ and ask for the absolute minimum j of J in Ω^*. Since $\Omega \subset \Omega^*$, $j \le i = \inf_\Omega I[x, u]$. Note that, by Carathéodory's theorem, the orientor field of this new problem,

$$\hat{Q}_H(t) = \{(z^0, z) \mid z^0 \ge H^*(t, p, u^*), z = \tilde{f}(t, p, u^*)\},$$

is just co $Q_H(t)$, the convex hull of the original orientor field. It is easy to check that the assumption of the growth condition (γ_H) for f implies that $\tilde{f}$ satisfies the analogous growth condition (γ_{H^*}). The central result is the following:

Theorem 1.   Let $\{x(t), p(t), u^*(t)\}$ be a generalized solution in the sense described above. Then there exists a usual solution $\{\bar{x}, \bar{u}\} \in \Omega_K$ such that $\eta(x) = \eta(\bar{x})$.

We indicate a proof of this theorem as follows. If $r(t, s)$ is the resolvent kernel associated with the kernel $h(t, s)$, we may represent the solution of (2) corresponding to u^* as

$$x(t) = [x(t_1) + G[u^*](t)] - \int_{t_1}^{t} r(t, s)G[u^*](s)\, ds,$$

where

$$G[u^*](t) = \int_{t_1}^{t} g(t, s)\tilde{f}(s, u^*(s))\, ds.$$

Then,

$$x(t_2) = x(t_1) + \sum_{l=1}^{n+1} \left| \int_{t_1}^{t_2} g(t_2, s)p^l(s)f(s, u^{(l)}(s))\, ds \right.$$
$$\left. - \int_{t_1}^{t_2} p^l(s)R(t_2, s)f(s, u^{(l)}(s))\, ds \right|.$$

By a Lyapunov-type theorem on vector integrals (proved by Cesari [4]), there exists a decomposition of $[t_1, t_2]$ into $n + 1$ disjoint measurable sets $E_1, \ldots, E_{n+1}$ such that

$$x(t_2) = x(t_1) + \sum_{l=1}^{n+1} \int_{E_l} [g(t_2, s) + R(t_2, s)]f(s, u^{(l)}(s))\, ds.$$

The ordinary control function $\bar{u}$ defined by $\bar{u}(t) = u^l(t)$, $t \in E_l$, $l = 1, \ldots, n + 1$, generates a trajectory $\bar{x}$ with the same ends as the (generalized) trajectory x. This result leads immediately to the following Neustadt-type existence theorem.

Theorem 2. Let the sets A, B, and M_0 be as described above and suppose:

(i) The functions $f = (f_1, \ldots, f_n)$ and H are continuous on M_0 and f satisfies (γ_H);

(ii) The sets $Q_H(t)$ are closed;

(iii) The functions h and g are continuous on Δ.

Let Ω be a nonempty closed class of admissible pairs such that $H(t, u(t))$ is L-integrable and $\int_{t_1}^{t_2} H(t, u(t))\, dt \leq M_1$ for some $M_1 \geq 0$. Then the cost functional $I(x, u) = e(\eta(x))$ has an absolute minimum in Ω.

To establish this theorem, one replaces $Q_H(t)$ with $\hat{Q}_H(t) = \text{co } Q_H(t)$ as described above, and considers the corresponding optimization problem in the class of generalized trajectories Ω^* with cost functional $J[x, p, u^*]$. We may then apply the existence theorem of [1] to obtain a generalized optimal solution $\{x_0, p_0, u_0{}^*\} \in \Omega^*$. Theorem 1 now guarantees the existence of a usual solution $\{\bar{x}, \bar{u}\} \in \Omega$ that is optimal.

These techniques, together with the McShane–Warfield lemma, also afford a result on the existence of bang–bang controllers for problems in which the control domain is a fixed compact set. Specifically, we can establish the following:

Theorem 3.   Let $A = [t_0, T] \times Z$, $Z \subset E^n$, and let $U \subset E^m$ be a fixed compact set. Let g and h be continuous for $t_0 \leq s \leq t \leq T$ and zero for $t_0 \leq t < s \leq T$, and let the system be monitored by

$$x(t) = x(t_1) + \int_{t_1}^{t} [h(t, s)x(s) + g(t, s)u(s)] \, ds, \qquad t_1 \leq t \leq t_2. \tag{3}$$

If $\{x, u\}$ is an admissible pair, i.e., the pair $\{x, u\}$ satisfies (3) as well as $u(t) \in U$, $t \in [t_1, t_2]$ almost everywhere, and $(t, x(t)) \in A$, then there exists another admissible pair $\{\tilde{x}, \tilde{u}\}$ defined on $[t_1, t_2]$, such that $\tilde{x}(t_1) = x(t_1)$, $\tilde{x}(t_2) = x(t_2)$, $\tilde{u}(t) \in \partial U$, $t \in [t_1, t_2]$.

REFERENCES

[1] T. S. Angell, On the optimal control of systems governed by non-linear Volterra equations, *J. Optimization Theory Appl.* (to appear).

[2] L. Cesari, Existence theorems for weak and usual optimal solutions in Lagrange problems with unilateral constraints I, II, *Trans. Amer. Math. Soc.* **124** (1966), 369–412, 413–430.

[3] L. Cesari, Closure theorems for orientor fields and weak convergence, *Arch. Rational Mech. Anal.* **55** (1974), 332–356.

[4] L. Cesari, Convexity of the range of certain integrals, *SIAM J. Control* **13** (2) (1975), 666–676.

[5] R. V. Gamkrelidze, On some extremal problems in the theory of differential equations with applications to the theory of optimal control, *SIAM J. Control* **3** (1965), 106–128.

[6] L. Neustadt, The existence of optimal control in the absence of convexity, *J. Math. Anal. Appl.* **7** (1963), 110–117.

[7] M. B. Suryanaryana, Linear control problems with total differential equations without convexity, *Trans. Amer. Math. Soc.* **200** (1974), 233–249.

[8] J. Warga, "Optimal Control of Differential and Functional Equations." Academic Press, New York, 1972.

Noncontrollability of Linear Time-Invariant Systems Using Multiple One-Dimensional Linear Delay Feedbacks

A. K. CHOUDHURY

Department of Electrical Engineering
Howard University, Washington, D.C.

1. Introduction

In this chapter we shall prove that the system

$$\dot{x}(t) = Ax(t) + \sum_{i=1}^{m} b_i c_i^{\mathrm{T}} x(t - ih), \qquad t > 0, \tag{1}$$

has the property that if

$$d^{\mathrm{T}} x(t_1) = 0, \qquad t_1 > 0,$$

for all solutions corresponding to different initial functions belonging to $C[-mh, 0]$, then $d^{\mathrm{T}} = 0$. This property is called pointwise completeness or pointwise nondegeneracy. The physical interpretation of this result is that for the control system

$$\dot{x}(t) = Ax(t) + Bu(t),$$

where A and B are $n \times n$ matrices and $u(t)^{\mathrm{T}} = (u_1(t), u_2(t), u_3(t), \ldots, u_m(t))$, there is no state feedback control of the form

$$u_i(t) = c_i^{\mathrm{T}} x(t - ih), \qquad i = 1, 2, 3, \ldots, m,$$

that ensures the vanishing of the output $d^{\mathrm{T}} x(t_1)$. The case with $b_1 = b_2 = b_3 = \cdots = b_m = b$ has recently been proved by Asner and Halanay [1], generalizing the method of Popov [4], who proved that the system

$$\dot{x}(t) = Ax(t) + bc^{\mathrm{T}} x(t - h)$$

is pointwise complete.

2. Preliminaries

Definition 1. The delay–differential system

$$\dot{x}(t) = Ax(t) + B_1 x(t - h) + B_2 x(t - 2h) + \cdots + B_m x(t - mh) \qquad (2)$$

is called pointwise degenerate at t_1 with respect to the n-vector $d^T \neq 0$, *if for all initial functions* $\varphi \in C[-mh, 0]$, the corresponding solution satisfies the relation

$$d^T x(t; \varphi) = 0, \qquad t \geq t_1 > 0.$$

If the system is not pointwise degenerate, then it is called pointwise complete. The largest set of real numbers at which the system is pointwise degenerate with respect to d is called the degeneracy set for d.

We introduce the following notation:

$$p_i(s) = (sI - A)^{-1} B_i \, \Delta(s) = G_i(s) \, \Delta(s), \qquad i = 1, 2, 3, \ldots, m,$$
$$\Delta(s) = \det (sI - A),$$

$q_0(s) = I$, the $n \times n$ identity matrix,

$$q_1(s) = q_0(s)p_1(s),$$
$$q_2(s) = q_1(s)p_1(s) + q_0(s)p_2(s) \, \Delta(s),$$
$$q_3(s) = q_2(s)p_1(s) + q_1(s)p_2(s) \, \Delta(s) + q_0(s)p_3(s)(\Delta(s))^2, \tag{3}$$

and in general,

$$q_k(s) = q_{k-1}(s)p_1(s) + q_{k-2}(s)p_2(s) \, \Delta(s) + q_{k-3}(s)p_3(s)(\Delta(s))^2 + \cdots$$
$$+ \, q_{k-m}(s)p_m(s)(\Delta(s))^{m-1}, \quad k = 1, 2, 3, 4, \ldots, r, \tag{4}$$

where r is a positive integer. $q_k(s) = 0$, whenever k is a negative integer, and

$$p_{ki}(s) = q_{k-1}(s)p_i(s) + q_{k-2}(s)p_{i+1}\Delta(s) + \cdots$$
$$+ \, q_{k-m}(s)p_{i+m-1}(s)(\Delta(s))^{m-1}, \qquad i = 1, 2, 3, \ldots, m, \quad k = 1, 2, 3, \ldots.$$
$$\tag{5}$$

We observe that

$$p_{k1}(s) = q_{k-1}(s)p_1(s) + q_{k-2}(s)p_2(s) \, \Delta(s) + \cdots + q_{k-m}(s)p_m(s)(\Delta(s))^{m-1}$$
$$= q_k(s).$$

3. Representation for Solution of Eq. (2) for a Special Class of Initial Functions

Consider the initial function $g(\cdot) \in C[-mh, 0]$ given by

$$g(t) = (\Delta(D))^k \phi(t), \quad D = d/dt,$$

where k is a positive integer $> n$, and $\phi(\cdot)$ is a function belonging to $C^{nk}[-mh, 0]$ with the property

$$D^r \phi(-ih) = 0, \quad i = 0, 1, 2, 3, \ldots, m; \quad r = 0, 1, 2, 3, \ldots, nk.$$

Then the representation for the solution of Eq. (2) in the first interval for the initial function $g(\cdot)$ defined above is given by

$$x(t; g) = \sum_{i=1}^{m} p_{1i}(D)(\Delta(D))^{k-1} \phi(t - ih)$$

$[p_{1i}(D)$ as defined in (5)]. This follows from the fact that $\sum_{i=1}^{m} p_{1i}(D) \times (\Delta(D))^{k-1} \phi(t - ih)$ satisfies the equation

$$(DI - A)x(t) = \sum_{i=1}^{m} B_i (\Delta(D))^k \phi(t - ih)$$

[using Eq. (3)] in the first interval, and, due to the fact that $D^r \phi(-ih) = 0$, $i = 0, 1, 2, \ldots, m$, $r = 0, 1, 2, 3, \ldots, nk$, the continuity condition of the solution at the origin is satisfied. Similarly, the representation for the solution of Eq. (2) in the second interval is given by

$$x(t; g) = p_1 \left(\sum_{i=1}^{m} p_{1i}(D)(\Delta(D))^{k-2} \phi(t - (i + 1)h) \right)$$

$$+ \sum_{i=2}^{m} p_{1i}(D)(\Delta(D))^{k-1} \phi(t - ih)$$

$$= \sum_{i=1}^{m} p_{2i}(D)(\Delta(D))^{k-2} \phi(t - (i + 1)h), \quad h \le t \le 2h, \quad (6)$$

where $p_{2i}(D)$, $i = 1, 2, 3, \ldots, m$, are defined in Eq. (5), and in general the representation for the solution of Eq. (2) in the rth interval is given by

$$x(t; g) = \sum_{i=1}^{m} p_{ri}(D)(\Delta(D))^{k-r} \phi(t - (i + (r - 1))h), \quad (r - 1)h \le t \le rh. \quad (7)$$

4. Necessary Conditions of Pointwise Degeneracy

The necessary condition that the system (2) is pointwise degenerate with respect to a nonzero n-vector d at the point t_1, $(r - 1)h \leq t_1 \leq rh$, is given by

$$d^T p_{ri}(D) = 0, \qquad i = 1, 2, 3, \ldots, m,$$

where $p_{ri}(D)$ is defined in Eq. (5).

Proof. We observe that $p_{ri}(D)$ is an $n \times n$ matrix all of whose elements are polynomials in D of degree less than or equal to $(rn - 1)$. Hence, from Eq. (7) $d^T x(t; g)$ in the rth interval for the above special class of initial functions can be expressed in the form

$$d^T x(t; g) = \sum_{i=1}^{m} \sum_{j=0}^{kn-1} S_{jri}^T D^j \phi(t - (i + (r - 1))h), \qquad (r - 1)h < t < rh, \qquad (8)$$

where S_{jri} is an n-dimensional column vector and S_{jri}^T denotes its transpose. Let us choose the initial function such that

$$D^j \phi(t_0 - (i + (r - 1))h) = S_{jri},$$
$$j = 0, 1, 2, \ldots, kn - 1, \quad i = 1, 2, \ldots, m, \quad t_1 < t_0 < rh, \quad t_1 \geq (r - 1)h.$$

Hence from Eq. (8), it follows that

$$d^T x(t_0; g) = \sum_{i=1}^{m} \sum_{j=0}^{kn-1} (S_{jri}^T)(S_{jri}),$$

which contradicts the fact the system (2) is pointwise degenerate at t_1. Hence $s_{jri} = 0, j = 0, 1, 2, \ldots, nk - 1, i = 1, 2, \ldots, m$, which proves the result.

5. Consequences of $B_i = b_i c_i^T$ on the Necessary Conditions of Pointwise Degeneracy

Theorem 5.1. Suppose that $B_i = b_i c_i^T$, $i = 1, 2, 3, \ldots, m$; then $d^T p_{ri}(D) = 0, r > 2, i = 1, 2, 3, \ldots, m$, implies that $d^T p_{2i}(D) = 0, i = 1, 2, 3, \ldots, m$, where $p_{ri}(D), p_{2i}(D)$ are defined in Eq. (5).

Proof. Using Eq. (5) we observe that $d^{\mathrm{T}}p_{ri}(D) = 0$, $i = 1, 2, \ldots, m$, can be expressed as a set of m equations as follows:

$$d^{\mathrm{T}}q_{r-1}(D)p(D)b_1 c_1{}^{\mathrm{T}}\Delta(D) + \cdots + d^{\mathrm{T}}q_{r-m}(D)p(D)b_m c_m{}^{\mathrm{T}}(\Delta(D))^m = 0, \qquad (9)$$

$$d^{\mathrm{T}}q_{r-1}(D)p(D)b_2 c_2{}^{\mathrm{T}}(\Delta(D))^2 + \cdots$$
$$+ d^{\mathrm{T}}q_{r-m+1}(D)p(D)b_m c_m{}^{\mathrm{T}}(\Delta(D))^m = 0, \quad (10)$$
$$\vdots$$
$$d^{\mathrm{T}}q_{r-1}(D)p(D)b_{m-1}c_{m-1}^{\mathrm{T}}(\Delta(D))^{m-1} + d^{\mathrm{T}}q_{r-2}(D)p(D)b_m c_m{}^{\mathrm{T}}(\Delta(D))^m = 0, \quad (11)$$
$$d^{\mathrm{T}}q_{r-1}(D)p(D)b_m c_m{}^{\mathrm{T}}(\Delta(D))^m = 0, \quad (12)$$

where $p(D) = (DI - A)^{-1}$, $\Delta(D) = \det(DI - A)$.

Case 1. Let $c_1, c_2, \ldots, c_m$ be linearly independent. Then it follows from Eqs. (9)–(12) that

$$
\begin{aligned}
d^{\mathrm{T}}q_{r-1}(D)p(D)b_i &= 0, & i &= m, (m-1), (m-2), \ldots, 3, 2, 1, \\
d^{\mathrm{T}}q_{r-2}(D)p(D)b_i &= 0, & i &= m, (m-1), \ldots, 3, 2, \\
&\;\;\vdots & & \\
d^{\mathrm{T}}q_{r-(m-1)}(D)p(D)b_i &= 0, & i &= m, (m-1), \\
d^{\mathrm{T}}q_{r-m}(D)p(D)b_i &= 0, & i &= m,
\end{aligned}
\tag{13}
$$

as $d^{\mathrm{T}}q_{r-j}(D)p(D)b_i$, $j = 1, 2, 3, \ldots, m$, $i = m, (m-1), \ldots, 3, 2, 1$, are scalar quantities. Let us define the following variables:

$$c_{ij} = c_i{}^{\mathrm{T}}p(D)b_j(\Delta(D))^{i-1}, \qquad i = 1, 2, 3, \ldots, m, \quad j = 1, 2, 3, \ldots, m \tag{14}$$

$$y_j^i = d^{\mathrm{T}}q_{r-2-j}(D)p(D)b_i(\Delta(D))^{i-1}, \qquad i = 1, 2, 3, \ldots, m, \quad j = 0, 1, 2, 3, \ldots, m \tag{15}$$

We observe from Eqs. (4) and (13)–(15) that

$$y_0^i = (y_i{}^1 c_{1i} + y_i{}^2 c_{2i} + \cdots + y_i{}^m c_{mi})(\Delta(D))^{i-1}, \qquad i = 1, 2, \ldots, m, \tag{16}$$

and

$$
\begin{aligned}
y_0{}^1 c_{1i} + y_0{}^2 c_{2i} + \cdots + y_0{}^m c_{mi} &= 0, & i &= m, m-1, \ldots, 3, 2, 1, & (17) \\
y_1{}^1 c_{1i} + y_1{}^2 c_{2i} + \cdots + y_1{}^m c_{mi} &= 0, & i &= m, m-1, \ldots, 3, 2, & (18)
\end{aligned}
$$
$$\vdots$$
$$y_{m-2}^1 c_{1i} + y_{m-2}^2 c_{2i} + \cdots + y_{m-2}^m c_{mi} = 0, \qquad i = m, m-1, \tag{19}$$
$$y_{m-1}^1 c_{1m} + y_{m-1}^2 c_{2m} + \cdots + y_{m-1}^m c_{mm} = 0. \tag{20}$$

If $\mathrm{rank}(c_{ij}) = n$, it follows from Eq. (17) that $y_0^i = 0$, $i = 1, 2, 3, \ldots, m$, showing that Eq. (13) is satisfied if r is replaced by $(r-1)$. Hence

$d^T p_{ri}(D) = 0$ will imply that $d^T p_{(r-1)i}(D) = 0, i = 1, 2, 3, \ldots, m$, and the process can be continued until $d^T p_{2i}(D) = 0$, $i = 1$, 2, 3, $\ldots$, m. If rank (c_{ij}) is less than n, then using Eqs. (16)–(20) the above theorem can be proved.

Case 2. If $c_1, c_2, \ldots, c_m$ are not all independent, then also the above can be proved, but the proof involves a little more algebra.

Theorem 5.2. If the system (1) is pointwise degenerate with respect to the n-vector $d \neq 0$, and if $d^T p_{2i}(D) = 0$, $i = 1$, 2, 3, $\ldots$, m, then the degeneracy set is $[h, \infty)$.

Proof. $d^T p_{2i}(D) = 0$, $i = 1, 2, 3, \ldots, m$, implies that

$$d^T(G_1(\sigma)G_i(\sigma) + G_{i+1}(\sigma)) = 0, \qquad i = 1, 2, \ldots, m, \quad B_{m+i} = 0, \quad (21)$$

where $G_i(\sigma)$ are defined in Eq. (3). Observing that $\sigma G_i(\sigma) = AG_i(\sigma) + B_i$, $G_i(\sigma) \to 0$ as $\sigma \to \infty$, and multiplying Eq. (21) successively by powers of σ and taking the limit as $\sigma \to \infty$, we have the following relations:

$$d^T(A^j B_{i+1} + \beta_j B_i) = 0, \qquad \beta_j = A^{j-1}B_1 + \beta_{j-1}A, \qquad \beta_0 = \beta_{-1} = 0, \quad (22)$$
$$i = 1, 2, 3, \ldots, m, \quad j = 0, 1, 2, 3, \ldots, m, \ldots, \infty.$$

Differentiating Eq. (1) successively and using Eq. (22), we see that $y(t) = d^T x(t; g)$ satisfies the following equations for $t > h$:

$$y^j(t) = d^T x^j(t) = d^T A^j x(t) + d^T \beta_j x(t - h), \qquad t > h, \quad j = 1, 2, \ldots, \infty. \quad (23)$$

Hence from Eq. (23), it follows that for $j > 2n$, there exists a nonzero j-vector $E^T = (e_1, e_2, e_3, \ldots, e_j)$ such that

$$e_j y^j + e_{j-1} y^{j-1} + \cdots + e_1 y = 0, \qquad t > h, \quad j > 2n. \quad (24)$$

Now $y(t)$ satisfies the ordinary differential equation (24) everywhere for $t > h$ and $y(t) = 0$, for $t > (r-1)h$ [system (1) is degenerate]; hence it follows that $y(t) = 0$, for $t \geq h$. But this is impossible (see [2, 4]). Hence the system (2) is pointwise complete.

REFERENCES

[1] B. A. Asner and A. Halanay, Non-controllability of time-invariant linear systems using one-dimensional linear delay feedback, *Rev. Roumaine Sci. Tech.*, *Ser.*, *Electrotech. Energ.* **18** (1973), 283–291.

[2] A. K. Choudhury, Necessary and sufficient conditions of pointwise completeness of linear time-invariant delay-differential systems, *Internat. J. Control* **16** (6) (1972), 1073–1082.

[3] A. K. Choudhury, Algebraic and transfer-function criteria of fixed-time controllability of delay-differential systems, *Internat. J. Control* **16** (6) (1972), 1083–1100.

[4] V. M. Popov, Pointwise degeneracy of linear, time-invariant delay-differential equations *J. Differential Equations* **11** (1972), 541–561.

A Perturbation Method for the Solution of an Optimal Control Problem Involving Bang–Bang Control

MARVIN I. FREEDMAN and JAMES L. KAPLAN
Department of Mathematics
Boston University, Boston, Massachusetts

In the newly emerging field of perturbation analysis for optimal control one typically looks at a system of differential equations with data depending on a small parameter ε, which arises as the Euler–Lagrange equation of some underlying optimal control problem. (See, for example, the papers by Kelly [1], Kokotovic and Sannuti [2], O'Malley [3], and Hadlock [4]). In the regular case the data depend on ε in a smooth way, while in the singular case the order of the system is reduced upon setting $\varepsilon = 0$. The objective of such studies has generally been twofold. First, one develops formal techniques for determining asymptotic expansions in powers of ε for the state variables, the adjoint response (costate variables), and possibly for the optimal control variable. (In the singular case, so-called boundary layer expansions in stretched variables are required in addition to the usual time variable.) Second, one demonstrates that the asymptotic series found are uniformly valid asymptotic expansions.

In [5] we treat such a problem for the system

$$\dot{x} = f(x, \varepsilon) + cu, \tag{1a}$$

$$\dot{\lambda} = g(x, \lambda, \varepsilon), \tag{1b}$$

on the interval $[0, T]$, together with the boundary conditions

$$x(0, \varepsilon) = a(\varepsilon), \tag{1c}$$

$$\lambda(T, \varepsilon) = b(\varepsilon). \tag{1d}$$

In the above $\cdot$ denotes d/dt, ε is a small, positive, real parameter, $x, \lambda, f, g, c \in R^n$, and u is a measurable function on $[0, T]$ with values in $[-1, 1]$. We may imagine system (1) arising as the result of an application of the maximal principal to some underlying optimal control problem in the variable x. The variable λ may be thought of as the "costate variable" corresponding to x. The function $u(t, \varepsilon)$ appearing in (1a) will be viewed as the optimal choice for the underlying control problem; that is, it represents the choice of admissible controller that maximizes the appropriate Hamiltonian function. We will suppose that $u(t, \varepsilon)$ takes the form

$$u(t, \varepsilon) = \operatorname{sgn}(c \cdot \lambda(t, \varepsilon)) = c \cdot \lambda(t, \varepsilon)/|c(t) \cdot \lambda(t, \varepsilon)|. \tag{1e}$$

Because of space limitations, we will demonstrate our ideas here by applying them to a specific optimal control problem. For the complete analysis, including asymptotic validity of the procedure illustrated, see [5].

Consider the problem of controlling a particle whose trajectory is determined by

$$\ddot{z}(t) + \varepsilon z^2(t) = u(t), \qquad |u(t)| \le 1, \quad t \in [0, 3],$$

starting at the initial point

$$z(0) = \tfrac{1}{2} + \varepsilon, \qquad \dot{z}(0) = 1 + 2\varepsilon.$$

Say, for instance, that we wish to maximize the functional

$$c(u) = \dot{z}(3, \varepsilon) - z(3, \varepsilon).$$

In system notation, if we let $x(t) = z(t)$, $y(t)$, $\dot{z}(t)$ we obtain

$$\dot{x} = y, \tag{2a}$$
$$\dot{y} = -\varepsilon x^2 + u, \qquad |u(t)| \le 1, \tag{2b}$$
$$x(0) = \tfrac{1}{2} + \varepsilon, \tag{3a}$$
$$y(0) = 1 + 2\varepsilon. \tag{3b}$$

The Hamiltonian function corresponding to system (2) is given by

$$H(x, y, \lambda, v, u, \varepsilon) = \lambda y + v(-\varepsilon x^2 + u),$$

from which the corresponding equations for the costate variables λ and v are determined. They are

$$\dot{\lambda} = 2\varepsilon x, \tag{4a}$$
$$\dot{v} = -\lambda. \tag{4b}$$

The costate variables must satisfy the boundary conditions

$$\lambda(3) = -1, \tag{5a}$$
$$v(3) = 1, \tag{5b}$$

which are independent of ε. When $\varepsilon = 0$, we may easily obtain an explicit solution to the problem (2)–(5), which we will denote by $x_0(t)$, $y_0(t)$, $\lambda_0(t)$, $v_0(t)$, $u_0(t)$. It is given by

$$x_0(t) = \begin{cases} -\frac{1}{2}t^2 + t + \frac{1}{2}, & 0 \le t \le 2, \\ \frac{1}{2}t^2 - 3t + \frac{9}{2}, & 2 \le t \le 3, \end{cases}$$

$$y_0(t) = \begin{cases} -t + 1, & 0 \le t \le 2, \\ t - 3, & 2 \le t \le 3, \end{cases}$$

$$\lambda_0(t) = -1,$$

$$v_0(t) = t - 2,$$

$$u_0(t) = \operatorname{sgn} v_0(t) = v_0(t)/|v_0(t)| = \begin{cases} -1, & 0 \le t \le 2, \\ +1, & 2 \le t \le 3. \end{cases}$$

[We recognize that this is also the solution of the problem of bringing to rest at the origin in minimum time a trolley that starts at $(z(0), \dot{z}(0)) = (\frac{1}{2}, 1)$ and moves along a horizontal track with negligible friction.]

Now, let $t(\varepsilon)$ denote the switch time of $u(t, \varepsilon)$ for the perturbed equations (5)–(8). Let

$$t = h(\tau, \varepsilon) = \frac{1}{2}\tau(3 - \tau)t(\varepsilon) + \tau(\tau - 2).$$

This Lagrange interpolating polynomial will freeze the switch point of the control at $\tau = 2$, while leaving the initial and terminal times fixed. Now define $X(\tau, \varepsilon) = x(h(\tau, \varepsilon), \varepsilon)$, and define $Y(\tau, \varepsilon)$, $\Lambda(\tau, \varepsilon)$, $N(\tau, \varepsilon)$, $U(\tau, \varepsilon)$ similarly. Thus

$$U(\tau, \varepsilon) = \begin{cases} -1, & 0 \le \tau \le 2, \\ +1, & 2 \le \tau \le 3. \end{cases}$$

Substituting these new variables into (5)–(8) results in

$$\begin{aligned}
dX/d\tau &= [\tfrac{1}{2}(3 - 2\tau)t(\varepsilon) + (2\tau - 2)]Y(\tau, \varepsilon), \\
dY/d\tau &= [\tfrac{1}{2}(3 - 2\tau)t(\varepsilon) + (2\tau - 2)][-\varepsilon X^2(\tau, \varepsilon) + U(\tau, \varepsilon)], \\
d\Lambda/d\tau &= [\tfrac{1}{2}(3 - 2\tau)t(\varepsilon) + (2\tau - 2)][2\varepsilon X(\tau, \varepsilon)N(\tau, \varepsilon)], \\
dN/d\tau &= [\tfrac{1}{2}(3 - 2\tau)t(\varepsilon) + (2\tau - 2)]\Lambda(\tau, \varepsilon).
\end{aligned} \tag{6}$$

Observe that $X(0, \varepsilon) = x(h(0, \varepsilon)) = x(0, \varepsilon) = \frac{1}{2} + \varepsilon$, while $Y(0, \varepsilon) = y(0, \varepsilon) = 1 + 2\varepsilon$. Thus, our original initial conditions are preserved in the new variable τ. In addition, we have right boundary conditions of the form $\Lambda(3) = -1$, $N(3) = 1$. Now let us suppose that each of the variables appearing in the transformed system possesses an asymptotic series expansion in powers of ε.

Thus

$$X(\tau, \varepsilon) = X_0(\tau) + \varepsilon X_1(\tau) + \varepsilon^2 X_2(\tau) + \cdots, \qquad t(\varepsilon) = t_0 + \varepsilon t_1 + \varepsilon^2 t_2 + \cdots,$$

with similar expressions for $Y(\tau, \varepsilon)$, $\Lambda(\tau, \varepsilon)$, and $N(\tau, \varepsilon)$. Observe that we must have $X_0(\tau) = x_0(\tau)$, $Y_0(\tau) = y_0(\tau)$, $\Lambda_0(\tau) = \lambda_0(\tau)$, $N_0(\tau) = v_0(\tau)$, and $t_0 = 2$.

Next, let us insert the assumed series expansions into (6). We then differentiate the resulting equations with respect to ε, and set $\varepsilon = 0$. This yields

$$dX_1/d\tau = Y_1(\tau) + [\tfrac{1}{2}(3 - 2\tau)t_1]Y_0(\tau), \tag{7a}$$

$$dY_1/d\tau = -X_0^2(\tau) + [\tfrac{1}{2}(3 - 2\tau)t_1]U_0(\tau), \tag{7b}$$

$$d\Lambda_1/d\tau = 2X_0(\tau)N_0(\tau), \tag{7c}$$

$$dN_1/d\tau = -\Lambda_1(\tau) - [\tfrac{1}{2}(3 - 2\tau)t_1]\Lambda_0(\tau). \tag{7d}$$

The appropriate boundary conditions for the first two equations in (7) are given by

$$X_1(0) = 1, \qquad Y_1(0) = 2. \tag{8}$$

For the second pair of equations, we must have

$$\Lambda_1(3) = 0, \qquad N_1(3) = 0. \tag{9}$$

In addition, since $U(\tau, \varepsilon) = \operatorname{sgn} N(\tau, \varepsilon)$, we know that we must have $N(2, \varepsilon) = 0$. We may now observe that system (7) may be solved by quadrature. Moreover, if we solve (7d), we can set $N_1(2) = 0$ to determine the value of t_1. Thus we can determine the coefficients in the asymptotic series expansion of the switch time. Using this value, the solution X_1, Y_1, Λ_1, N_1, U_1 is known completely. We may then differentiate (7) once again with respect to ε and set $\varepsilon = 0$, to determine the next terms in the expansions.

The actual value computed for t_1 for the problem treated here is $t_1 = 76.6 \ldots$. Thus we have

$$t(\varepsilon) = 2 + 76.6\varepsilon + O(\varepsilon^2),$$

and the relationship $t = h(\tau, \varepsilon)$ takes the form

$$t = \tau + \varepsilon[38.3\tau(3 - \tau)] + O(\varepsilon^2).$$

The expansions (in the variable τ) found as indicated above will be uniformly valid on $0 \le \tau \le 3$.

We conclude by remarking that the technique we have demonstrated for a control problem involving one switch is indicative of a more general technique that is applicable to a wide variety of practical control problems involving multiple switches. This will necessitate the use of a higher degree Lagrange interpolation polynomial. Other bang–bang control problems can be treated similarly. In the case of a singular bang–bang control problem, one expects boundary layer behavior at the switch points as well as at the initial and terminal times. For the minimum-time problem, it is necessary to choose a Lagrange interpolation polynomial that "freezes" the terminal time as well as the switch time. These ideas will be explored in detail in forthcoming papers.

REFERENCES

[1] H. J. Kelley, Singular perturbations for a Mayer variational problem, "Advances in Control Systems" (C. T. Leondes, ed.), Vol. IX. Academic Press, New York, 1972.
[2] P. V. Kokotovic and P. Sannuti, Singular perturbation methods for reducing the model order in optimal control design, *IEEE Trans. Automatic Control* **13** (1968), 377–384.
[3] R. E. O'Malley, Jr., Boundary layer methods for certain nonlinear singularly perturbed optimal control problems, *J. Math. Anal. Appl.* **45** (1974), 468–484.
[4] C. R. Hadlock, Existence and dependence on a parameter of solutions of a nonlinear two point boundary value problem, *J. Differential Equations* **14** (1973), 498–517.
[5] M. I. Freedman and J. L. Kaplan, Perturbation analysis of an optimal control problem involving bang–bang controls (Submitted for publication). Also available as *B.U. Tech. Rep.* '74–**8**.

Sufficient Conditions for a Relaxed Optimal Control Problem

RUSSELL D. RUPP
Department of Mathematics
State University of New York, Albany, New York

Introduction

A sufficiency theorem is established for the problem of minimizing a functional of the form

$$F(C) = g(b) + \int_{T^1}^{T^2} M[t, L(t, x(t), u)] \, dt$$

over a class of terminally and differentiably admissible generalized arcs (generalized arc = boundary parameters + trajectory + relaxed control).

Description of Results

This sufficiency theorem applies to deterministic as well as relaxed control problems and includes several known sufficiency results as special cases. Consequently, the proof, which is by a new type of expansion argument, demonstrates that useful sufficiency results can be derived without highly sophisticated mathematics. Hypotheses are made with respect to transversality, the costate (or Euler–Lagrange) equations, and a strengthened maximum principle. Several illustrative examples are given and reference is made to some of the many places in the literature where derivation of a problem's extremals using necessary conditions (the maximum principle) is tantamount to verifying these sufficiency hypotheses.

The details of proof and the application of this fixed-time-domain result to minimum-time problems will be given in later work.

REFERENCES

[1] L. Cesari, Existence theorems for weak and usual optimal solutions in Lagrange problems with unilateral constraints. II. Existence theorems for weak solutions, *Trans. Amer. Math. Soc.* **124** (1966), 413–430.

[2] M. Hestenes, "Calculus of Variations and Optimal Control Theory." Wiley, New York, 1966.

[3] E. McShane, Relaxed controls and variational problems, *SIAM J. Control* **5** (1967), 438–485.

[4] W. Reid, Lagrange multipliers revisited, *in* "Control Theory and the Calculus of Variations" (A. Balakrishnan, ed.). Academic Press, New York, 1969.

[5] R. Rupp, A nonlinear optimal control minimization technique, *Trans. Amer. Math. Soc.* **178** (1973), 357–381.

Author Index

Subject Index